Framing the World

Framing the World

Classical Influences on Sixteenth-Century Geographical Thought

Margaret Small

THE BOYDELL PRESS

First published 2020
The Boydell Press, Woodbridge

ISBN 978-1-78327-520-5

The Boydell Press is an imprint of Boydell & Brewer Ltd
PO Box 9, Woodbridge, Suffolk IP12 3DF, UK
and of Boydell & Brewer Inc.
668 Mt Hope Avenue, Rochester, NY 14620–2731, USA
website: www.boydellandbrewer.com

A catalogue record of this publication is available from the British Library

To my parents, Alastair and Carola Small

Contents

List of Illustrations

The author and publisher are grateful to all the institutions and individuals listed for permission to reproduce the materials in which they hold copyright. Every effort has been made to trace the copyright holders; apologies are offered for any omission, and the publisher will be pleased to add any necessary acknowledgement in subsequent editions.

Preface and Acknowledgements

This book has been the result of many of years of research and study. It took its present form in the West Midlands – first at the University of Keele where I began to teach the subject of Europe and the Wider World in the early modern period to students, but mostly it was formed in the University of Birmingham where I owe a great deal of gratitude to my colleagues. It would be difficult to mention them all, but I cannot refrain from thanking the following in particular: Alex Gadja, Sadia Qureshi and David Gange who all read portions of the book, and Richard Cust, Simone Laqua-O'Donnell, Jonathan Willis and Elaine Fulton. Tara Hamling has also read large sections and given me huge encouragement and advice. I also owe thanks to Sheila Christie at Cape Breton University and Madeleine Hron at Wilfrid Laurier for their support and advice on specific chapters. I must acknowledge a special debt of gratitude to Catherine Delano-Smith, editor of *Imago Mundi* who has been a good friend and counsellor, and to Surekha Davies who gave invaluable advice in helping the book take its present shape.

The book has had a long conception. I have always been fascinated by the interplay between history and geography and by the history of geographical ideas. At the University of Alberta, I read a joint honours degree in classics and history. The combined studies helped me to realise how important was the role played by ancient texts in the history of geography, and I was particularly fortunate to be able to take a course on early modern history of geography with Lesley Cormack, an expert in the field. I wrote an honours thesis on geographical boundaries in Herodotus which complemented my work in early modern geography. This led me on to study for a Master's degree on the geography of Herodotus and Pre-Socratic influences on his conceptualisation of the world under Gordon Shrimpton at the University of Victoria. The thesis confirmed to me the important role philosophy played in the history of geography. At Oxford I was able to develop these interests further in my doctoral thesis, *Boundaries and Balance, Classical Influence on Sixteenth-century Geographical Thought* (2004) supervised by Nicholas Davidson. After I completed my Doctorate I went to the National Maritime Museum as a postdoctoral fellow where I began a project on Neoplatonic influences on English exploration in the sixteenth century. This project reaffirmed to me the importance of theoretical and philosophical ideas in the history of geography, and the way that theoretical ideas could provide a springboard into action. Without their philosophical framework for the world the Renaissance navigators would not have explored where they did.

The book could not have been written without the friendly help I have received in many libraries. I am especially grateful to the librarians in the Bodleian Map room, in particular Jenny Coles, Nigel James and Nick Millea, and to the staff in the rare books room at the British Library. I am also grateful to the personnel at the Library of San Marco in Venice, the Biblioteca Ambrosiana in Milan and Trinity College Library in Dublin who helped me in the hunt for elusive information about scholarly networks and readers of geographical information.

The book has involved texts in a variety of languages. Numerous people have helped me with the translations: Paul Brownutt with Spanish, Michela Okada and Caterina Belo with Portuguese, Jonathan Kirkpatrick with Greek and Latin, Caterina Bruschi with Italian, and my parents Alastair and Carola Small with Italian, French, Latin and Greek.

Finally I would like to thank my family, in particular my husband, Joel Lewis, who has been an enormous support, and has learnt a great deal about early modern geography over the course of my writing this book, and my parents, Alastair and Carola Small, who have read the book in its entirety, been immensely patient with my linguistic queries and instilled an interest in history and classics from childhood.

Abbreviations

DK	H. Diels and W. Kranz, *Die Fragmente der Vorsokratiker griechisch und deutsch* (Berlin: Weidmannsche Werlagsbuchhandlung, 1960, first published 1903; 3rd edition).
F.Gr.Hist	F. Jacoby, *Die Fragmente der griechischen Historiker*, 3 vols. (Berlin: Weidmann, 1926).
M –	H. J. Mette, *Sphairopoiia: Untersuchungen zur Kosmologie des Krates von Pergamon: mit einem Anhang, Texte* (Munich: Beck, 1936).
NV	G. B. Ramusio, *Navigazioni e viaggi*, Marica Milanesi ed. (Turin: Giulio Einaudi, 1978–1984, first published 1550–1559).
PN	R. Hakluyt., ed. *The Principal Navigations, Voiages, Traffiques and Discoveries of the English Nation,* 12 vols. (Glasgow: Maclehose and Sons, 1903, first published 1598–1600. Reprint, 2nd).
RE	*Paulys Realencyclopädie der classischen Altertumswissenschaft* (Stuttgart, 1937).

Introduction

> This is clear, said the mariners, that beyond this Cape [Bojador] there is no race of men nor place of inhabitants; nor is the land less sandy than the deserts of Libya, where there is no water, no tree, no green herb; and the sea is so shallow that a whole league from land is only a fathom deep. The currents are so terrible that no ship having once passed the Cape will ever be able to return. Therefore our forefathers never attempted to pass it... Now what sort of a ship's captain would he be who, with such doubts placed before him by those to whom he might reasonably yield credence and authority, and with such certain prospect of death before his eyes, could venture the trial of such a bold feat as that?[1]

The long sixteenth century saw the greatest single shift in European geographical knowledge and understanding that has occurred in written history. This book demonstrates how over this key period western Europeans moved from seeing the northern temperate regions of the world as the only knowable space to thinking of the world as a global unit in which God made all parts of the sphere for human exploitation and to interact with one another. In the fifth century BC the ancient Greeks had developed the idea that only parts of the world were inhabitable. This 'inhabitable world', termed *oikoumene,* in which boundaries of climate and Ocean separated the knowable regions of the earth from the rest of the sphere,

[1] Isto he claro, deziam os mareantes, que despoís deste cabo nom ha hi gente nem povoraçom algũa; a terra nom he menos areosa que os desertos de Libya, onde nom ha augua, nem arvor, nem herva verde; e o mar he tam baixo, que a hũa legoa de terra nom ha de fundo mais que hũa braça. As correntes som tamanhas, que navyo que la passe, jamais nunca podera tornar. E por tanto os nossos antecessores nunca se antremeterom de o passar ... Hora qual pensaaes que avya de seer o capitam do navyo, a que posessem semelhantes duvydas dyante, e mais per homeês a que era razom de dar fe e autoridade em taaes lugares, que ousasse de tomar tal atrevimento, sob tam certa sperança de morte como lhe ante os olhos apresentavam? G. E. de Azurara, *Crónica do descobrimento e conquista da guiné* (Paris: J. P. Aillaud, 1841), Cap. VIII, 50.

became the primary model in geographical thought for more than two thousand years. The quotation above from Gomes Eanes de Zurara's *Chronicle of Guinea* written about 1450, at the very cusp of the period of European expansion, captures the fear with which the world beyond the *oikoumene* was viewed. According to Zurara, the mariners were certain that there was no point in venturing beyond Cape Bojador since neither sea nor land could support human life. Humans were confined to the *oikoumene*. In the late fifteenth and sixteenth centuries, however, the world came to be seen as a truly global space and the traditional boundaries of classical and medieval geographical thought were undermined. Nonetheless, ideas of inhabitability continued to dominate geographical literature throughout the sixteenth century. This book demonstrates that the dominant framework for the new geography remained the *oikoumene* and that far from disappearing from geographical thought the *oikoumene* reached out to cover the sphere. Humanist geographers from every country re-reading the classics in the light of the geographical evidence that was reaching them from the newly encountered territories were the ones that fuelled this re-interpretation. Through their works the traditional boundaries were transformed, enabling the whole world to be seen as a globally connected space designed by God for humans.[2]

It is argued here that the European re-imagining of the world as a globally connected inhabitable space was dependent on Greek and Roman theories and concepts. Despite a sixteenth-century emphasis on eye-witness knowledge, the coincidence of the Renaissance and the period of European exploration ensured that a re-interpretation of classical ideas directed the way in which the Europeans re-framed the world. This book proposes that the geographical re-interpretation of the classics and new understanding had huge implications for early modern Europe. Europeans came to see the entire sphere as subject to human mastery, and

2 On the transformation in geographical thought and the use of classical knowledge see in particular D. E. Cosgrove, *Apollo's Eye: A Cartographic Genealogy of the Earth in the Western Imagination* (London: Johns Hopkins University Press, 2001); W. G. L. Randles, 'Classical Models of World Geography and their Transformation Following the Discovery of America,' in *The Classical Tradition and the Americas: European Images of the Americas and the Classical Tradition,* ed. by Wolfgang Haase and Meyer Reinhold, vol. 1 (Berlin: 1994); W. G. L. Randles, *De la terre plate au globe terrestre: une mutation épistémologique rapide (1480–1520)* (Paris: Armand Colin, 1980); G. Atkinson, *Les nouveaux horizons de la Renaissance française* (Paris: Droz, 1935); J.-M. Besse, *Les Grandeurs de la terre; aspects du savoir géographique à la Renaissance* (Lyon: ENS, 2003); N. Broc, *La géographie de la Renaissance (1420–1620)* (Paris: Bibliothèque nationale, 1980); F. Lestringant, *Mapping the Renaissance World: The Geographical Imagination in the Age of Discovery* (Berkeley: University of California Press, 1994); S. Davies, *Renaissance Ethnography and the Invention of the Human: New Worlds, Maps and Monsters* (Cambridge: Cambridge University Press, 2016); Z. Shalev and C. Burnett, 'Ptolemy's 'Geography' in the Renaissance,' Warburg Institute Colloquia, 17 (London: Warburg Institute, 2011).

furthermore, came to believe it was a God-given duty to interact with one another to gain access to the different resources of each region to the best of their ability. As the book concludes, many of the hallmarks of the early modern world could not have developed without the paradigm shift in geographical thought. The great global trading companies of the late sixteenth and seventeenth centuries and even the vast European colonial empires relied on this shift in understanding from the idea of bounded *oikoumenai* (inhabitable zones) to an *oikoumene* that reached out across an unsegregated terrestrial sphere. Two-thousand-year-old theories informed the interpretation of the new empirical data, and the Greco-Roman ideas that balance and reciprocity, boundaries and limits were identifiable in all aspects of the natural world remained key to the new geography.

In order to move to the idea of a globally inter-dependent world, however, people had to challenge the traditional theory of the *oikoumene*. The boundaries of the *oikoumene* were never universally accepted. Both classical writers such as Polybius and Posidonius and medieval authors such as Sacrobosco and Roger Bacon challenged various aspects of the classical *oikoumene*, and its limits were already beginning to crumble by the early fifteenth century.[3] Were it not for a number of classical and medieval writers who discussed inhabitable torrid zones, neither the Portuguese nor Columbus would ever have sailed south and west.[4] Nonetheless until the arrival of the Portuguese in India and the Spanish in Africa, the concept of a single, bounded *oikoumene* remained the dominant paradigm for understanding human geography. Even in the aftermath of the fifteenth-century voyages of discovery round the coast of Africa to India and across the Atlantic to the newly encountered lands of the Americas, the idea of an *oikoumene* was not entirely destroyed; rather it was transformed. For much of the sixteenth century, the expanding margins of the *oikoumene* acted as a means of controlling the unknown space of the Americas, Southeast Asia and sub-Saharan Africa.

In the aftermath of the discovery of the New World, and of the circumnavigations of Africa and later the globe, the world in a sense lost its shape. Nobody knew the precise configuration of lands, or even the number of continents, and yet these were suddenly relevant since the transatlantic and sub-Saharan African voyages had incontrovertibly proved that the old idea of a confined *oikoumene* acting almost as a knowable island in the space of an unknowable world was erroneous. Nobody could experience the world in its entirety. Those interested in both the exploration and the nature of the world therefore studied and used the written word to gain an understanding of the earth's space and to create conceptual constructs of the shape and nature of the world. On one level the

[3] A. Hiatt, 'Blank Spaces on the Earth,' *The Yale Journal of Criticism,* 15 (2002): 233–36.

[4] N. Wey Gómez, *The Tropics of Empire: Why Columbus Sailed South to the Indies* (Cambridge, Mass.: MIT Press, 2008).

history of geography in Western Europe in the aftermath of the fifteenth- and early sixteenth-century voyages of discovery is the story of devising means for a mental appropriation and control of unknown space. Although over the course of a long fifteenth century Europeans reached China, Japan, the Philippines, and mainland southeast Asia, circumnavigated Africa, encountered and rounded South America for the first time, and sailed as far north as the Arctic circle in search of northeast and northwest passages, there were still vast areas of the world about which they had little empirical information. Yet by the end of the century they had developed theoretical frameworks which brought the whole world into their mental horizon.

It is no coincidence that the classics – at times consciously, at others less explicitly – influenced these theoretical frameworks. The classical revival of the fifteenth and sixteenth centuries brought to light a vast array of Greek and Roman texts which had been forgotten for hundreds of years. Some of these contained essentially new ideas and concepts which helped to spur on scholarship while others simply reinforced existing theories. The great respect for classical knowledge held by the sixteenth-century humanist-trained authors – key players, as this study will show, in creating and transmitting the new geographical understanding – ensured that Greek and Roman theories played a fundamental role in the way that geography became re-interpreted in the light of the Spanish and Portuguese voyages of exploration in the late fifteenth century, and the subsequent French, Dutch and English ones of the sixteenth century.

The sixteenth century was a period of transition. This was not the period of absolute innovation of the fifteenth century when cosmographers at the Portuguese and Spanish courts were making the case for crossing Cape Bojador at the entry to the tropics, and traversing the Atlantic to find sea-routes to India. Wey-Gomez rightly points out the 'fundamental inventiveness of Columbus's India enterprise' when new readings of virtually every existent genre of literature enabled Columbus and his supporters to make the case for crossing the Atlantic to Asia.[5] Nor was this the period of what Ayesha Ramachandran has recently called 'the worldmakers', where the inhabitability of the world had begun to be taken for granted, and Europeans were devising means to take mental possession of a whole world that they could never see in its entirety.[6] Rather this was a period when European geographical knowledge was in a state of significant flux. As will be demonstrated, it was only towards the end of the sixteenth century that the *oikoumene* really came to be seen as coinciding with the sphere. Until that point geographers and explorers alike continued to struggle with whether there were climatic boundaries to human inhabitation. They

5 *Ibid.*, 358.

6 A. Ramachandran, *The Worldmakers: Global Imagining in early modern Europe* (Chicago: The University of Chicago Press, 2015).

thought of the knowable world as confined by monsters and marvels populating the regions of the world furthest from Europe. From the time of the earliest Greek geographical writings, the inner limits of the *oikoumene* had been defined as the realms of the monstrous who populated the *eschatai*, the farthest reaches, which lay in regions of relatively extreme climates, bordering on the *eremoi* (wastelands) at the outermost borders of the knowable world.[7] Sixteenth-century geographical texts and travellers' tales are notable for the continuing presence of such marvels and monsters. They remained a means of managing space, of separating the known from the unknown and to a certain extent of distinguishing the knowable from the unknowable. They play a major role in the history of space in western experience, and, as this book concludes, it is only when they largely disappeared from the pages of western European geographical texts in the late sixteenth century that the idea that the *oikoumene* truly coincided with the sphere – and that all the world was not only knowable, but indeed intended to be known by humans – really came to the fore.

The disappearance of the monsters was the final stage in the construction of the new geographical model, however. This book proposes that the earlier part of the process of reshaping the *oikoumene* was founded on a reinterpretation and re-application of ideas that the natural world and the ordered *kosmos* were governed by laws of balance and proportionality and limit and transgression. The ancient Greek geography which became the foundations of western European geographical thought was founded on the idea that the effect of these laws could be discerned in the structure and geography of the physical world. An understanding that these laws were fundamental to nature was integral to the way in which the sixteenth-century humanists redefined the *oikoumene*. This book traces the way in which the *oikoumene* was conceived in the ancient world in accordance with these principles, and then how each element of the bounded, inhabitable world, was reinterpreted in the sixteenth-century Age of Exploration from the challenging of the boundaries of climate, *eremoi* and monsters, through the reimagining of the ocean from a confine to a means of communication, to the interpreting of relatively unexplored landmasses as proportionate and harmonic counterparts to the old world. It required the reconsideration of each of these aspects of traditional geography to take place for the sixteenth-century Europeans really to be able to claim the entire earth as the domain of humans. The sixteenth century was the period in which Europeans came to see the whole world as having the potential for conquest, colonisation, exploitation and global trade, thereby laying the groundwork of the modern globalised world. This transition, however, was not made as part of a dramatic overthrowing of traditional sources of knowledge. Rather the classical authorities played an integral role in the

[7] J. Romm, *The Edges of the Earth in Ancient Thought: Geography, Exploration, and Fiction* (Princeton: Princeton University Press, 1992).

reinterpretation of the *oikoumene* in the sixteenth century. If this new understanding of the world heralded modernity, then the roots of modernity can be firmly traced to theories of the Greeks and Romans.

As early as 1884 William Tillinghast, in his contribution to Justin Winsor's vast compendium on the history of America, raised the importance of the role that the classics played in European geographical interpretation of the New World.[8] Yet in the ensuing century remarkably few scholars contributed anything further to the discussion of the role of the classics in sixteenth-century geographical understanding.[9] Historians of ethnography such as Lewis Hankes, Sabine McCormack and Anthony Pagden have demonstrated the vast importance that classical writings, in particular those of Aristotle, played in European interpretations of the nature of New World peoples. Yet this focus on the role of the classics in forming the European perception of newly encountered inhabitants of the expanding *oikoumene* has not created a corresponding focus on how Europeans used the classics to create a new understanding of the physical world.[10] Similarly, historians of cartography have long appreciated the influence of the Hellenistic author, Ptolemy, on European

8 W. H. Tillinghast, 'The Geographical Knowledge of the Ancients Considered in Relation to the Discovery of America,' in *Narrative and Critical History of America*, ed. by Justin Winsor, vol. 1 (Boston: 1884, repr. New York, 1967).

9 John Elliott drew attention to the importance of the classics in geographical thought, but geography only played a small part in his argument. J. H. Elliott, *The Old World and the New*, 1st Canto edn (Cambridge: Cambridge University Press, 1970). The situation began to change in 1976 when Fredi Chiappelli edited an important two-volume work: *First Images of America: The Impact of the New World on the Old* (Berkeley: University of California Press, 1976). This work contained several essays relating explicitly to the impact of Renaissance ideas and Greek and Roman thought on the European discovery and interpretation of the nature of the Americas.

10 L. Hanke, *All the Peoples of the World Are Men*, James Ford Bell Lectures, 8 (Minnesota: University of Minnesota Press, 1970); L. Hanke, *The First Social Experiments in America: a Study in the Development of Spanish Indian Policy in the Sixteenth Century* (Cambridge, MA.: Harvard University Press, 1935); L. Hanke, *Aristotle and the American Indians: A Study in Race Prejudice in the Modern World* (Chicago: Henry Regnery, 1959); L. Hanke, *All Mankind Is One: A Study of the Disputation between Bartolomé De Las Casas and Juan Ginés De Sepúlveda in 1550 on the Intellectual and Religious Capacity of the American Indians*, 1st edn (DeKalb, IL: Northern Illinois University Press, 1994; first edn 1974); A. Pagden, *European Encounters with the New World* (New Haven and London: Yale University Press, 1993); A. Pagden, *Peoples and Empires: Europeans and the Rest of the World, from Antiquity to the Present*, paperback edn (London: Phoenix Press, 2002); A. Pagden, *The Fall of Natural Man: The American Indian and the Origins of Comparative Ethnology*, 1st edn, repr. with corrections and additions (Cambridge: Cambridge University Press, 1982). S. MacCormack, *On the Wings of Time: Rome, the Incas, Spain, and Peru* (Princeton: Princeton University Press, 2007); S. MacCormack, *Religion in the Andes: Vision and Imagination in Early Colonial Peru* (Princeton: Princeton University Press, 1991).

mapping in the early modern period, but they too paid little attention to the wider corpus of classical authors.[11] Until the quincentennial of Columbus's first trans-atlantic voyage, the influence of the classics on sixteenth-century geographical understanding remained rather neglected.

In the early 1990s, however, a number of books brought the mental world of the European geographical writers and explorers back into focus. Authors such as Valerie Flint and William and Carla Rahn Phillips began to draw attention to the way in which explorers' (in this case, Columbus's) preconceptions and imagination helped them to interpret the land which they encountered.[12] In doing so they drew attention to the role that the classics played as an educational and intellectual background for his discoveries and understanding of the world he encountered. Meanwhile, also around the celebrations of the five-hundredth anniversary of Columbus, Anthony Grafton published his volume accompanying an exhibition at the New York Public Library, entitled *New Worlds, Ancient Texts*.[13] In this book Grafton threw the attention back onto the classics, demonstrating ways in which the classics, rather than contradicting the new European discovery of the Americas, were part of the process of understanding them. His work, like the present book, placed the focus on textual contributions to the history of geography. At about the same time, Wolfgang Haase and Meyers Reinhold edited a volume on *The Classical Tradition and the Americas* which drew together a variety of articles dealing with the way European understanding of the Americas was filtered through their classical knowledge in the early modern period.[14] Since the publication of these two works, classicists, historians and geographers have paid far more attention to the role that the classics have played in geographical understanding in the sixteenth century, broadening the scope to other regions

11 See for example W. G. L. Randles, 'The Recovery of Ptolemy's Geography in Renaissance Italy and Its Impact in Spain and Portugal in the Period of the Discoveries,' in *Geography, Cartogaphy and Nautical Science in the Renaissance: the Impact of the Great Discoveries* (Aldershot: Ashgate, 2000); P. G. Dalché, *La géographie de Ptolémée en Occident* (Turnhout, Belgium: Brepols, 2009); O. A. W. Dilke and M. Dilke, 'The Adjustment of Ptolemaic Atlases to Feature the New World,' in *The Classical Tradition and the Americas: European Images of the Americas and the Classical Tradition,* ed. by Wolfgang Haase and Meyer Reinhold, vol. 1 (Berlin: 1994); S. Y. J. Edgerton, 'From Mental Matrix to "Mappamundi" to Christian Empire: The Heritage of Ptolemaic Cartography in the Renaissance,' in *Art and Cartography*, ed. by David Woodward (Chicago: Chicago University Press, 1987).

12 V. Flint, *The Imaginative Landscape of Christopher Columbus* (Princeton Princeton University Press, 1992); W. D. P. Junior and C. R. Philips, *The Worlds of Christopher Columbus* (Cambridge: Cambridge University Press, 1992).

13 A. Grafton, A. Shelford, and N. Siraisi, *New Worlds, Ancient Texts: The Power of Tradition and the Shock of Discovery* (Cambridge, Massachusetts: Harvard University Press, 1992).

14 W. Haase and M. Reinhold, *The Classical Tradition and the Americas.*

of the world rather than strictly focussing on the Americas.[15] While historians began to look backwards, classicists also took a longer view. Perhaps the most important book for the present work was written by the classicist, James Romm, also in 1992. In his *Edges of the Earth in Ancient Thought* he examines the way in which the boundaries of the *oikoumene* were conceived and defined in the classical world, demonstrating in his concluding chapter how the Greek and Roman ideas about the edges of the earth developed into a literary tradition which survived into the Renaissance.[16]

Historians of cartography have also begun to examine in much more detail the influence of the written word on cartography, and indeed the interplay between text and image, bringing to the fore the kind of ideas and theoretical traditions that formed the background to cartography. The collection of essays on cartography in the Renaissance in the multi-volume *History of Cartography* does much to demonstrate this change in emphasis in the history of cartography.[17] More recently, works such as Sean Roberts' *Printing a Mediterranean World: Florence, Constantinople and the Renaissance of Geography* and Surekha Davies's *Renaissance Ethnography and the Invention of the Human: New Worlds, Maps and Monsters*, have shown, in very different ways, how the dichotomy that was once demonstrated between the history of geography and the history of cartography has been overstated.[18] As the collection of essays on *Mapping Medieval Geographies* edited by Keith Lilley has demonstrated, the idea of mapping had far wider implications than simply cartographical representation, and, as I am arguing here, textual geography could provide a means

15 See the work of historians A. Hiatt, *Terra Incognita; Mapping the Antipodes before 1600* (London: British Library, 2008); N. Wey Gómez, *Tropics of Empire*. The classicist James Romm has also provided insights from a more classical angle: J. Romm, *Edges*; J. Romm, 'New World and 'Novos Orbes': Seneca in the Renaissance Debate over Ancient Knowledge of the Americas,' in *The Classical Tradition and the Americas,* 77–116; J. Romm, 'Mythes, cartes et histoire: L'utilisation par Ortelius de L'Atlantide dans le développement de la théorie de la dérive des continents,' in *Abraham Ortelius 1527–1598: Cartographe et Humaniste* (Turnhout: Brepols, 1998), 104–16. For the purpose of the present study, W. G. L. Randles and the geographer Denis Cosgrove have contributed particularly important discussions of the classical tradition and the change in perception from what Randles called a 'flat earth' to a terrestrial globe in the early modern period. But as their titles suggest their foci are on the epistemology and genealogies of the change in perception of the earth, rather than on the theoretical underpinnings which allowed for the conceptualisation, appropriation and reuse of space. D. E. Cosgrove, *Apollo's Eye*; W. G. L. Randles, *De la Terre plate*.

16 J. Romm, *Edges*.

17 D. Wodward ed., *Cartography in the European Renaissance*, *The History of Cartography,* vol. 3, parts i and ii (Chicago: University of Chicago Press, 2007).

18 S. Davies, *Renaissance Ethnography*; S. E. Roberts, *Printing a Mediterranean World: Florence, Constantinople, and the Renaissance of Geography* (Cambridge: Harvard University Press, 2013).

of representing the world.[19] Nonetheless there was a difference between cartography and textual geography. The majority of the works referred to in this book either had no or very few maps, or had maps which were not republished with the texts when the texts were published in different languages and contexts. This does not mean that maps did not have a widespread readership. Genevieve Carlton's work, *Worldly Consumers, the Demand for Maps in Renaissance Italy* has shown just how extensive a market there was for maps in sixteenth-century Italy, but the textual and cartographical traditions, although related, were not synonymous.[20] This book focusses on the textual. In doing so, it draws on the work of various literary scholars such as Tom Conley and Mary Fuller and John Gillies who have looked at the impact of voyages in print and the way in which the new geography became part of the Renaissance literary understanding; it is perhaps even more influenced by writers such as Frank Lestringant who have used literary tools to focus on single writers and thereby trace developments on geographical understanding in the sixteenth century.[21]

The last quarter of a century has seen an ever-increasing interest in connected histories, in Global History and the History of the Atlantic World. Early modern scholars having largely turned their back on the fifteenth- and sixteenth-century European voyages of exploration throughout most of the post-imperial era of the latter half of the twentieth century have begun to refocus on these expeditions. Colonial and postcolonial periods have long received attention, but these voyages constituted the pre-amble to the true colonial period. Writers such as Felipe Fernandez-Armesto, Joa Pau Rubies, Nicolas Wey-Gomez, Christine Johnson and Surekha Davies have helped clarify the intellectual environment in which these voyages took place and the reception of them back in Europe. The Age of Reconnaissance, as Parry once termed it, did not spring out of the blue, and nor were its discoveries immediately accepted back in Europe. Instead they were filtered through existing beliefs about climate, ethnicity and environment.[22]

19 K. D. Lilley, ed., *Mapping Medieval Geographies: Cartography and Geographical Thought in the Latin West and Beyond: 300–1600* (Cambridge: Cambridge University Press, 2014).

20 G. Carlton, *Worldly Consumers: The Demand for Maps in Renaissance Italy* (Chicago: The University of Chicago Press, 2015).

21 J. Gillies, *Shakespeare and the Geography of Difference* (Cambridge: Cambridge University Press, 1994); T. Conley, 'Early modern Literature and Cartography: an Overview,' in *Cartography in the European Renaissance: the History of Cartography*, ed. by David Woodward, vol. 3.1 (Chicago: University of Chicago Press, 2007); T. Conley, *The Self-Made Map: Cartographic Writing in early modern France* (Minneapolis, Minn.: University of Minnesota Press, 2010); M. C. Fuller, *Voyages in Print: English Travel to America, 1576–1624* (Cambridge: Cambridge University Press, 1995); F. Lestringant, *André Thevet: Cosmographe Des Derniers Valois* (Geneva: Droz, 1991); F. Lestringant, *Mapping*.

22 J. P. Rubiés, *Travel and Ethnology in the Renaissance: South India through European*

The most recent theme in scholarship on geographical thought has been to see the use of the classics in the fifteenth and sixteenth centuries as part of a long tradition which reaches back at least to the twelfth century.[23] There was certainly not a complete break with the medieval past, and indeed medieval texts such as *The Book of Sir John Mandeville* or the *Cosmography* of Peter D'Ailly provided some of the geographical imagination which enabled explorers such as Columbus and Vasco da Gama to break out of the mental horizons of the *oikoumene* and conceive of a connected world. This emphasis on continuity has, however, obscured the fact of change in both thought and knowledge in the fifteenth and sixteenth centuries, attributable in large part to the widespread reintroduction of Greek and Roman texts.[24] This book also shows how these ideas, once harnessed, helped fashion the idea of a globally-connected world, thereby tracing the incipient theories of globalisation back to the early modern period.

Numerous factors came together in the sixteenth century to refashion the world into a global, interconnected space. Explorers, merchants, philosophers, theologians, politicians and others all had to grapple in different ways with the changing conception of space, and presented very different attitudes to this space. Some of those at the forefront of exploration leave very little indication that they thought much about the bigger picture of how their voyages were

Eyes, 1250–1625 (Cambridge: Cambridge University Press, 2000); J. P. Rubies, 'New Worlds and Renaissance Ethnology,' in *Travellers and Cosmographers: Studies in the History of early modern Travel and Ethnology*, Variorum Collected Studies (Aldershot: Ashgate, 2007); C. R. Johnson, *The German Discovery of the World: Renaissance Encounters with the Strange and Marvelous* (Charlottesville, Va.: University of Virginia Press, 2008); F. Fernández-Armesto, *Before Columbus: Exploration and Colonisation from the Mediterranean to the Atlantic 1229–1492* (Basingstoke: Macmillan Education, 1987); J. H. Parry, *Age of Reconnaissance*, 2nd edn (London: Praeger, 1969, first edn 1963).

23 See for example, F. Fernández-Armesto, *Before Columbus*; P. G. Dalché, *L'espace géographique au Moyen Âge* (Florence: SISMEL – Edizioni del Galluzzo, 2013); P. G. Dalché, 'Principes et modes de la representation de l'espace géographique durant le haut moyen âge,' in *L'espace géographique Au Moyen Âge* (Florence: Edizioni del Galuzzo, 2013, first published 2003); K. D. Lilley, *Mapping Medieval Geographies*; A. Hiatt, *Terra Incognita*.

24 For an excellent work showing the both continuity and change with late medieval geography see M. O'Doherty, *The Indies and the Medieval West: Thought, Report, Imagination* (Turnhout: Brepols, 2013). O'Doherty demonstrates how works such as the *Book of Sir John Mandeville* began to develop a Christian appropriation of global space is a precursor to the connected world of the sixteenth century, but also how these works stopped short of a theoretically completely connected space so that, for instance, Lamoreans in t*he Book of Sir Mandeville* resist 'the appropriation of their space and consequent absorption into a global, interconnected, common Christian world' (p. 302).

reshaping the world. This book does not attempt to cover every angle of the European discovery and appropriation of the world in the sixteenth century. Rather, it focusses on the construction of a new geography by an academic elite, almost all of whom had a university education. I argue that humanists laid the geographical understanding that eventually formed the backbone of European concepts about the nature of a global world and its universal suitability for human endeavour. Humanists' writings are therefore at the core of this book. Even where individual travel narratives by those who had received less formal education, outside humanist circles, have been used, these are normally the ones filtered through the understanding of the educated. The great sixteenth-century travel compilations of Giovanni Battista Ramusio and Richard Hakluyt, for example, are vital to the present study. These authors assembled the narratives of individual travellers and presented them in edited versions which helped support their ideas of world geography. While many of the travel narratives they published were written by those who lacked humanist training or indeed a formal education, Ramusio and Hakluyt themselves were both university graduates, and published works on classical authors as well as on contemporary geography and exploration.[25] They used their humanist skills to edit and present their texts. They did not insert classical allusions into the narratives which they published, but they framed their organisation of the narratives and indeed their own discourses in terms of their classical learning. Both authors, for example, framed their discussions of their third volumes on the Americas in terms of Plato's *Timaeus* and Atlantis.[26] André Thevet is among the few examples of geographical writers studied in this work who did not have a high level of classical education and humanist training, but even he fits the pattern since he used Mathurin Héret, a medical doctor and classicist who had published translations of Greek texts, to provide classical references and present an educated veneer to his work.[27] Other humanist-trained writers such as Jean Bodin, Giovanni Botero and Louis le Roy who might nowadays be thought of more as philosophers than as geographers also find a place in this study. Their understanding of humans and their relationship to space and the geographical environment demonstrates clearly the way the understanding of the *oikoumene* and the relationship of humans to the globe was changing. They contributed to the geographical discourse of the era.

[25] L. V. Ryan, 'Richard Hakluyt's Voyage into Aristotle,' *The Sixteenth Century Journal*, 12, no. 3 (1981); G. B. Parks, 'Ramusio's Literary History,' *Studies in Philology*, 52 (1955).

[26] G. B. Ramusio, 'Discorso di messer Gio. Battista Ramusio sopra il terzo volume delle Navigazioni e viaggi nella parte del mondo nuovo,' *NV*, vol. 5; R. Hakluyt, 'The Epistle Dedicatorie in the Third Volume of the Second Edition', *PN*, vol. 1, lxxv.

[27] Héret's involvement in the publication of Thevet's *Singularitez de la France Antartique* was so extensive that he gained rights for all royalties for the work to be paid to him. F. Lestringant, *Mapping*, 62–63.

The preponderance of works used in this book are those known to have had an international readership (and frequently were widely translated) and were therefore undoubtedly ones that were part of an international community of knowledge. Perhaps the least widely read was Richard Hakluyt who, because he was at the forefront of English geographical education, has sometimes been seen to have been disproportionately important. His key work, *The Principal Navigations, Voyages, Traffiques and Discoveries of the English Nation* went through only two significant (and very different) editions in the sixteenth century and was never a best-seller. His writings, however, were known by such influential continental scholars as Gerard Mercator, Abraham Ortelius and Theodor de Bry so that, even if they were never translated in their entirety, they contributed to the European network of geographical ideas. The focus on the printed word has its limitations. Many of the most innovative Spanish cosmographers find no place in this study because their works remained unprinted. Likewise John Dee, sometimes called the most important geographer of Tudor England, also receives little attention since the majority of his geographical writings were unpublished.[28] Such geographers were undoubtedly influential. John Dee's networks reached half across the continent and his teaching informed many.[29] This book, however, focusses on those humanists who used print to disseminate their ideas and create a public understanding of the nature and structure of the world. They were the people who framed the new geographical paradigm. Although individual sixteenth-century authors put their own interpretations onto classical theories, and emphasised the importance of different classical writers in their works, this book shows that the overall effect across Europe of the humanist reframing of geographical knowledge was the creation of a new geography which over the course of the sixteenth century came to emphasise the global nature of the knowable, inhabitable world.

The book is divided into six substantive chapters. The first examines the way a concatenation of events – the European reconnaissance, the Renaissance, and the sixteenth-century reformations – were fundamental in creating new, overarching theories about the shape and nature of the world. It begins by introducing the concepts of descriptive and mathematical geography and explores why descriptive geography was so important in the reconceptualisation of space in the sixteenth century. It then examines the theoretical distinctions between geography and cosmography in the works of the Greek writers, and demonstrates why the concept of the *oikoumene* was a fundamental aspect of the original idea of geography. The focus of this chapter is

[28] E. G. R. Taylor, *Tudor Geography 1485–1583* (London: Methuen, 1930).

[29] W. Sherman, *John Dee: The Politics of Reading and Writing in the English Renaissance* (Amherst: University of Massachussetts Press, 1995).

the way in which Europeans humanists, writing for a literate public, created a new theory of the earth which was nonetheless heavily dependent on classical understanding. The idea of an entirely habitable world was a theoretical concept, at times supported by empirical evidence, at times almost undermined by it. Just as the idea of the limited *oikoumene* had been spread by the writers in the ancient world, so the idea of the unbounded *oikoumene* with all the implications it held for Europeans in terms of global trade and colonisation was one interpreted and spread by the written word in the sixteenth century.

The second chapter argues that the search for validation in the classical authorities by humanist-trained geographical writers and travellers in the sixteenth century was so great that the early modern authors revived classical geographical opinions from every genre of writing in order to develop their new concept of global human geography. The idea that humans could inhabit every region of the world was not common among the traditional corpus of classical material but could be discerned in certain classical authors. It shows how the search for authority in the classics changed which authors were considered authorities in geographical matters.

The third chapter then addresses the way in which the traditional corpus of classical authors developed and maintained the idea that the limits of the *oikoumene* were marked by a frigid zone to the north and torrid zone to the south in which the climates were respectively too cold and too hot to sustain human life, and addresses the way in which these zones were transformed in the sixteenth-century humanist construction of space. It is proposed that rather than acting as uninhabitable zones they came to be seen as playing a key role in enabling different parts of the world to communicate with one another. A classical geography of limit and transgression was reimagined as one of connectedness and communication. It looks at the reevaluation of authority which still saw the Greeks and Romans as the arbiters of knowledge, but enabled non-canonical classical authors to become important in revising attitudes to the climatic zones which had once been thought to define the edges of the *oikoumene*. In the new geography, the arctic and equatorial regions were believed to play key roles in allowing the different parts of the world to interact with one another. It is argued here that without this shift in mind-set about the frigid and torrid zones, it would have been impossible for Europeans to begin to think of the world as a single, exploitable, inhabitable global unit.

The fourth chapter interrogates Ocean, the other major conceptual confine of the classical *oikoumene* and examines the way in which Ocean was redefined in the fifteenth and sixteenth centuries. It begins with an overview of the three different theories of Ocean which existed in the classical world. It then looks at how the understanding of Ocean was altered in a Christian Renaissance context. It proposes that despite the overarching emphasis on Ocean as a confine throughout classical and medieval times, there was much in the enduring idea of

Ocean which could enable sixteenth-century writers and travellers to reenvisage it as a means of communication. Far from limiting human inhabitation, Ocean provided the means for humans to have global interchange and exchange – it became the pathways of the *oikoumene*. The chapter then examines how Ocean was also thought to define the nature and extent of the landmasses because it surrounded all the different parts of the world. Owing to their belief in the classical Ocean, humanists came to believe that the Americas must be a continent in their own right, entirely separated from Eurasia and Africa by Ocean, but also connected with them for trade purposes by the same water that separated them. The chapter concludes by showing how the Christian and classical contexts worked together in the reenvisioning of the nature of ocean as a means of connecting all parts of the world. To do so, however, required humanists not only to divest it of earlier religious connotations but also to fit it into their new conception of a world designed by God for humans to have full dominion over. Rather than being the place where nature no longer worked according to law, it had to be seen as part of nature and was expressly created by God to link all parts of the earth.

The fifth chapter continues this examination of the way in which humanists used classical theory to try to understand the shape of the physical landmasses which were now brought into communication with one another by Ocean. It examines how ideas of symmetry, balance and proportionality, which were all-pervasive in the way in which Greeks and Romans understood and rationalised the nature and structure of the world, allowed humanist geographers, proto-ethnographers and travellers to assimilate the discoveries and organise them into a comprehensible format. As this chapter demonstrates, ideas of balance, symmetry and proportionality were particularly evident in the way in which geographers discussed the proportions and arrangements of the continents. In the sixteenth century empirical evidence had proved that the physical arrangement of the land and sea was not such as had been conceived in earlier periods; instead nobody knew the full extent of continents, their precise locations or their exact relationship with the seas. It argues that Renaissance geographers inferred the structure of the world from their understanding that a fundamental aspect of the laws of nature was a need to maintain all aspects of the created world in balance and equilibrium. Such equilibrium required the world to be made in opposites or counterparts with the balance maintained between these pairs. A balance could be inferred between land and sea, or between one landmass and another, which allowed Renaissance geographers to hypothesise about the physical structure of the world they were now discovering. The chapter begins by showing how the Greeks constructed a geography based on polarity between opposites or analogy between counterparts and how this created a way of understanding the physical geography of the world that endured until the sixteenth century. The chapter then continues with an examination of the terminology applied to the western continent, looking particularly at the meanings of the phrases *novus mundus* and *novus orbis* which were frequently applied to the Americas. It contends that these terms implied an extraordinary degree of separation for the western continent, at the very time that geographers were beginning to argue that the world was a single unit. Despite the superficial separation indicated by such terminology, however, the underlying principle of a world acting as a unit was maintained. The chapter proposes that at the core of the ideas about the nature and proportion of the world was a sense that this was a world made by God for human domination.

The sixth chapter focusses particularly on the perceived relationship between the newly discovered western landmass and the Old World. It looks at the way in which sixteenth-century descriptive geographers used Platonic and Neoplatonic ideas, and particularly the idea of *harmonia mundi* (harmony of the world), to understand the nature of the world in the sixteenth century with particular reference to the New World. Throughout the sixteenth century European geographers and explorers sought justification for the European reconnaissance in the works of Plato, and particularly in one of his more abstract works, the *Timaeus*. They connected the idea of the best possible world which Plato had referred to in the *Timaeus* with the idea of a world made by God for humans. Although Plato was concerned with the world of ideas rather than the changing physical world, his ideas of proportionality and hidden affinities reinterpreted the ideas of *harmonia mundi*, Atlantis, and the 'best possible world' to understand the New World, and its relationship to the Old one. The chapter shows how a geography based on ideas of *harmonia mundi* and hidden affinities became an important way of understanding the relationship between the Americas and Europe in the sixteenth century... The New World was on one level entirely distinct, with a new order of animals, products and peoples, but on another level it was the complementary counterpart to the Old World and all parts of it were knowable to the one common humanity.

The final chapter examines how, in the ongoing process of European exploration and expansion in the sixteenth century, the Europeans retained a geography of centre and periphery despite theoretically believing that the entire world was made for human dominion and exchange. The ideas of balance and proportionality had given the lands a theoretical shape, but within those outlines of land and sea the newly encountered territories reached out, vast and un-subjugated by Europeans. Europeans did not know the nature and topography of all the lands that were now believed to be part of the *oikoumene*, nor did they know the character of their inhabitants. While there were a few European forays into the interior of Africa, Asia and the Americas, these were insufficient to provide much information about these lands and their inhabitants. As a result, throughout much of the sixteenth century there remained the idea that there was a European norm at the heart of the world, and zones of inhabitation surrounding this, so that in the regions furthest from Europe all was monstrous and abnormal. This theory of centre and periphery was underpinned by classical humoral theory and the Greek and Roman idea that inhabitants' characteristics reflected the characteristics of their environment, so that the more extreme the environment the more extreme its population. Europe, at the centre of the world, had the most equitable environment and population and the least exotic resources. Further afield all of these became proportionately more extreme so that in the reaches of the world furthest from Europe there were monstrous peoples: Amazons, cannibals, sciapods and dog-heads among others. The presumed existence of such realms of monstrous races, derived almost unaltered from the classical world, created mental horizons to the world, in a sense taking the place of the torrid and frigid zones – but, as this chapter will demonstrate, such horizons were movable. As exploration took place, and people discovered that the regions on the peripheries of their knowledge were not inevitably inhabited by monstrous peoples, the horizons receded. The chapter proposes that the general disappearance of uniform locations of monstrous peoples from the humanist texts by the end of the century allowed for the final acceptance that the globe and the *oikoumene* were synonymous space.

The conclusion considers how the general belief in a world subject to human dominion, and the redeployment of classical theories in a Christian context where God made all parts of the world to be connected, both served to create a geographical understanding that paved the way for globalisation. The adaptation of geography in the sixteenth century required a total reinterpretation of the nature of the world, from a confined and restricted space to a space of expansion and exchange, but Greek and Roman geographical thought and writings, which had been integral to the creation of the geography of limit, was no less important in creating the new geography of globalised communication.

1

Renaissance, Discovery, and the Written Word: Influences on Sixteenth-Century Geography

In the fifteenth century, Portuguese ships edged southwards along the African coast, crossing the equator and circumnavigating the Cape of Good Hope, eventually arriving in India. In 1492, notoriously, Columbus and his crew sailed across the Atlantic Ocean hoping to find an easier, westwards route to the Orient, instead finding outposts of a New World. With these two initiatives, the Iberians irrevocably destroyed the framework of the world within which medieval and classical western European geographical theories had evolved. A geography of boundaries and limits began to be transformed into one of connectivity.

Although there were always writers who had questioned the widely believed classical theories about the edges of the earth, the Spanish and Portuguese voyages of exploration round Africa and the Americas in the last quarter of the fifteenth century were the chief catalyst for the creation of a new widespread approach to world geography.[1] In the ensuing century a tussle took place to create a new understanding of the world that took account of a vast wealth of information about previously unimagined peoples and places, while not entirely countering the received wisdom of the ancients. By the end of the sixteenth century a new

[1] On the questioning of traditional geography see in particular D. Abulafia, *The Discovery of Mankind: Atlantic Encounters in the Age of Columbus* (New Haven and London: Yale University Press, 2008); N. Wey Gómez, *Tropics of Empire*; A. Hiatt, *Terra Incognita*. As recent scholarship has pointed out, medieval geographical thought has often been written about in a teleological way, looking at an evolution of thought culminating in the age of discovery, and not considering the wider context within which individual medieval geographical authors were writing. This book is not intending to suggest that there was one uniform medieval geography. Works such as N. Lozovsky, *The Earth Is Our Book: Geographical Knowledge in the Latin West ca. 400–1000* (Ann Arbor: University of Michigan Press, 2000) and K. D. Lilley, *Mapping Medieval Geographies: Cartography and Geographical Thought in the Latin West and Beyond: 300–1600* have undermined such approaches. It does, however, propose that the Iberian voyages necessitated a widespread recalculation of geographical thought.

geographical framework became widely accepted throughout western Europe. It was one which remained firmly founded on classical theories, but which also argued that the entire world was made by God for human inhabitation and that Ocean was designed to connect all parts of the globe. The *oikoumene* or inhabitable world now embraced the sphere.

This chapter will offer a brief examination of the way in which the coinciding events of the European explorations, the Renaissance, and the sixteenth-century reformations were fundamental in creating new, overarching theories about the shape and nature of the world. Starting with definitions of descriptive and mathematical geography, the chapter explores why descriptive geography, which has a human focus, was so important in the reconceptualisation of space in the sixteenth century. It examines the theoretical distinctions between geography and cosmography in the Greek writers, and demonstrates why the concept of the bounded *oikoumene* was a fundamental aspect of the original Greek idea of geography. The chapter then offers a brief discussion of the limits of the *oikoumene*, arguing that there was a time-lag between discovering that these limits extended further than previously believed, and the acceptance of these discoveries and their geographical implications into western European society. As the chapter argues, the increasing popularity of print in the sixteenth century, which helped new ideas to be more rapidly disseminated, somewhat mitigated this time-lag. The chapter continues by demonstrating how the majority of these authors focussed on textual geography and either ignored or downplayed the importance of cartography in the creation of this printed geography. Just as authors in the ancient world had spread the idea of the limited *oikoumene*, so writers interpreted and disseminated the idea of the unbounded *oikoumene* in the sixteenth century.

For the purposes of this work, I am concerned with descriptive rather than mathematical geography, and particularly descriptive geography produced by scholarly elites. By descriptive geography I mean 'the natural and political description of other lands, both exotic and well known'.[2] In terms of both the classical world and the sixteenth century it often involved a verbal and literary approach that used quite a different methodology from the more scientific, mathematical geography.[3] It was intended to be empirical but, as this book will demonstrate, theoretical interpretations often influenced the empirical observations. Even today geography is a discipline that is difficult to define, straddling Arts and Science faculties in universities and involving anything related to the study of the earth, from its physical make-up and the creation of its landforms to humans and their relationship with their natural environments. In the sixteenth

2 L. B. Cormack, *Charting an Empire: Geography at the English Universities 1580–1620* (Chicago: Chicago University Press, 1997), 15.

3 D. Dueck, & K. Brodersen, ' Introduction,' in *Geography in Classical Antiquity*, Key Themes in Ancient History (Cambridge: Cambridge University Press, 2012), 5.

century it was no easier to describe. Philosophical discourses, travel reports, religious dialogues, navigational manuals, astronomical and mathematical treatises were among the many subjects which contributed to the sixteenth-century understanding of geography.

The ancient Greeks coined the word *geographia*, but the term and to a certain extent the concept had fallen somewhat into abeyance by the sixteenth century.[4] More often writers thought in terms either of chorography, involving the study of particular regions, or of cosmography which at least until the sixteenth century involved the study of the entire *kosmos* or universe and not merely the earth. This cosmographical influence played a significant role in geographical understanding because it prescribed to a large extent how much of the earth could be directly known to humans through the senses rather than through reason. Cosmography dealt with the regions of the earth beyond the inhabitable region and therefore, by definition, the world of reason rather than of the senses, whereas geography was concerned with the *oikoumene* – the knowable, visible world. As Strabo said, 'Geography seeks to mark off the sector of the temperate zones being inhabited by us from the other'; and even Ptolemy, who was deeply interested in the relationship of the world to the universe, declared that 'Geography is the representation of the diagram of the whole known part of the earth together with the portions that have generally been combined with it', laying emphasis on the known world which he then placed it into the wider context of the spheres.[5]

In theory, therefore, geography was completely distinct from cosmography. The two terms, however, became ever more closely entangled over the course of the fifteenth and sixteenth centuries as the *oikoumene* expanded to cover more and more of the sphere. Cosmographers ostensibly 'discussed the evidence for the spherical form of the elements; the borders of the habitable world; the locations of climate zones; the relationship between the spheres of earth and water; and the existence of antipodes, people dwelling on the opposite side of the earth'.[6] But as the *oikoumene* expanded in the aftermath of the Portuguese and Spanish voyages across the ocean and into the torrid zone, many of these

4 The term *geographia* seems to have been coined by Eratosthenes in the third century BC (preserved in Strabo) although the earliest extant citation is from the first century BC. See Eratosthenes and D. W. Roller, *Eratosthenes' Geography* (Princeton, NJ: Princeton University Press, 2010), 1. As Dalché has recently shown, the loss of a concept of geography in the Middle Ages can be overstated (P. G. Dalché, *L'espace Géographique*); nonetheless, both the terminology and the idea of geography became much more widespread in the sixteenth century.

5 ζητεῖ γὰρ ἡ γεωγραφία τῆς ἑτέρας τῶν εὐκράτων ἀφορίσαι τὸ οἰκούμενον ὑφ'ἡμῶν τμῆμα (Strabo, 2.3.1) Ἡ γεωγραφία μίμησις ἐστι διαγραφῆς τοῦ κατειλημμένου τῆς γῆς μέρους ὅλου μετὰ τῶν ὡς ἐπίπαν αὐτῷ συνημμένων (Ptoelmy, *Geographia*, 1.1).

6 Klaus A. Vogel, 'Cosmography,' in *The Cambridge History of Science*, ed. by

subjects became part of the inhabitable world, and so the concern of geography. As a result, the terms and concepts were continuously evolving throughout the sixteenth century, and for much of the period the distinction between the two remained murky. Indeed many of the more famous geographical works of the sixteenth century were titled cosmographies although they were much more geographies than cosmographies. The German geographer and theologian, Sebastian Münster, for instance, called his study of the earth and its inhabitants a cosmography even though it focussed solely on the known, inhabited world.[7] As late as 1575 André Thevet called his work a *Cosmographie Universelle* even though he paid very little attention to the relationship of the earth to the heavens and was rather more interested in providing descriptions of lands, flora, fauna and peoples.[8] By the end of the sixteenth century, however, exploration pushed the boundaries of the world back sufficiently for the whole earth to be perceived as inhabitable, and cosmography and geography became entirely separate subjects. The term cosmography came to be applied to the study of the heavens. Johannes Kepler, for instance, famously used the words *Mysterium Cosmographicum* to discuss the relationships between the planets. As the knowable world of the human *oikoumene* came to coincide with the sphere, the study of the inhabitants, features and physical make-up of the earth, which is the subject of this book, became exclusively the study of geography. This clarity of terms evolved over the course of the sixteenth century, as the ideas of cosmography and geography came under scrutiny in the late European voyages of exploration and colonisation.

Meanwhile the discipline of geography itself evolved and divided. Mathematical geography and cartography, which had been a rather neglected subject in western Europe until the late fifteenth and sixteenth centuries, became an important sub-discipline.[9] Ptolemaic mathematical geography and the need to create practical navigational advice and to provide precise locations on a longitudinal and latitudinal grid spurred geographers throughout Europe to turn their attention to mathematics. Many of the most famous geographers of the late fifteenth and sixteenth centuries such as Peter Apian, Pedro Nunes, Gemma Frisius and Oronce

Katherine Park and Lorraine Daston (Cambridge: Cambridge University Press, 2006), 470.

7 S. Münster, *Cosmographiae Universalis Lib Vi* (Basel: Henric Petri, 1544).

8 A. Thevet, *Cosmographie Universelle illustrée de diverses figures des choses plus remarquables veuës par l'Auteur, et incogneuës de noz Anciens & Modernes* (Paris: Pierre L'Huillier, 1575).

9 As Harvey says in the introduction to medieval maps in the *History of Cartography* 'few maps were made in medieval Europe': P. D. A. Harvey, 'Medieval Maps: An Introduction', in *The History of Cartography*, ed. by J. B. Harley and D. Woodward, vol. 1 (Chicago: University of Chicago Press, 1987), 283. There were a few such as Roger Bacon and Robert Grosseteste (discussed below) who worked on mathematical geography, but they were aberrations rather than the norm.

Fine focussed on mathematical geography. Their work became incorporated into sixteenth-century educational systems.[10] They wrote, however, a very different genre of geography from that of the people who form the core of this study. There were few writers like Mercator whose works crossed the mathematical and descriptive divide; for the most part there was a distinct division between those who adopted a more mathematical approach and those who used the written word to describe the attributes of peoples, animals and places. These descriptive writers were an eclectic group of people whose only shared interest was in the visible and discernible aspects of the new lands and territories. Many of them would not, indeed, have described themselves as geographers. They were virtually all humanists drawn from across Europe with differing access to new geographical material, but with a common educational background in classical writings. Men such as Peter Martyr and Jose de Acosta in Spain, João de Barros and Antonio Galvão in Portugal, Sebastian Münster in Germany, Giovanni Battista Ramusio in Venice, André Thevet in France, Gerard Mercator and Theodore de Bry in the Low Countries, and Richard Hakluyt in England drew on vastly different contemporary sources of information. But they all drew on classical geographical theories of balance and symmetry, proportionality and harmony to fill in the void in knowledge which was apparent in the aftermath of the Iberian voyages round Africa and across the Atlantic and later the Pacific. They were interested in the visible and inhabitable world, but this does not mean that they had empirical information on all the parts of the earth that were now deemed to be inhabitable. Together, humanist geographers created a new written, descriptive, geography which was rooted in the rediscovery of theoretical ideas from the Greeks and Romans, made newly familiar to western Europeans through the Renaissance emphasis on the *studia humanitatis*. Their works had a widespread and often popular readership.

Changing the perception of the *oikoumene* from an enclosed to an all-encompassing space was not an easy task. In destroying the idea of the bounded *oikoumene*, the sixteenth century saw the total overthrow of one of the most closely held tenets of European geography. Although the earth was known to be a sphere, for two thousand years until the Iberian expansion most western Europeans had accepted the Greek theoretical notion that the habitable world covered only a small portion of the earth's surface culminating in the frigid and torrid zones and the encircling impassable Ocean.[11]

Throughout classical and medieval times, the distinction between the *oikoumene* and the sphere was important. Indeed the focus among pre-sixteenth

[10] L. B. Cormack, *Charting an Empire*, 108–9; Isabelle Pantin, 'Oronce Fine's Role as Royal Lecturer', in *The Worlds of Oronce Fine: Mathematics, Instruments and Print in Renaissance France*, ed. by Alexander Marr (Donington: Shaun Tyas, 2009).

[11] J. Romm, *Edges*.

century writers on the *oikoumene* may be partly why a myth has developed that before the period of Columbus Europeans thought the world was flat. This was simply not the case. The idea that the majority of medieval writers believed in a flat earth is a nineteenth-century fabrication superimposed onto medieval thought.[12] In both classical and medieval times the world was commonly thought of as a sphere. Pythagoras, living in Greece in the sixth century BC, was the first to propose that the earth was spherical. His arguments were not founded on physical evidence, however, but on the idea that the sphere was the most perfect form. Because he believed in the perfection of the earth, he thought that the earth must hold such a form. It was not until the fourth century BC that the sphericity of the earth became more generally accepted when Aristotle presented the earliest clearly reasoned arguments for the earth's sphericity.[13] He argued that

> the form [of the earth] must necessarily be spherical since each of the constituent parts has weight until [it gets near to] the centre and the smaller part being thrust by the larger part is unable to expand, but rather is compressed and converges one with another (i.e. part by part) until it comes to the centre.[14]

He continued:

> either therefore [the earth] is spherical, or it is at least spherical by nature. And it is necessary to say that everything is such as it wants to be by nature and as it is, but not such as it is by force and contrary to nature.[15]

Thereafter the concept of a spherical earth was never abandoned in western Europe, but the majority of ancient and medieval maps which showed a circular landmass surrounded by Ocean had the effect of reinforcing the idea that the *oikoumene* was distinct from the rest of the sphere.[16]

12 J. Russell, *Inventing the Flat Earth: Columbus and His Modern Historians* (New York: Praeger, 1991); L. B. Cormack, 'Flat Earth or Round Sphere: Misconceptions of the Earth and the Fifteenth-Century Transformation of the World,' *Ecumene*, 1 (1994).

13 Herodotus, writing in the fifth century BC, derided those who depicted the world (γῆς περιόδους) as perfectly circular as if drawn by a pair of compasses and surrounded by Ocean, but it is unclear about what he believed was the precise shape of the world. Herodotus, 4.36.

14 Σχῆμα δ' ἔχειν σφαιροειδὲς ἀναγκαῖον αὐτήν· ἕκαστον γὰρ τῶν μορίων βάρος ἔχει μέχρι πρὸς τὸ μέσον, καὶ τὸ ἔλαττον ὑπὸ τοῦ μείζονος ὠθούμενον οὐχ οἷόν τε κυμαίνειν, ἀλλὰ συμπιέζεσθαι μᾶλλον καὶ συγχωρεῖν ἕτερον ἑτέρῳ, ἕως ἂν ἔλθῃ ἐπὶ τὸ μέσον Aristotle, *De Caelo*, 2.14; 297a 10–14.

15 ἢ οὖν ἐστι σφαιροειδής, ἢ φύσει γε σφαιροειδής. δεῖ δ' ἕκαστον λέγειν τοιοῦτον εἶναι ὃ φύσει βούλεται εἶναι καὶ ὑπάρχειν, ἀλλὰ μὴ ὃ βίᾳ καὶ παρὰ φύσιν. Aristotle, *De Caelo*, 2.14; 297b 20–24. He then adduced arguments from the senses including the perception of eclipses and of the horizon.

16 Agrippa's map no longer exists and there are many arguments about it, but it seems to have been in the tradition of earlier maps in depicting the *oikoumene*. J. J. Tierney, 'The

Medieval *T-O* maps, for instance, give a clear visual representation of this emphasis on *oikoumene*; rather than trying to show the whole globe (an impossibility without a good mathematical system such as Mercator's for projecting the three-dimensional sphere onto a plane surface), they focussed on the world as it was known (and knowable) to humans. In so doing they created a visual representation which ignored all the rest of the globe outside the inhabitable world. As Alfred Hiatt has shown, a *terra incognita* lying across the southern equator was certainly part of the classical and even medieval imagination.[17] There were even various classical terms for inhabitants of different zones and hemispheres, including *synoikoi, perioikoi, antoikoi, antichthones, antomoi* and *antipodes*, but though part of the geographical discourse they were neither the most relevant nor the most common aspect of geography.[18] Instead the main trends were to perpetuate the idea of the importance of the *oikoumene.*

The *oikoumene* was not always an entirely static concept in ancient geography, but by the end of the Roman period it came to have fairly defined margins. Homer, writing in the eight century BC, at a time when the Greeks still believed the surface of the earth was flat, had a very different idea of it from Herodotus, writing in the fourth century BC who in turn had a different notion of it from Martianus Capella. Yet even Martianus, writing more than a millennium after Homer first recorded the idea, still thought that there was only a portion of the earth about which humans could have eye-witness knowledge. Once the Greeks and Romans came to believe in the sphericity of the earth they simply superimposed the concept of a bounded *oikoumene* onto the landmasses of the globe, and the location and extent of the *oikoumene* altered as Greeks and Romans learnt more about the physical geography of their world. Marking the absolute limits of the knowable world were expanses of almost infinite seas and pathless deserts, of intolerable blistering heat and of freezing, unsurvivable cold.[19] No human could enter such a region and survive – the vastness of the Ocean or the extremity of the climate simply could not support life. Therefore, despite repeated references throughout ancient and medieval times to lands and peoples living in another separate *oikoumene*, few Europeans were prepared to test the theoretical climatic boundaries of their own *oikoumene* until the fifteenth century when enough scholars, and a few travellers, had given sufficient indication that the *oikoumene* might extend further than the ancient writers had suggested. Even then, few were willing to speculate about whether it might be possible to cross into the realms of

Map of Agrippa,' Proceedings of the Royal Irish Academy, Section C; Archaeology, Celtic Studies, History, Linguistics, Literature, 63 (1962), 151–66.

17 A. Hiatt, *Terra Incognita.*

18 *Ibid.*, 19; J. L. Berggren and J. Evans, *Geminos's Introduction and Study of a Hellenistic Survey of Astronomy* (Princeton: Princeton University Press, 2006), 210.

19 J. Romm, *Edges.*

the *antichthones* or *antipodes*. The *oikoumene* had become a defined, distinct unit on the surface of the sphere. These theoretical limits deterred human exploration considerably. By the late Middle Ages, a few individuals such as Roger Bacon, Sacrobosco, John Mandeville, and Marco Polo had written about the possibility of travel beyond the old *oikoumene*, but until the fifteenth-century Portuguese voyages, the northern temperate zone and Ocean had more or less defined the boundaries of the medieval expansion of Europe.[20]

By definition, any description of the world beyond the *oikoumene* was founded therefore on theoretical and mathematical ideas rather than on eye-witness reports since as soon as humans survived in a region and were able to provide reports of that area, it was clearly inhabitable and therefore technically part of the *oikoumene*. As recent scholarship has emphasised, the Iberian voyages did not come out of the blue, but it was not until Gil Eanes crossed the Tropic of Cancer in 1434, and Columbus the Atlantic in 1492, that the Iberians definitively proved that these confines had only been conceptual, and that previously unimagined, inhabitable lands lay beyond the margins in all directions.[21] These discoveries did not, however, immediately create an expectation that the whole world was habitable. Rather they ushered in a century where humans were testing the limits of the *oikoumene* geographically. For much of the sixteenth century the idea of an *oikoumene* remained a concept. As late as 1600 John Pory wrote in the introduction to his translation of Leo Africanus's *Description of Africa* that

> the greater part is comprized betweene the saide Tropiques vnder the Torrid or burnt Zone, for which onely cause the ancient writers supposed it to be vnhabitable and desert in so many places: which indeed is much rather to be ascribed to the waste wildernesse, the barren and sandie soile, and the scarcitie of waters and fountaines.[22]

Pory had been educated at Cambridge, but was also keenly interested in travel and a close associate of Richard Hakluyt at whose urging he translated the work. He is the epitome of the humanists who were creating the new geography –

20 On Bacon, Grosseteste et al., see chapter 3 below.

21 N. Wey Gómez, *Tropics of Empire*; S. Davies, *Renaissance Ethnography*. Patricia Seed argues that Gil Eanes's crossing was not so much a conceptual achievement but a feat of navigation. The winds, seas and sands around Cape Bojador are difficult to navigate. Crossing it required revolutions in navigational learning and there were at least twelve failed attempts. Seed's argument is undoubtedly true, but the difficulty in crossing merely reinforced the conceptual boundaries of the *oikoumene* which needed to be transformed. P. Seed, 'Navigating the Mid-Atlantic, or What Gil Eanes Achieved', in *The Atlantic in Global History, 1500–2000*, ed. by Jorge Cañizares-Esguerra and Erik R Seema (London: Pearson, 2007).

22 A. Leo and J. Pory, *A Geographical Historie of Africa*, in Early English Books, 1475–1640, 262:12 (George Bishop: London, 1600), 2.

fascinated by the travel reports and new knowledge, but deeply immersed in the classics and still influenced by the ancient descriptions.[23] It was a century in which the geographical margins of the inhabitable world were steadily pushed back, as explorers and theorists tested the potential for human dominion over the whole earth, but there was no certainty that there were 'no lands uninhabitable, no Sea innavigable', as Robert Thorne an Henrician merchant and theorist optimistically declared in 1527.[24]

The most popular geographical books printed around 1500 gave the appearance of dealing with the whole world, but most of them showed little or no interest in the new discoveries.[25] Jacques Signot, for instance, wrote the most popular French world geography of the sixteenth century but neglected to include the New World.[26] Likewise Johann Boemus's international bestseller, *Omnium Gentium Mores, Leges et Ritus*, first published in 1520 and then translated and published all across Europe from the Czech Republic to England resolutely ignored the New World while dealing with Europe Asia and Africa.[27] Their subject was the essentially unaltered geographical legacy of the ancient world, with the tripartite, narrowly defined *oikoumene*. Works on the New World were coming into print, but they had a surprisingly limited circulation and were generally published in

23 Pory did not simply translate Leo Africanus but augmented it with information from other ancient and modern writers. On Pory and his relationship with Hakluyt, his translation and augmentation see W. S. Powell, *John Pory 1572–1636: the Life and Letters of a Man of Many Parts* (Chapel Hill: The University of North Carolina Press, 1977), 5–15.

24 R. Thorne, 'The Booke Made by the Right Worshipful M. Robert Thorne in the Yeere 1527 in Sivil to Dr. Ley, Lord Ambassadour for King Henry the Eight, to Charles the Emperour, Being an Information of the Parts of the World, Discovered by Him and the King of Portingal: And Also of the Way to the Moluccas by the North', *PN*, vol. 2, 178.

25 Perhaps the most typical is J. Boemus, *Omnium Gentium Mores Leges Et Ritus Ex Multis Clarissimia Rerum Scriptoribus, a I.B.A. Sacerdote Tentenicae Militiae Devoto Nuper Collectos: & in Libros Tris Distinctos Aphricam, Asiam, Europam, Optime Lector Lege* (Augsburg: Augustae Vindelicorum, 1520). This was widely read and republished as late as 1582 in Leiden, but ignored the New World entirely. There were later editions as well, but these were augmented with excerpts from literature about the New World. The work was originally published as *Omnium gentium mores, leges et ritus*. It was translated into seven languages in the course of the sixteenth century, and was printed all over Europe (A. Grafton, A. Shelford, and N. Siraisi, *New Worlds*, 13). See also R. Hirsch, 'Printed Reports on the Early Discoveries and Their Reception', in *First Images of America: The Impact of the New World on the Old*, ed. by Fredi Chiappelli, vol. 2 (Berkeley: University of California Press, 1976), 537–58.

26 It was first published in 1539 and was republished as late as 1590. J. Signot, *La Division du monde, contenant la déclaration des provinces et régions d'Asie, Europe et Aphricque* (Paris: Lotrian, 1539); G. Atkinson, *Nouvelles Horizons*, 24.

27 The book was translated from Latin into French, Italian, Czech, German English and Spanish by the end of the century and went through multiple editions, proving its popularity.

separate publications, most often in Italy or Germany.[28] It was not until the 1560s that it became unacceptable to publish a new book on world geography in France that did not include the New World, and England was even slower to recognise the importance of the Americas. It was not until 1553 that the first book in English on the New World was published. Nonetheless, though some countries lagged behind others in the dissemination of knowledge, by the mid-sixteenth century western European geographers became eager for news of any non-Christian country, aware that numerous gaps in geographical knowledge remained to be filled. The world had, in a sense, lost its cohesion. It was accepted that it was a sphere, but within that sphere no-one now knew the extent of the continents, the pattern of land and sea, or the relationship of humans to the vast expanses of newly discovered territory.

Over the course of the late fifteenth and sixteenth centuries Europeans explored enough of the new coastlines to provide a new geography, but there was a time-lag of approximately a hundred years after the breaking of the old boundaries before a new conceptualisation of the world as consisting of wholly inhabitable counterpart landmasses of Eurasia and Africa in the eastern hemisphere and the Americas in the western one became universally accepted in the literature. Although the printing press helped to spread empirical information and ideas about the whole world more widely and rapidly, they still took time to receive widespread acceptance. The discoveries of previously unknown uninhabited landmasses and transoceanic sea-routes involved a complete paradigm shift. As a result it is not surprising that the new ideas about the shape and structure of the world took time to percolate. As the communication theorist Everett Rogers showed in his work on diffusion of innovation, new innovations require adopters, communication channels, time and a social system in order to become accepted. According to Rogers, most adopters rate an innovation on factors such as how compatible it is with the previous system, how testable it is, how observable it is, its complexity and its relative advantage.[29] Some adopters are quicker to take on board new concepts while others need more convincing, but all take into account these matters. In terms of all these factors, the idea of wholly inhabitable globe was a testing concept, and it is hardly surprising that it took nearly a century to become widely accepted in Europe. The humanists, with their training in classical geography, were strangely in a better position to adapt the new discoveries to the previous system since their training in the classics had exposed them not only to the framework of classical geography but also to the underlying theories which could be compatible with the new concepts. In the immediate aftermath of Iberian

[28] R. Hirsch, *Printed Reports*; C. R. Johnson, 'German Cosmographers and the Naming of America', *Past and Present*, 191 (2006); C. R. Johnson, *German Discovery*.

[29] E. M. Rogers, *Diffusion of Innovations*, 5th edn (New York: Free Press, 2003; first published 1963), 15–17.

discoveries, even the majority of humanist geographers were loath to query the supposed facts of classical geography, but over the century they became the adopters and disseminators of the new theories.

Few of the sixteenth century geographical writers were travellers themselves. Instead they were collectors, compilers and above all scholars who with varying degrees of success tried to reconcile the classical works in which they had been educated with contemporary information.

The Iberian discoveries of sea-routes to Asia and the Americas coincided directly with the Renaissance. Humanist scholars of the fifteenth and sixteenth centuries read geographical texts such as Ptolemy's *Cosmographia* and Strabo's *Geographia* which, although technically extant, had ceased to circulate in western Europe with greater attention than they had enjoyed even in classical times. As Anthony Grafton has said, these ancient texts, newly coming to light again, had 'the crisp immediacy of new books'.[30] As a result, a new geography that took no account of the classics would have been inconceivable, but this automatically brought sixteenth-century writers into conflict with the old idea of the *oikoumene.* Even those geographers and explorers who were more receptive to the ideas of unencountered lands and wholly navigable seas pillaged the ancient authorities for any detail that could be made to fit with their newly expanded knowledge of the world. Columbus for instance, although clearly open to new ideas, noted that Seneca the younger had referred to another *orbs* lying to the west in his Medea (and included both the quotation and a translation in his book of Prophecies justifying his expeditions). The passage reads: 'The time will come in a number of years, when Oceanus will unfasten the bounds, and a huge land will stretch out, and Typhis the pilot will discover new worlds, so the remotest land will no longer be Thule.'[31] Similarly, Humphrey Gilbert, an English explorer and statesmen, justified his search for the northwest passage partly on the basis that the Americas were Plato's Atlantis, and also found references to sea voyages from Asia to Europe in the classical authors. Jose de Acosta, a Spanish Jesuit missionary to Peru, likewise referred to both Atlantis and classical voyages to Asia.[32] All three,

30 A. Grafton, A. Shelford, and N. Siraisi, *New Worlds*, 29, 45.

31 Venient annis secula series, quibus Oceanus vincula rerum Laxet, & ingens pateat tellus Tiphisque novos detegat orbes, nec sit terries ultima Thule. C. Columbus et al., *The Libro De Las Profecías of Christopher Columbus: An En Face Edition* (Gainesville: University of Florida Press, 1991), 224. Curiously, as Moretti points out, it is the only secular quotation that Columbus uses. G. Moretti, *Gli antipodi: avventure letterarie di un mito scientifico* (Parma: Nuova Pratiche Editrice, 1994), 123.

32 On Seneca, see for example J. de Acosta, *Historia Natural Y Moral De Las Indias* (Madrid, 1894, repr. from the first edn, Seville, 1590), vol. 1, 45.; C. Colon, *Historia del Almirante Don Cristobal Colon*, 2 vols, vol. 2, Colección de libros raros o curiosos que tratan de America (Madrid: Tomas Minues, 1892), vol. 2, 32; D. Clay, 'Columbus' Senecan Prophecy', *American Journal of Philology*, 113 (1992), 117–20, G. Moretti,

of different ethnicities and backgrounds, but with a similar interest in risking their lives on the basis of new geographical understanding, still set their ideas of world geography into a classical context.

As it became increasingly evident that many of the regions of Spanish, Portuguese, French and English discoveries were entirely outside the knowledge of the ancient writers, humanists ceased to use the works of classical authors so extensively for specific data, yet they continued to look to them for theories and ideas. Louis le Roy, writing *On the Vicissitutes and Variety of Things* towards the end of the sixteenth century, still wrote of the rediscovery of the classics as the foremost achievement of his time, saying 'in addition to the restoration of the classics which has almost been accomplished, the discovery of many and beautiful things which are of benefit, not only for the necessities of life but also for its enjoyment and enrichment has been left to this age.'[33] Nearly a century after Columbus, classical theories about the nature and structure of the world were still relevant for geographers trying to describe the world with its vast unexplored territories and unknown pattern of land and sea.

Yet the late Renaissance writers of the sixteenth century used the classical theories in a very different way from how they had been used in the ancient world. For the ancient writers, concepts of boundaries and balance had created a framework for their world which, in a sense, imposed limits on the possibility of European expansion, by emphasising that the *oikoumene* was distinct from the whole earth. They had talked of Ocean as a confine, and had described boundaries created by monstrous peoples or marked by desert lands which separated out the

---

'Nec Sit Terris Ultima Thule (La profezia di Seneca sulla scoperta del Nuovo Mondo)', *Columbeis* 1 (1986): 95–106; J. Romm, *Seneca in the Renaissance*, 77–113. On the voyages from Asia to Europe, see J. de Acosta, *Historia*, vol. 1, 45, H. Gilbert, 'A Discourse Written by Sir Humphrey Gilbert Knight, to Prove a Passage by the Northwest to Cathaia and the East Indies', *PN*, vol. 7, 172. On Atlantis see J. de Acosta, *Historia*, vol. 1, 75–79, 103.; G. Benzoni, *La historia del mondo nuovo laqual tratta dell'isole & mari nuovamente ritrovati & delle nuove città da lui proprio vedute, per acqua & per terra in quattordeci anni* (Venice: Francesco I Rampazetto, 1565), 111; H. Gilbert, 'Discourse to Prove Passage,' *PN*, vol. 7, 160. Ortelius labelled the Americas as Atlantis, ensuring that the two would remain equated in people's minds. See J. Romm, *Mythes, Cartes Et Histoire*, 104–16; H. Cook, 'Ancient Wisdom, the Golden Age, and Atlantis: The New World in Sixteenth-Century Cosmography', *Terrae Incognitae*, 10 (1978): 25–43.

33 Outre la restitution presqu'accomplie du sçavoir ancien, l'invention de plusieurs belles chose nouvelles servantes non seulement à la necessité: mais aussi au plaisir & ornement de la vie, a esté reservae à cest aage. L. le Roy, *De la vicissitude ou variété des choses en l'univers, et concurrence des armes et des lettres par les premieres et plus illustres nations du monde, depuis le temps où à commencé la civilité, et memoire humane iusques à present,* 2nd edn (Paris: Pierre l'Huilier, 1576), 99v. The edition is very inconsistent in spelling and the use of accents; I have quoted the text as is throughout.

oikoumene from the globe. The Renaissance reformulation of classical theory reframed the geography of the world to emphasise its unity. All the mechanisms of balance, reciprocity, symmetry, harmony and proportionality which they used to determine the shape and structure of the unexplored regions and the nature of their inhabitants were inherent properties of nature and applied universally to the whole world. In the sixteenth-century perspective of a world designed by God for human dominion, it was necessary for the *oikoumene* to coincide with the globe, for such dominion to be a possibility.

In the classical writings it was largely nature that was responsible for maintaining balance, but nature in the classical world was a complicated phenomenon. There were writers who from the fifth century BC onwards spoke of *pronoia,* roughly translatable as divine providence, an all-pervading feature of the *kosmos.* Such writers provided an argument from design and generally linked nature with the divine. Herodotus for instance spoke of *pronoia* when discussing how carnivorous animals had fewer offspring than herbivores. In Herodotus there was no particular human element to *pronoia*, but he did have the idea that the gods were responsible for setting everything in order and that nature was part of this order.[34] This order, then, came to have a human element to it. Xenophon in a famous passage talks about the limits of the movements of the sun being made for humans, and talks about how an unstinting supply of water, vital for all aspects of human undertaking and life, shows design at work. The concept of a world designed for humans was there in the classical writers for the Renaissance writers to draw on. *Pronoia* became a common feature of stoic philosophy in particular – perhaps the most famous expression of it being in Cicero's *On the Nature of the Gods*, where he explicitly stated: 'Therefore I say that the world and all its parts were established in the beginning and administered for all time by the providence of the gods', and that 'god is neither obedient or subject to any nature; therefore he himself rules all nature'.[35] Others are less clear. In an interesting passage in Strabo, he appears to talk about *phusis* (nature) and *pronoia* as having distinct roles. It is providence, however, which is responsible for making the world the best suited for humans. Strabo, in a passage very similar to one which Mercator wrote a millennium and half later, argued that 'providence [*pronoia*] formed many eminences and cavities in the earth, so that these cavities should receive the whole or a great part of the water covering the land beneath it: and at the

[34] Herodotus, 3.108; 2.53. For a discussion of the earliest ideas of *pronoia* demonstrated in Herodotus and Xenophon, see R. Parker, 'The Origins of Pronoia: A Mystery', in *Apodosis: Essays Presented to Dr. W. W. Cruickshank to Mark His Eightieth Birthday* (London: St. Paul's School, 1992).

[35] Cicero, 'Dico igtur providentia deorum mundum et omnes mundi partes et initio constitutas esse et omni tempore administrari ... [N]ulli igitur est naturae oboediens aut subiectus deus; omnem ergo regit ipse naturam'. *De Natura Deorum,* 2.75–77.

eminences the earth should rise and conceal the water beneath it, except so much as was necessary for the use of the human race and the animals and plants around it.[36] Others such as the Epicureans ignored any argument from design altogether, but clearly there was sufficient in the classical writings and even in the more strictly geographical texts, about design and natural law for the Renaissance writers to draw upon when reframing their ideas about balance, natural law and a world made for humans in terms of a sixteenth-century Christian perspective.

In the sixteenth century, different writers use the terms Nature, God and Providence and it is impossible to sort out any consistent theory about who or what maintained the balance. What is clear is that all sixteenth-century writers saw the unity of the world, underpinned by the properties of balance, proportion and harmony, as evidence of God's creation of it for humankind. It was God's intention that Ocean, rather than constituting a boundary, should act as a means of communication to unite his whole world. It was his purpose that no one land should have everything so that humans would have to interact with all parts of his creation. He designed the world as the best possible world, but by that was meant the best possible world for humans. As André Thevet said,

> Also man was so created by God that he could live in whatever part of the world, whether it is hot or cold or temperate… Experience informs us, as we have said several times, how large the world is, and how ample and how suited the world is to all creatures and this it has done as much by continual travelling on the sea as by long (or far off) journeys on the land.[37]

The world was for humans to define, to exploit and to have dominion over. This book draws largely on descriptive geographical texts with an international circulation in the sixteenth century. They were either known to have been closely read or widely cited by authors outside their own country, or they were published or printed in more than one country in the sixteenth century. The international nature of their circulation shows their relevance for arguments about the way in which European-wide discourses on the nature of world geography was reshaped by an understanding of the classics. They were written or published by people who were interested in putting geographical knowledge

[36] Strabo 17.1.3.6. M. Hatzimichali, 'Strabo's Philosophy and Stoicism', in *The Routledge Companion to Strabo*, ed. by Daniela Dueck (London: Routledge, 2017), 14–16.

[37] Aussi a esté l'homme ainsi crée de Dieu qu'il pourra vivre en quelque partie de la terre, soit chaude, froide, ou temperée. … L'experience d'avantage (comme plusieurs fois nous avons dit) nous certifie, combien le monde est ample, & accommodable à toutes creatures, & ce tant par continuelle navigation sus la mer, comme par loingtains voyages sur la terre. A. Thevet, *les Singularitez de la France antarctique, autrement nommée amérique, & de plusieurs terres et isles découvertes de nostre temps* (Paris: Maurice de La Port, 1558), 37r-v.

into the public sphere, often in defiance of others' vested interests or censorship policies. Although printing became increasingly widespread in the sixteenth-century, there was still a large amount of geographical material that circulated only in manuscript form. This naturally reached a more restricted audience than printed material. Much of the vital information regarding the Spanish and Portuguese voyages, for example, was never published. Some works were censored or self-censored and other forms of material were more frequently constructed for private than for public record. Ships' journals and logs, for instance, were occasionally published, but more often they remained as manuscripts. Once a work was printed, however, it unsurprisingly had a far wider circulation than manuscripts, since every manuscript had to be produced individually whereas the print runs for books in the sixteenth century were normally in the region of a thousand.[38] Ptolemy's *Geographia* provides a clear example of the differing availability and circulation of print and manuscript works. Although it was one of the most influential geographical texts in the Renaissance, it exists in only a few dozen manuscripts.[39] It reached a scholarly but circumscribed audience while it remained in manuscript. Once it was printed in 1475, however, it permeated all levels of literate society throughout Europe. By 1600 there had been thirty-one editions printed in various different European countries in either Latin or Italian.[40] Several of these had run to more than one printing.

Printed books also reached a less controlled market. Once a book had been published it could circulate across Europe with fair ease. For instance the Italian geographer Giovanni Battista Ramusio, who compiled one of the most important travel collections of the sixteenth century, obtained an account of the Coronado expedition through illicit means – possibly on account of his friendship with the Spanish ambassador to Venice.[41] After he published it in his *Navigazioni e viaggi* it was readily available to other scholars. The English geographer, Richard Hakluyt, then made a translation of the Ramusio version

[38] C. B. Schmitt et al., eds, *The Cambridge History of Renaissance Philosophy* (Cambridge: Cambridge University Press, 1988), 28.

[39] B. Weiss, 'The Geography in Print: 1475–1530', in *Ptolemy's Geography in the Renaissance*, ed. by Zur Shalev and Charles Burnett, vol. 17, Warburg Institute Colloquia (London: The Warburg Institute, 2011), 91–120.

[40] J. J. S. Goss, *The Mapmaker's Art: A History of Cartography* (London: Studio Editions, 1993), 123.

[41] G. B. Parks, 'Literary History', 129; V. di Coronado, 'Relazione che mandò Francesco Vazquez di Coronado, capitano generale della gente che fu mandata in nome di sua Maestà al paese nuovamente scoperto: quel che successe nel viaggio, dalli ventidue d'aprile di questo anno MDXL, che partì da Culiacan per innanzi, e di quel che trovò nel paese dove andava', *NV*, vol. 6.

to be published in his *Principal Navigations*.[42] The original manuscript account has not survived, and had it not been for Ramusio's version, little would have been known about the expedition, and it would have had little impact on general European geographic thought. By publishing the document, Ramusio ensured not only that it survived, but that the material reached an unrestricted audience.

By contrast, André Thevet owned a manuscript copy of the Huguenot René Goulaine de Laudonnière's manuscript testimony of his attempt to found a French protestant colony in Florida, for over twenty years. He never published it, perhaps because he pretended that the information he included in his *Singularitez* and later his *Cosmographie* was his own rather than another's eye-witness report. Unlike Ramusio and Hakluyt, he never published other people's travelogues. He let Richard Hakluyt see the Laudonnière manuscript, however, and was incensed when Hakluyt took it and published it.[43] Their quarrel was about the need to make geographical information known. Hakluyt thought that Thevet was wrong to hold onto the narrative rather than adding it to the public pool of information about the world. Throughout his life, Hakluyt exerted himself in tracking down, editing, translating and publishing material. He believed absolutely in the authority of the printed word and did his best to place accurate versions into the public sphere. He even published material which showed the fallacy of some of his own earlier writing.[44] The idea of wilfully withholding material for personal reasons would have been anathema to him. All over Europe there were networks of 'translators, diplomats and spies' who channelled works from one country to another in order to put them into print.[45] Their works may frequently have served as propaganda, but they also widened geographic knowledge.

Even when authors were interested in publication, it was not always easy either to exchange or publish geographical information in the sixteenth century

42 V. di Coronado, 'An Extract of the Letter of Captain Vasquez di Coronado', *PN*, vol. 9, 116-18; M. Small, 'A World Seen through Another's Eyes: Hakluyt, Ramusio and the *Navigazioni e viaggi*', in *Richard Hakluyt and Collected Travel Writing in Early Modern Europe*, ed. by Daniel Carey and Claire Jowett (London: Hakluyt Society and Ashgate, 2011).

43 See R. Schlesinger and A. P. Stabler, eds, *André Thevet's North America: A Sixteenth-Century View* (Kingston and Montreal: McGill University Press, 1986), xxiv; see also B. Bucher, *Icon and Conquest: A Structural Analysis of the Illustrations of De Bry's Great Voyages*, trans. by Basia Miller Gulati (Chicago: Chicago University Press, 1981), 8. Bucher points out that Basanier, the Parisian editor of Laudonniere's diary of the Huguenot expedition to Florida, declared that Hakluyt was to be commended for having rescued the diary from 'the tomb'. Hakluyt was behind not only the English publication but also the French one. P. Mancall, *Hakluyt's Promise: An Elizabethan's Obsession for an English America* (New Haven: Yale University Press, 2007), 170.

44 M. Small, *A World Seen through Another's Eyes*.

45 L. F. Parmelee, 'French Propaganda in Late Elizabethan England', *Sixteenth-Century Journal,* 25 (1994), 860.

despite a proliferation of geographical writing and an eager readership for some of the more lurid travellers' tales. Some countries felt a greater need to censor geographical publications while others, in contrast, placed great importance on printed publication. The contribution to geographical discourse in the sixteenth century did not equate with the contribution to exploration, which creates a slightly peculiar slant to the source material used in this book.

Vested interests and censorship policies meant that some of the most theoretically important and empirically based works of the period did not play a significant role in reshaping European understanding of the form and nature of the world in the sixteenth century. Despite both the Portuguese and the Spanish being at the forefront of sixteenth-century exploration and expansion, for example, the Iberian contribution to a European-wide discourse of descriptive world geography is comparatively small. The Spanish government strictly controlled the publication of geographical works, fearing that they could offer their national rivals insight into vulnerabilities and the importance of the empire.[46] They thought that geographical knowledge was a tool of power, and that dissemination of geographical knowledge about their empire threatened their control over these regions and their assets. Many Spanish works were therefore not even printed in their own country, but instead circulated solely in manuscript form and so cannot be said to have contributed greatly to a public, let alone a European-wide, conception of geography. The Portuguese likewise prohibited the publication of large amounts of geographical material so successfully that in many ways much of the eastern Portuguese empire remained remarkably unknown to western Europeans.

Nonetheless, there were both Portuguese and Spanish writings which had a notable impact on European geographical thought. There were three ways in which Portuguese and Spanish works entered the European sphere. Some, such as Columbus's letter, Peter Martyr's *Decades*, and Acosta's *Natural and Moral History*, were allowed to be published in full. The Low Countries were particularly useful in ensuring printed Spanish works circulated beyond Iberian borders. Others were first printed outside Spain such as Girava's *Dos Libros De Cosmographia* which he printed in Milan where he was living at the time. Likewise, Juan Gonzalez de Mendoza's *Historia de las cosas mas notables, ritos y costumbres del gran Reyno de la China,* although published in Spanish in 1585, was first published in Rome, at the request of Pope Gregory XIII.[47] Once printed

46 M. M. Portuondo, *Secret Science: Spanish Cosmography and the New World* (Chicago: University of Chicago Press, 2013); J. B. Harley, 'Silences and Secrecy: The Hidden Agenda of Cartography in Early Modern Europe', *Imago Mundi; The International Journal for the History of Cartography*, 40 (1988), 61.

47 B. Schmidt, *Innocence Abroad: The Dutch Imagination and the New World, 1570–1670* (Cambridge: Cambridge University Press, 2001), 23.

it had a rapid retransmission, being republished in both Valencia and Madrid, also in 1585, and was translated into Italian, English, French, German, Latin and Dutch by 1595, becoming by far the most influential book on China written in that period. The rapid republication of the *Historia... de la China* is indicative of the kind of demand there could be for Iberian geographical material.

The third method of evading Spanish censorship was the exploitation of diplomatic contacts. Ramusio was arguably the single most important scholar to make Iberian descriptive geographical treatises available to the rest of Europe. His friend Andrea Navagero went as an envoy to Spain, and while he was there he succeeded with some difficulty in obtaining a copy of Oviedo's first *Decade* as well as the first three *Decades* of Peter Martyr's *De Orbe Novo*. Even he, however, had considerable difficulty tracking down many works relating to the New World. In 1525 he wrote to Ramusio saying, 'I can find no books here on the Indies; but in time I will send enough matter to weary you. I have means of learning everything from Peter Martyr, who is my close friend, as well as from the president of the Council of the Indies and from many other members of the council.'[48] Ramusio's position as secretary to the Venetian council and assistant keeper of the library of San Marco put him in a better position than many to obtain Iberian material, since he had contacts with both Venetian and foreign travellers and diplomats. He even succeeded in obtaining some official travel reports through the Spanish ambassador to Venice.[49]

Sometimes even Ramusio was only able to obtain partial versions of texts in which the most financially or politically valuable information was suppressed. An interesting case in point concerns the Portuguese factor to Malabar, Duarte Barbosa, who wrote a first-hand account of the journey to India, and of the country. His book would have been particularly valuable to Europeans because Barbosa was fluent in the Malayalam language and was able to obtain detailed

[48] A. Navagero, *Andreae Nagerii Opera omnia quae quidem magna adhibita diligentia colligi potuerant*, ed. by Gaetano Cristoforo Volpi and Giovanni Antonio Volpi (Padua: J. Coominus, 1718), 300–1. Delle cose de *las Indias* qui non si truova niente di stampato, ma io con tempo vi manderò tante cose, che vi stancherò. Io ho modo d'intender il tutto, sì per M. Pietro Martire, che è amiccissimo mio, come per via del Presidente del Consiglo *de las Indias,* e di monlti altri di detto Consiglio. ... Vi scriverò anche di *Panama* che mi chiedete, ma ora non penso; ne di giorno in giorno resterò di scriver circa tal material quell che s'intenderà di momento.

[49] G. B. Parks, 'Literary History', 130. See also my forthcoming article on Ramusio's literary network: 'From Colleagues to Friends: a sixteenth-century Venetian literary network'. S. Grande, 'Le relazione geografiche fra P. Bembo, G. Fracastoro, G. B. Ramusio, G. Gastaldi', *Memorie della società geografica italiana*, 12 (1956); M. Donattini, 'Giovanni Battista Ramusio e le sue 'Navigationi', appunti per una biografia', *Critica Storica, Rivista Trimestrale*, 17 (1980–1981); T. Veneri, 'Giovanni Battista Ramusio, molto piu di uno spettatore. Le quinte delle 'Navigationi et viaggi', *Italica*, 89, no. 2 (2012).

information on much of south India, but it was subject to the Portuguese efforts to control the dissemination of information.[50] Although his description of the voyage from the Cape of Good Hope to India was translated from Portuguese into Castilian in 1524, it remained confined to the Iberian Peninsula for several decades. Clearly the ban on free exchange of geographical information had some effect.[51] Ramusio eventually succeeded in publishing a partial version of Barbosa's book in 1550, but his description of the problems he had in obtaining a copy of it, or of Tomé Pires: *Suma Oriental que trata do Mar Roxo até os Chins*, one of the most detailed descriptions of east Asia which he believed was written a short time later, shows the problems of extracting geographical information from Spain and Portugal. He claimed he had searched diligently for complete copies of Pires's and Barbosa's works, but to his extreme frustration was unable to obtain them. He published the limited sections that he had, but the complete versions which would have done much to augment European geographical knowledge remained beyond his grasp. In the case of Pires's *Suma Oriental*, the Portuguese had ensured that all the information about the Moluccas was censored and not even Ramusio with all his contacts could get hold of the data.[52] As he declares, he would

> have published them and brought them to light much more willingly and quickly, for no other reason, and for no other purpose of our own (as we have said in several places in the present volume), than to do a kindness to students who take pleasure in such reading.[53]

Similarly as we have seen, even though he corresponded directly with the historian of the New World, Fernando Gonzalez de Oviedo, Ramusio had

50 D. F. Lach, *Asia in the Making of Europe,* 3 vols, vol. 1 (Chicago: University of Chicago Press, 1965), 186.

51 J. P. Rubiés, *Travel and Ethnology*, 3; D. F. Lach, *Asia*, 186.

52 T. Pires, 'The Suma Oriental of Tomé Pires: An Account of the East, from the Red Sea to Japan, Written in Malacca and India in 1512–1515, and the Book of Francisco Rodrigues, Rutter of a Voyage in the Red Sea, Nautical Rules, Almanack and Maps, Written and Drawn in the East before 1515', ed. by Armando Cortesão (London: The Hakluyt society, 1944), lxvi–lxvii.

53 che molto piú volentieri e presto gli arebbon publicati e messi in luce, non ad altro fine né per altro nostro proposito (come in piú luoghi del presente volume abbiamo detto) che per far cosa grata agli studiosi che si dilettano di tal lezione. G. B. Ramusio, 'Discorso sopra il libro di Odoardo Barbosa e sopra il sommario dell Indie Orientali', *NV*, vol. 2, 542. Ramusio can be somewhat disingenuous about how much he is solely catering for the scholars. As Marica Milanesi has observed, he also had an interest in Venice's fortune and invested in the spice trade himself. M. Milanesi, *Tolomeo sostituito: studi di storia delle conoscenze geografiche nel XVI secolo* (Milan: Unicopli, 1984), 41.

difficulty in obtaining a copy of the Spanish historian's *De la natural historia de las Indias* and was able to publish only the first of three *Decades* of that work.[54] That said, there may have been an additional reason for Ramusio's difficulties in obtaining that particular book. Butzer suggests that the advocate of Indian rights, Bartolomé de Las Casas, had quarrelled with Oviedo and had helped to block the publication of the last volumes of his work even in Spain.[55] By the second half of the sixteenth century the enforcement of the censorship restrictions became laxer, although there was never the free exchange of geographical information that there was between other countries.[56] Nonetheless, through permitted publication, diplomatic interventions and personal contacts a large number of Iberian documents reached a wider audience and contributed to western European geographical knowledge in general.

The Iberian countries may have been the most successful at censoring, but they were certainly not the only ones to try to do so. Henry VIII, for instance, largely for political and religious reasons, but also partly in an attempt to save the English printing industry from foreign competition, placed a ban on the import of foreign books in 1534, which probably helped slow the dissemination of the new geography in England. English printers did not publish a work on the New World until the 1550s and foreign books were blacklisted.[57] Later the import ban was repealed, but the English government still attempted to impose restrictions. It is probable that the reason that there was no publication of Francis Drake's circumnavigation of the world until 1587 was that the crown deemed the information so important that it placed a ban on its publication.[58] Similarly, some of John Dee's more propagandistic documents seem to have circulated only in manuscript because either he or the crown hoped to retain some element of control over their readership. Many of his works, hermeneutic, geographical and political, were not deemed suitable for a 'wider reading public'.[59] Nor was it only governments who tried to control the flow of geographical information. Even people who had happily communicated with each other in one set of circumstances saw the

54 G. B. Ramusio, 'Discorso Sopra Il Terzo Volume', *NV,* vol. 5, 9–10. On Ramusio and Oviedo see A. Gerbi and J. Moyle, *Nature in the New World: From Christopher Columbus to Gonzalo Fernandez De Oviedo*, 1st edn (Pittsburgh: University of Pittsburgh Press, 1985), 165–73.

55 G. B. Ramusio, 'Discorso Sopra Il Terzo Volume', *NV*, vol. 5, 9–10. See K. W. Butzer, 'From Columbus to Acosta: Science, Discovery and the New World,' *Annals of the Association of American Geographers* 82 (1992): 549.

56 J. B. Harley, 'Silences and Secrecy,' passim.

57 J. Parker, *Books to Build an Empire; a Bibliographical History of English Overseas Interests to 1620* (Amsterdam: Nico Israel, 1965), 10.

58 *Ibid.*, 107.

59 W. Sherman, *Politics of Reading*, 117; K. Macmillan and J. Abeles, eds, *The Limits of the British Empire* (Westport, CT: Praefer, 2004), 3.

value in keeping information secret in another. Richard Hakluyt, for instance, although he was friendly with Ortelius, was not entirely pleased when he felt that the purpose of Ortelius's visit to England in 1577 was to 'prye and looke into the secrets of Frobisher's voyage'.[60] It was one thing to share information but another to damage the prospects of one's own country or merchants, and Ortelius had readily admitted to Hakluyt that if the Netherlands had not been in a state of civil war, they would have been searching for a northwest passage themselves.[61] Companies too tried to protect their own interests. In the seventeenth century the Dutch East India Company achieved some success in restricting the dissemination of its materials.[62] Few countries or companies, however, were as uniform or as effective in their publication embargos as Spain and Portugal.

Despite all these barriers to the creation of a shared geography, the new geography that did develop was dependent on pooled information from the new reports arriving from voyages of exploration, from the new interpretations made and shared across Europe and from the common starting point of the classics.

The authors whose works did obtain an international readership were a disparate group. Some were travellers whose written reports gave an empirical underpinning to the new geography; others were published in the works of compilers of travel narratives such as Giovanni Battista Ramusio, Richard Hakluyt and Theodore de Bry, whose compilations were never translated in entirety but who introduced travel narratives from other regions of Europe to reading publics that reached throughout Italy, England, the Low Countries and beyond. Still others wrote more conventionally formatted single-authored cosmographies, some in the vernacular, others in Latin. What all these writers and editors had in common was that they were almost all educated as humanists, and because throughout western Europe the key interpreters and disseminators of geography shared a common education in the classics, sixteenth-century geographical theory was founded as much on a reinterpretation of Greek and Roman ideas as on new empirical information.

The shared sixteenth-century descriptive geography as presented by these humanists was highly dependent on text. As a result, maps, although appearing alongside many descriptive geographical texts, do not hold a prominent role in this study. Many of the authors discussed in this book either saw maps as irrelevant or thought them inadequate in describing a world in which geographical knowledge was in a constant state of flux. Maps have an essentially static nature, which was problematic for keeping place with the flow of discovery. Giovanni Battista Ramusio, as I have argued elsewhere, chose the format of the travel

[60] R. Hakluyt, 'Discourse of Western Planting', in *Hakluyt Society Extra Series*, ed. by David Beers Quinn and Alison M. Quinn (London 1584 facsimile edition, 1993), 76.

[61] *Ibid.*

[62] J. B. Harley, 'Silences and Secrecy', passim.

compilation because it allowed narratives about new regions to be added to the original compilation as they became available, and so meant that subsequent editions could keep pace with the process of discovery, enabling the compiler and the reader to keep abreast of the constantly expanding state of knowledge about the world.[63] The book could therefore ostensibly remain useful and relevant up to and after the whole world was made known to Europeans. This could not be the case for any large scale map made to accompany the text. Maps could only show the state of knowledge at the time they were made, and could be outdated easily. This is not to deny that Ramusio saw value in maps. Indeed he stated that part of the purpose of making his compilation was so that others would have a source book from which, in combination with existing portolan maps, they could make accurate depictions of the world. But he saw this as a separate function to his own work.[64] He himself included only thirteen maps in his *Navigazioni e viaggi*, and even these were not necessarily part of his original plan and may have been included largely to please his scholarly network.[65]

Richard Hakluyt, who followed Ramusio's model of constructing a geography through compiling travel narratives, was even more grudging in his use of cartographic material. He even went so far as to state that

> nowe, because peradventure it would bee expected as necessarie, that the ***descriptions*** of so many parts of the world would farre more easily be conceived of the Readers, by adding Geographicall, and Hydrographicall tables thereunto... I have contented my selfe with inserting into the worke one of the best generall mappes of the world.[66]

Hakluyt was clearly aware of the power of the image in creating an impression

63 M. Small, 'Displacing Ptolemy? The Textual Geographies of Ramusio's Navigazioni e viaggi', in *Mapping Medieval Geographies: Cartography and Geographical Thought in the Latin West and Beyond, 300–1600*, ed. by Keith D. Lilley (Cambridge: Cambridge University Press, 2014).

64 G. B. Ramusio, 'All'eccellentiss. M. Ieronimo Fracastor. Gio. Battista Ramusio', *NV*, vol. 1, 4–5; J. R. Barnes, 'Giovanni Battista Ramusio and the History of Discoveries: An Analysis of Ramusio's Commentary, Cartography and Imagery in *Delle Navigationi Et Viaggi*' (University of Texas, 2007).

65 G. B. Ramusio, 'Discorso Sopra Il Terzo Volume', 16. He worked closely with the cartographer Gastaldi who made the few maps that were included in the compilation. But the majority of Gastaldi's maps were printed separately from the *Navigationi e viaggi*, although Gastaldi undoubtedly used the compilation as source material. D. Cosgrove, 'Mapping New Worlds: Culture and Cartography in Sixteenth-Century Venice,' *Imago Mundi: The International Journal for the History of Cartography*, 44 (1992), 74.

66 R. Hakluyt, 'To the Favourable Reader', in *The Principall Navigations, Voiages and Discoveries of the English Nation* ed. Richard Hakluyt, David B. Quinn, and R. A. Skelton, vol. 1 (London: 1589 facsimile edition 1965), xlvii–l.

of the geography of the world, but though he felt a need to acknowledge his lack of illustrative material, he did not include any more. Maps would have been too expensive, and concentrating on supplying up-to-date maps would have taken Hakluyt's time and attention away from his primary interest of printing as much geographical documentation as possible. He also faced the problem that they would become quickly outdated in a way his compilation would not. For Hakluyt, as for Ramusio, the text, not the map, was at the heart of geography.[67]

Hakluyt, when he used maps at all, largely used maps by British cartographers. He reluctantly included a map drawn by Robert Thorne early in the sixteenth century, and also used maps by Michael Lok and Edward Wright; but although he modelled his text on Ramusio's work, he did not reissue any of the maps drawn by Gastaldi which accompanied Ramusio's texts.

The travel compiler Theodore de Bry is a notable exception to what has just been said: he was not primarily a cartographer but his travel compilations are filled with engraved maps and images taken directly from other sources. For example, he issued a map of Virginia, which was largely based on a watercolour map by John White, to which he added information derived from Ralph Lane's voyages. But for de Bry, maps were part and parcel of other images. The de Bry family made their name as illustrators, and in many ways images, including maps, are the focus of his book. Although he became famous for his geographical material, for him (as Michael van Groesen has shown) his publications were very much a means of making money rather than a driving passion; he did, however, make great effort to make the images complement the text.[68] He would take existing images, such as the woodcuts made by the French colonist le Moyne de Morgues, to accompany his eye-witness account of the short-lived French colony in Florida,[69] tidy them up, in many ways classicise them, and re-etch them onto copperplate.[70] The engravings made them costly books: volume IV alone cost 4 guilders at the Golden Compass in Antwerp; a complete Latin edition of de Bry's *America* cost forty-two guilders, a considerable sum.[71] Yet, despite their cost,

67 R. A. Skelton, 'Hakluyt's Maps', in *Hakluyt Handbook*, ed. David B. Quinn, vol. 144, 2nd (London: Hakluyt Society), 66; J. G. Kohl, *A Descriptive Catalogue of Those Maps, Charts and Survey Relating to America: Which Are Mentioned in Vol. III of Hakluyt's Great Work* (Washington: Henry Polkinhorn, 1857), 5; D. B. Quinn, 'Introduction', in R. Hakluyt, *Principall Navigations*, xlviii.

68 M. v. Groesen, 'The De Bry Collection of Voyages (1590–1634): Early America Reconsidered', *Journal of Early Modern History* 12, no. 1 (2008).

69 See P. Hulton, ed, *The Work of Jacques le Moyne de Morgues: A Huguenot Artist in France, Florida and England* (London: British Museum Publications, 1977). The second volume is a collection of drawings and engravings by him.

70 B. Bucher, *Icon and Conquest*, 4.

71 M. v. Groesen, *The Representations of the Overseas World in the De Bry Collection of Voyages (1590-1634)* (Leiden: Brill, 2008), 313.

most purchasers seemed to have bought the entire series, and the Latin edition, at any rate, reached a fairly large number of moderately wealthy readers (both Protestant and Catholic). They interested merchants, nobles, clerics and scholars alike but all needed a relatively large income to be able to afford them.[72]

De Bry was not alone in providing his own maps and images when he republished an existing narrative. When authors from one country translated and reissued works previously published in another, they rarely reused each other's illustrative material exactly, so that although the travel narratives received an audience throughout much of western Europe, the illustrations including the maps seldom did. Although usually coproduced with geographical texts, they were seen as a separate commodity from geographical writings. Few geographical works were entirely without cartographical illustration, but these maps were not seen as integral to the texts. It was not just the travel compilers who altered the illustrations. Sebastian Münster's *Cosmographiae Universalis Libri VI* of 1544, for example, is peppered with illustrative material, and his revised edition of 1550 contains even more. When François Belleforest translated and expanded his book in French, however, he did not use most of Münster's maps. Similarly Richard Eden omitted them from his partial English translation.[73]

Such omissions show that it was not merely the concern that maps could become easily outdated that prevented them from being republished. Some of Münster's maps were still fairly up-to-date by the time that Belleforest was producing his text. It was also that the humanist scholars had a concept of geography that was much broader than the location and depiction of physical features. They took over the concept of the *oikoumene* – the world inhabited by human beings as the proper field of geographical study from authors such as Ptolemy and Strabo. For these humanists the inhabitable world was the relevant world, even if it was a vastly larger world than had previously been imagined. They were interested in physical and mathematical geography only as it related to human beings, and placed the texts rather than the maps at the forefront of publication.[74]

In a world where photographic images of the earth from space are now readily available, where GPS and computerised mapping have become the norm, and longitude and latitude are ascertained at the touch of a button, it is hard to imagine the difficulties in transmitting geographical information about an entirely

[72] *Ibid.*, 318–9.

[73] F. de Belleforest, *La cosmographie universelle de tout le monde, augmentée, ornée et enrichie* (Paris: Nicolas Chesneaus, 1575).

[74] As Surekha Davies's recent work has shown, it is possible to overstate this divide since cartography did find a place for the human and indeed incorporated a vast range of ethnographic depiction and understanding in early modern maps. The maps could provide information on peoples and their relationship to place, bringing to the fore the apparent relationship between humans, latitude, environment and the products of certain regions. S. Davies, *Renaissance Ethnography*.

unknown area; yet this was the task of the Renaissance geographer far more than the cartographer. The historian of cartography Johannes Keuning was not unusual in thinking that although the geographical horizons greatly expanded in the fifteenth and sixteenth centuries, 'these events did not lead... to the creation of a geographical literature of great significance'.[75] He was mistaken. There was a symbiotic relationship between maps and texts, but, I would argue, the geographical imagination of the sixteenth century began with the writers, and the new mapping of the world began with the written word. Through rereading classical authors and reconceptualising classical theories, and combining the ancient ideas with eye-witness material recorded by travellers to the previously unknown lands, sixteenth-century writers were able to provide a new, shared, descriptive geography which could be transmitted independently from the cartographic image.

In order to create this shared geographical understanding, geographical writers had to cross religious and political divides. Many of the key geographical texts of the period were written with political or religious aims. Münster and Hakluyt had a geographical understanding which was inseparable from their faith, but also had strong nationalistic inclinations. Münster, for instance, thought that the study of geography with its investigation into earthly, human things could help provide an understanding of things divine. He believed that in studying the earth, the geographer could learn about the work of God's hands and so, by knowing more about God's creation, could indirectly know more about its maker. He explicitly set out show how God's creation was eternal whereas human fabrication was transitory. At the same time he set the German lands at the heart of his cosmography.[76] Hakluyt, likewise, was full of evangelical protestant fervour, but also nationalistic pride as can be seen from the very title of his work: *The Principal Navigations... of the English Nation.*

Martyr and Thevet were both employed as cosmographers royal and promoted their countries' interest through their geographical writing. Paradoxically, however, in an era of constant religious turmoil, and of growing national divisions, geographers whose works had extremely strong implications for nationalism and religious propaganda formed communities of scholarship which extended all across Europe, creating a network of scholars who frequently ignored both national and confessional divides in order to exchange and

75 J. Keuning, 'The History of Geographical Map Projections until 1600', *Imago Mundi: The International Journal for the History of Cartography* 12 (1955), 1.

76 The Latin version of his Cosmography became the main European geographical best-seller of the first half of the sixteenth century. It was translated in whole or in part into seven different vernacular languages and was printed more than thirty times by the end of the century. Despite making scant reference to the New World, it became the compulsory text for all Western European geographers. L. B. Cormack, *Charting an Empire*, 113, 53.

accumulate information. Governments sought jealously to control the dissemination of geographical information for nationalistic reasons, but did not succeed in preventing a trans-European exchange of information, which is reflected in the source material used for this book.

Protestants and Catholics communicated directly with one other. Gerard Mercator, for instance, probably remained a Catholic his entire life (although not an entirely orthodox one; he was even imprisoned for heresy).[77] He was a deeply devout man, and for him as for so many Renaissance geographers, his faith defined his geographical purpose. As he himself wrote: 'For this is what we strive for while we hand over the *Cosmography* that the infinite wisdom of God and his inexhaustible goodness may be seen from the wonderful harmony of all things directed to God's one end, and in the inscrutable providence of their composition.'[78] Nonetheless, Mercator looked far beyond his co-religionists, and had strong links with the Evangelical community in Louvain. Richard Hakluyt, a Church of England chaplain, was also numbered among his friends, as was the Familist Ortelius.[79] Hakluyt and Ortelius in turn had their own circles of friends and correspondents that extended across the confessional divide. Ortelius's circle of contributors spread half across Europe, straddling the religious divides, and

[77] N. Crane, *Mercator: The Man Who Mapped the Planet* (London: Weidenfeld and Nicholson, 2002), 130–41.

[78] Id enim molimur, dum Cosmographiam tradimus, ut ex mirabili omnium rerum in unum Dei finem concordia, & ex inperscrutabili in compositione providentia, infinita Dei sapientia, & inexhausta eius bonitas conspiciantur, quo ad venerandam eius maiestatem, colendam amplexandamque; eius divitem bonitatem feramur. G. Mercator, *Atlas Sive Cosmographicae Meditationes De Fabrica Mundi Et Fabricati Figura* (Amsterdam: Albertus Busius, 1595), a2.

[79] Ortelius's religious views are not entirely clear, but see G. Mangani, 'Abraham Ortelius and the Hermetic Meaning of the Cordiform Projection', *Imago Mundi; The International Journal for the History of Cartography*, 50 (1998); A. Hamilton, *The Family of Love* (Cambridge: James Clarke and Co. Ltd, 1981), 71–73. R. Boumans, 'The Religious Views of Abraham Ortelius', *Journal of the Warburg and Courtauld Institutes*, 17, no. 3/4 (1954); T. Meganck, *Erudite Eyes: Friendship, Art and Erudition in the Network of Abraham Ortelius (1527–1598)* (Leiden: Brill, 2017), 227, 71. A letter between Richard Hakluyt the elder and Ortelius can be found in Harry Ransom Center, The University of Texas in the Abraham Ortelius Correspondence, HRC 55, Kraus-57 and has been digitised online at https://hrc.contentdm.oclc.org/digital/collection/p15878coll99/id/156 accessed 07/01/2020. A letter from Ortelius to his nephew Jacob Cole also contains greetings for Richard Hakluyt the younger and discusses map-making for him. Abraham Ortelius Correspondence HRC 55, Kraus-59 https://hrc.contentdm.oclc.org/digital/collection/p15878coll99/id/145/rec/3. A letter from Mercator to Hakluyt can be found in E. G. R. Taylor, ed., *The Original Writings and Correspondence of the Two Richard Hakluyts*, 2 vols, vol. 1, 2nd series (London: Hakluyt Society, 1935), 160.

included writers from as far afield as Poland and Portugal.[80] Hakluyt's network of acquaintances was similarly eclectic. He was not merely acquainted with both Ortelius and Mercator, but also met André Thevet, and had contacts among the British merchant community in Spain who obtained documents for him.[81] One of his key contacts in France was the strongly Catholic explorer and merchant Étienne Bellenger.[82] Similarly, his fellow countryman John Dee studied with the Catholic Gemma Frisius, became close friends with both Ortelius and Mercator, and even named the Portuguese Catholic *converso*, Pedro Nunes, his literary executor.[83] Both Dee and Hakluyt were ardent patriots and proponents of English imperialism, yet in one-to-one relationships the shared intellectual interests of geography surmounted all other concerns. All these men were able to build extensive networks for sharing ideas that placed greater emphasis on the knowledge than on the religion or even country of origin of the correspondents.

Nor were they simply reliant on personal contacts for the exchange of knowledge. Virtually from the moment of schism with Rome, Protestant works were read with fervour by Catholics, and Catholic geographical works were printed in Protestant countries. Patriotic protestant Englishmen and Dutchmen avidly scoured the works of their hated enemies, the Spanish. Moreover there was no secrecy about this indebtedness to foreign or heretical authors. Geographers throughout western Europe obtained, translated and published material written by men of different nationalities and confessions. The Protestant and Catholic writers tended to have slightly different preconceptions in their approach to geography. The Catholics, as Büttner has pointed out, were more interested in God the Creator; the Protestants, in God as the Divine Providence.[84] But though the way Catholic and Protestant writers interpreted the world differed, the facts and details in their geographical works were still of value to readers and fellow geographers of either confession.

The Protestant Sebastian Münster wrote with an explicitly theological purpose, but believed that it was more important that his geography be accurate and comprehensive than that it be derived solely from writers sympathetic to the

80 J. Depuydt, 'Le cercle d'amis et de correspondants autour d'Abraham Ortelius', in *Abraham Ortelius, 1527–1598: Cartographe Et Humaniste* (Turnhout: Brepols, 1998), 117–20.

81 D. Beecher, 'The Legacy of John Frampton: Elizabethan Trader and Translator', *Renaissance Studies*, 20, no. 3 (2006), 322.

82 P. Mancall, *Hakluyt's Promise*, 110.

83 Pedro Nunes was from a Jewish converso family. M. van Durme, *Correspondance Mercatorienne* (Anvers: de Nederlandsche Boekhandel, 1959), 36–38 contains a letter from Dee to Mercator telling Mercator about his appointment of Pedro Nunes. It also talks about how Dee was suspected of heresy.

84 M. Büttner, 'The Significance of the Reformation for the Reorientation of Geography in Lutheran Germany', *History of Science*, 17 (1979), 151–69.

Reformation. For example, he unashamedly used the 1539 *Carta Marina* of Olaus Magnus despite it being a clear piece of Catholic propaganda.[85] Olaus's goals were as strongly theological as Münster's. In both his map and his later work, *A History of the Northern Peoples*, he wanted to demonstrate that the Scandinavian lands could and should be reclaimed for the Catholic faith. In this community of knowledge, however, readers such as Münster were able to distinguish the thesis of Olaus's works from the information they contained and scoured them for data.

The same was true of Catholics with Protestant material. Despite his secession from the Catholic church, Münster's own work was admired and reissued by Catholics. It crossed the linguistic and even the religious divide. Pietro Andréa Mattioli and Giacomo Gastaldi used Münster's text of Ptolemy (though not his maps) as the basis for their Italian edition in 1548.[86] Nearly thirty years later, the French Franciscan André Thevet wrote a eulogistic life of Münster, in which, after spending the first few paragraphs explaining that Münster was misguided in his beliefs, he insisted that the great geographer was nonetheless worthy of honour.[87] Meanwhile Thevet's compatriot and enemy, the devoutly Catholic François Belleforest, produced an expanded French version of the Cosmographia.[88] The international exchange of information and ideas across national and religious boundaries allowed for the relatively rapid creation of a new geography. The interchange of religious and political information allowed geographers to accumulate a shared pool of knowledge, but for the new ideas to become the mainstay of geographical understanding in western Europe the geographers relied on print to disseminate their information to a wider reading public. The printed descriptive geographies did not necessarily show what was at the forefront of geographical thought in the sixteenth century, but they clearly demonstrate how geography altered over the century (particularly since they often left out some of the most up-to-date mathematical geography). George Abbot, later Archbishop of Canterbury, published *A Brief Description of the whole Worlde* in 1599, intending to provide an introductory text for his students at University College, Oxford, which nicely sums up the state of accepted geographical knowledge in England expected of would-be scholars at the time.[89] To a degree, it shows how little the

[85] See my forthcoming article on the Language of the Carta Marina.

[86] Ptolemy, *La Geografia*, ed. by Sebastian Münster, Giacomo Gastaldi, and Pietro Andrea Mattiolo, trans. by Pietro Andrea Mattiolo (Venice: Gioa. Baptista Pedrezano, 1548). The text was edited by Giacomo Gastaldi from the 1540 edition of Sebastian Münster.

[87] A. Thevet, *Pourtraits et vies des hommes illustres grecz, latins et payens*, 2 vols, vol. 2 (Paris: La vesve I. Kervert et G. Chaudière, 1584), 559r–62v.

[88] F. de Belleforest, *Cosmographie*. Thevet extended his quarrel with Belleforest to saying that he should not have tampered with Münster's work: A. Thevet, *Pourtraits*, 560r.

[89] J. N. L. Baker, 'Academic Geography in the Seventeenth and Eighteenth Centuries',

discovery of whole new continents and peoples had penetrated into the popular consciousness by the end of the century. It opens with the statement:

> The olde knowne firme land was conteined onely in Asia, Europe, and Africa, Europe is deuided from Africa by the Mediterran sea, and from Asia by the riuer Tanais, whereby appeareth that the North-partes of Asia, and of Europe in olde time were but little knowne, and discouered. Africa is deuided from Europe by the Mediterran sea, from Asia by the riuer Nilus: and so Asia, by Tanais and Nilus, is seuered from Europe and Afrike.[90]

The first mention of the Americas appears four pages from the end when Abbot states that

> Although some dispute out of Plato, and the olde writers that there was not onely a gesse, but a kinde of knowledge in auntient time, that ... there was another large countrey lying to the West, yet he that shall aduisedly vse the coniectures made therevpon, may see that ... all antiquitie was vtterly ignorant of the newe founde countries towarde the West[91]

Abbot does not offer much more than the ancients, however. The last four pages provide little substantial information about the New World. Essentially Abbot, a century after Columbus, was still following the guidelines of classical and Biblical geography, yet his work proved popular, going through nine editions by 1636. As Parker has stated, 'like the translations from Münster and the classical geographers, it appears to have had far more appeal to the general public than well-founded accounts of recent geographical discoveries'.[92] Nonetheless, despite its traditionalist viewpoint, the book demonstrates the changes that had occurred in a century. In the first half of the century virtually no geographical works were published in England. By the end of the century, as Abbot's text indicates, they had become part of early modern education; they were used to show the errors of the ancient writers. Even in a work that essentially dwelt on the subjects of traditional geography, the New World had to be mentioned.

In the sixteenth century geography developed from a 'general interest in the world as part of the cosmos... into the discipline of geography, which was centred on political society and driven by concerns that were both intellectual and

---

in *The History of Geography: Papers by J. N. L. Baker* (Oxford: 1963), 16; G. Abbot, *A Brief Description of the Whole Worlde Whereby Are Particularly Described Al the Monarchies, Empires and Kingdomes of the Same with Their Severall Titles and Situations Thereunto Adjoining* (London: T. Iudson for Iohn Browne, 1599).

90 G. Abbot, *Brief Description*, 4.

91 *Ibid.*, 30.

92 J. Parker, *Books to Build an Empire*, 164.

pragmatic'.[93] The advantages of world-wide trade seemed obvious as gold, spices and luxury goods flowed in from east and west, while there were also economic and political benefits to be gained from colonisation. Giovanni Botero in Italy and Jean Bodin in France, among others, were using the latest geographical and ethnographical theories to support political arguments.[94] By that time the old geography of limit and boundaries had begun to vanish from European conceptions. Although writers such as Abbot still adhered to the classical paradigms, more up-to-date geographers were beginning to look forward to the products of discovery, arguing that it was divinely ordained that no region should have everything, and consequently that all the diverse parts of the world were necessarily dependent on one another for different products. The new geography held that this, the best possible world, was created in perfect balance and harmony by God for humans to hold dominion over in entirety. All parts of the world were designed to communicate and trade with one another. This new geography encompassed new racial theories, concepts of empire and ideas of globalisation which had long-lasting ramifications. Over time people became less aware of the classical foundations on which these theories were founded, to the extent that they have frequently been overlooked in the study of the history of geography to this day; and yet underlying them all were the ideas of boundaries, balance and harmony, derived from the classics, and reformulated in an era when all scholars were deeply immersed in the classical training of the Renaissance.

93 L. B. Cormack, *Charting an Empire*, 225.

94 See for example, G. Botero, *Della ragion di stato libri dieci con tre libri delle cause della grandeza e magnificenza della città* (Venice: i Gioliti, 1589); J. Bodin, *Methodus Ad Facile Historiarum Cognitionem* (Paris: M. Iuuenem, 1566).

2

The Classical Revival and the New Geography

As we have seen, in the late fifteenth and sixteenth centuries when the idea of an *oikoumene* distinct from the whole world still permeated geographical interpretation, the majority of scholars who responded to the new discoveries were immersed in the humanist idea that the Greek and Roman writers held the keys to knowledge of the world.[1] Many of the classical writings had been preserved in the Byzantine world, but had only reached western Europe in the fifteenth century, not long before the Portuguese and the Spanish began opening up the world to European eyes.[2] Simultaneous with the drive for geographical discovery was the eager searching for lost classical texts that characterised the fifteenth and sixteenth century renaissance. These ancient texts, newly coming to light again, had 'the crisp immediacy of new books'.[3] In Italy, at the forefront of the new interest in the *studiae humanitatis*, geography came to be considered an important area of study, to such an extent that in 1469, when Battista Guarini was composing his school syllabus in his *De ordine docendi et discendi*, he declared that all students should read the Roman geographers Pomponius Mela and Solinus, and the Greek geographer and astronomer Ptolemy in order to learn about geography and astrology, and that older students should read the Latin works of Aulus Gellius, Macrobius and Pliny the elder, all of which contain lengthy geographical references.[4] Some, such as the newly rediscovered geographical works of Strabo and Ptolemy, had an obvious relevance for humanist writers seeking to explain the shape and nature of the physical world, but others such as the philosophical writings of Plato and the Neoplatonists had a more surprising

1 Well into the seventeenth and even eighteenth centuries, classical geographical works were reprinted and used as basic texts for geographical study. See J. N. L. Baker, *Academic Geography*, 175.

2 A. Grafton, A. Shelford, and N. Siraisi, *New Worlds*, 45.

3 *Ibid.*, 29,45.

4 P. F. Grendler, *Schooling in Renaissance Italy, Literacy and Learning, 1300–1600* (Baltimore and London: John Hopkins University Press, 1989), 203.

influence. This chapter argues that the humanist-trained geographical writers' and travellers' search for validation in the classical authorities was so great that the early-modern authors revived classical geographical opinions from every genre of writing in order to develop their new concept of global human geography. In the process, the humanist geographers revolutionised the idea of who formed the key ancient authorities on geography.

The quest for geographical understanding, combined with the Renaissance revival of forgotten classical texts, and the rigour of Renaissance humanist training, meant that the sixteenth-century humanist geographers sought authority in Greek and Roman writings which were far outside the traditional corpus of classical works which had done much to create the ancient and medieval idea of a bounded *oikoumene*. As the historian of Renaissance humanism Paul Kristeller said, 'guided by their enthusiasm for everything ancient, and by their conscious program of imitating and reviving ancient learning and literature, Renaissance scholars had a much more comprehensive interest in ancient literature than either medieval or modern students'.[5] Although they read Virgil, Cicero, Aristotle and other now canonical authors by preference, Renaissance readers also avidly examined the works of later and minor authors, and to a large extent it was the more recently rediscovered authors and texts which formed the backbone of the new geography.

Sixteenth-century writers referred to a vast repertoire of classical works ranging from the full-length treatises of Ptolemy and Strabo to the works of Greek authors such as Anaximander, Thales, Ctesias, Megasthenes, Eratosthenes and Posidonius whose texts had long since disappeared, and existed only in fragmentary form.[6] While certain Latin writers such as Pliny, Solinus and Pomponius Mela featured prominently, the humanist geographers placed a particular new emphasis on Greek writings. Sometimes they cited authors in support of the newly evolving ideas of the shape and nature of the world; at other times they referred to them in order to demonstrate the inadequacy of the existing theories and the need for sixteenth-century scholars to create a new geographical framework. But either way their thought underpinned the new scholarship on the nature of the inhabitable world.

For the most part, sixteenth-century descriptive geographers were not interested in examining an evolution of geographical ideas in the ancient world, but rather, at least in terms of geography, tended to see the classical world as having

5 P. O. Kristeller, *Renaissance Thought: The Classic, Scholastic, and Humanist Strains; a Revised and Enlarged Edition of the Classics and Renaissance Thought* (New York: Harper, 1961), 7.

6 Guarino, for instance, took note of every reference to Posidonius in Strabo. E. B. Fryde, *Humanism and Renaissance Historiography* (Hambledon Press: London, 1983), 79–83.

a uniform authority.[7] The Portuguese pilot cited by Ramusio was a relative rarity, when, in his introduction to the voyage of Hanno, he described Polybius and Pliny, writing about voyages down the coast of Africa as far as the equator, and discussed how 'Ptolemy, who was a long time after Pliny and Pomponius Mela,' tried to assign degrees of latitude to these voyages.[8] This is not to say that they believed all Greek and Roman writers held exactly the same ideas about geography – they were far too familiar with the classical authors to hold such beliefs – but rather most humanist geographers took advantage of any lack of consensus to single out those ancient writers whose ideas conformed with their own, and uphold their authority regardless of when they were writing.

Robert Recorde, for instance, a mid-sixteenth-century English humanist, mathematician and geographer, cited Posidionius and Polybius alongside Eratosthenes and Ptolemy as authorities on the inhabitability of the equatorial zone. [9] Eratosthenes and Ptolemy lived some four hundred years apart from one another with Polybius and Posidonius falling in between, but Recorde listed them together, thinking little about their different historic contexts and the ways in which ideas might have changed over the course of nearly half a millennium. Indeed he seems to have thought very little about the authors themselves. Eratosthenes was a Greek philosopher and geographer whose works, like those of Posidionius, existed only in fragmentary form. Polybius was predominantly an untheoretical Hellenistic historian whose historical works had been recovered in western Europe in the fifteenth century but whose geographical treatise had also disappeared. Only the works of Ptolemy, the second-century Hellenistic geographer, were still extant. The style and form of his works differed greatly from what we know of Polybius. But none of that was relevant to Recorde. The important aspect for Recorde was that the extant fragments showed that a range of classical writers had believed in the inhabitability of the torrid zone and therefore provided evidence that not all ancient Greeks had held the misguided opinion that the *oikoumene* had hard and fast boundaries. He blamed the belief in the burning zone on the lack of ancient trade to the south, saying that, as a result, 'manye of the Latines as well as of the Grekes phantasied that there did dwell no inhabit-antes, neither could dwell there for the vehement heat: wherfore they called it the Burned Zone'. He argued, however, that reason, experience and the 'autority of many of their owne writers' proved that in fact the tropics could easily be

7 It is important to emphasise that this was in terms of geography because chronology was an increasingly important discipline in this period, gaining particular shape as a discipline from the work of Joseph Scaliger. A. Grafton, *Joseph Scaliger: A Study in the History of Classical Scholarship*, vol. 2 (Oxford: Clarendon Press, 1993).

8 'Ptolemeo, che fu molto tempo dapoi Pomponio Mella e Plinio'. G. B. Ramusio, *Discorso sopra la navigazione di Annone Cartaginese, fatto per un pilotto portoghese*, *NV*, vol. 1, 553.

9 R. Recorde, *The Castle of Knowledge* (London: Reyner Wolfe, 1556), 175.

entered.[10] Recorde was a humanist and a pragmatist and all his dialogues show the need for a balance between reason and authority when searching for knowledge.[11] Even he did not decry the authority of the classical authors altogether, however. Rather, by reasonable deduction, he placed the authority of one set of ancient writers in the place of another. Writers such as Polybius and Posidonius could be cited as great authorities despite the bulk of their relevant works having been lost, while other 'Grekes and Latines' could be decried as fantasists.

In addition to viewing all ancient writers as having the same potential but not uniform authority, sixteenth-century geographical writers placed little value on distinguishing the authority of different genres of ancient writing. Philosophical and geographical works of the entire classical period both played a vital role in shaping sixteenth-century European understanding about the shape and nature of the world. Admittedly, the distinction between geography and philosophy is itself a later one. Philosophy as both ancient and Renaissance writers would have defined it implied a love of knowledge. The limited modern implications of the term to denote 'A particular system of ideas or beliefs relating to the general scheme of existence and the universe' became common only in the eighteenth century, although there were occasionally uses of it earlier.[12] Nonetheless, although the word, as I am using it, would have been meaningless to both Renaissance and classical writers, it is useful for showing the way ideas which might be considered as belonging to an entirely different discipline could be used for geography. Geography, as we have seen, was not a precisely defined subject, spilling over into mathematics and navigation, travellers' tales, and anthropology, cartography and history. While both mathematical and descriptive geographers drew upon ideas from all disciplines to clarify their knowledge about the world, they were particularly indebted to what would now be known as philosophy.

In the ancient world anyone who wished to discuss the world beyond the *oikoumene* was forced to rely upon philosophical speculation to fill the void in scientific knowledge left by a lack of empirical evidence. The sixteenth-century writers, also short of observational material and often actively seeking compelling justifications for the funding of further voyages of exploration, picked up the classical practice of relying on philosophy to determine the nature of the unexplored regions of the earth. They drew on classical philosophical writings to supplement the clearly geographical works.

It is hardly surprising that, at a time when Aristotelian Natural Philosophy still dominated university curricula and when Aristotle continued to feature prominently in reading lists throughout the renaissance, some of his works, particularly

[10] *Ibid.*

[11] J. Fauvel, 'Platonic Rhetoric in Distance Learning: How Robert Record Taught the Home Learner', *For the Learning of Mathematics*, 9, no. 1 (1989): 3.

[12] '*philosophy, n.*' (OED Online) <http://www.oed.com/view/Entry/142505> .

the *Politics*, the *De Caelo* and the *Meteorologia*, played their part in the new geography. The majority of the writers who are at the core of this study about the construction of geography were university-trained academics for whom his work had formed the foundation of their advanced studies.[13] The *Meteorologia* in particular became a thesis that needed to be contended with as the bounds of the *oikoumene* were enlarged.[14] It had circulated widely throughout antiquity and numerous manuscript copies of medieval translations by Gerard of Cremona and, even more importantly, by William of Moerbeke existed. It was clearly a well-known treatise, but it contained strong endorsements of the zonal theory which had to be discredited for the idea of an *oikoumene* coterminous with the world to be credible.[15] The *De Mundo*, now known to be a later forgery but then thought to be Aristotelian, was also important, it too having circulated since the thirteenth century. One need only look at the repeated references to Aristotle throughout Jose de Acosta's and Gonzalo Fernández de Oviedo's natural histories of the New World, or the frequency with which less well-travelled writers from every country referred to him, to see that Aristotle remained a leading authority. He could not simply be disregarded but had to be definitively refuted when his theories were proved wrong.[16] Although his zonal theory of climates was hotly

13 D. Armitage, *The Ideological Origins of the British Empire*, Ideas in Context (Cambridge: Cambridge University Press, 2000), 73. See also J. McConica, 'Humanism and Aristotle in Tudor Oxford', *The English Historical Review*, 94, no. 371 (1979): 291–331; L. V. Ryan, 'Hakluyt's Voyage,' 81–84; N. Crane, *Mercator*, 41–42. Ramusio, who almost certainly attended the University of Padua, would certainly have studied these texts where they formed a set part of the curriculum. P. F. Grendler, *The Universities of the Italian Renaissance* (Baltimore: Johns Hopkins University Press, 2001), 269.

14 C. Martin, 'Experience of the New World and Aristotelian Revisions of the Earth's Climates During the Renaissance', *History of Meteorology*, 3 (2006),1–16.

15 On its circulation see P. L. Schoonheim, *Aristotle's 'Meteorology' in the Arabico-Latin Translation: Critical Edition of Both Texts, Introduction, Index and Registers (Aristoteles Semitico-Latinus)* (Leiden: Brill, 2000), xiii.

16 The references to Aristotle in Acosta are almost too numerous to count. He uses his authority on anything from the position of the earth in relation to the heavens to the extent of the waters. He also cites his authority on animals – for example, to discuss birds' toleration of water. He clearly read a range of texts from the *De Caelo* and *De Meteorologia* to *De Animalia.* See, for example, J. de Acosta, *Historia*, 7, 34, 426. Elsewhere he feels he has to dispute Aristotle expressly if Aristotle has made statements which Acosta feels have been disproved. See for instance his discussion of the inhabitability of the torrid zone (*ibid.*, 120). Acosta's familiarity with Aristotle is not surprising since he had studied at the University of Alcala where the third year of the degree was devoted to Aristotelian natural philosophy. C. M. Burgaleta, *José De Acosta, S. J., 1540–1600: His Life and Thought* (Chicago: Jesuit Way, 1999), 18. Indeed in some ways more surprising is his willingness to challenge Aristotle. Oviedo likewise filled his works with references to Aristotle. See e.g. G. F. de Oviedo, 'Della naturale e generale istoria dell'Indie a' tempi nostri ritrovate Libro Primo', *NV*, vol. 6,

contested early on in the Age of Reconnaissance, he remained an authority in other matters related to geography. He was still thought to provide insight into both the structure of the world and the societies it contained.[17]

More surprisingly, Platonic philosophy, despite its emphasis on the world of reason and intellect and not of the senses, became a key contributor to sixteenth-century geographical theory, and indeed exploration. Many of the same writers who referred to Aristotle also cited Plato: they clearly saw philosophy as providing insight into understanding the world.[18] Some (though not the majority) even valued Plato more highly than Aristotle. The Tudor explorer and courtier, Sir Humphrey Gilbert, for example, made repeated references to Plato's *Timaeus* and *Critias*, and found inspiration in Platonic and Neoplatonic philosophy in his *Discourse on a Northwest Passage*, but did not refer to Aristotle.[19] Mercator did not accept Plato's ideas of what was created and what was eternal as valid, but nonetheless believed that Plato's ideas of theology and harmony and the order of creation, although flawed, contained useful premises for understanding the nature of the world, whereas he largely rejected Aristotle. While he makes multiple references to Plato, there are only three references to Aristotle in his Atlas, and of these only his discussion of the best form of commonwealth is a truly affirmative citation.[20]

Nor were Aristotle and Plato the only classical philosophers to provide a foundation for European geographical thought in the sixteenth century. Many

361. Oviedo refers to Aristotle three times in reference to a pre-Columbian European knowledge of the New World. Oviedo was less of a scholar than Acosta, but, as Gerbi puts it, after years in Italy he came back to Spain 'above all, as a wholehearted disciple of humanism… filled with reverence of antiquity reborn and the new literature', and this is clearly to be seen in his approach to knowledge and his interpretation of the New World. His use of Aristotle is just one aspect of this. A. Gerbi and J. Moyle, *Nature in the New World*, 139.

17 See for example, J. Bodin, *Methodus*, 99; W. Cuningham, *The Cosmographical Glasse* (London: in officina Ioan Daij, 1559), 6; L. le Roy, *Vicissitude*, 23v. There has been a very great deal written on Aristotle and new world inhabitants. Leading the way was Lewis Hanke's ground-breaking book, *Aristotle and the American Indian.* Perhaps the most important was A. Pagden's *The Fall of Natural Man*. Less has been written on his role in the understanding of physical geography in the aftermath of the discoveries.

18 J. Bodin, *Methodus*, 122; W. Cuningham, *Cosmographical Glasse*, 6; L. le Roy, *Vicissitude*, 23v; J. de Acosta, *Historia*, chapters 12 and 22 passim; I. Barbaro, 'Di Messer Iosafa Barbaro, Gentiluomo Veneziano, Il viaggio della Tana e nella Persia', *NV*, vol. 3; G. B. Ramusio, 'Discorso Sopra Il Terzo Volume', vol. 5, 5; G. Fracastoro, 'Risposta dello Eccellentissimo Messer Ieronimo Fracastoro del crescimento Del Nilo a Messer Gio. Battista Ramusio', *NV*, vol. 2, 412, 13. Chapter 6 contains much more discussion about the role of Platonic philosophy in sixteenth-century geography.

19 G. Gascoigne, 'To the Reader', in *A Discourse of a Discoverie for a New Passage to Cataia, Reprinted in Voyages and Colonising Enterprises of Sir Humphrey Gilbert*, ed. by David B. Quinn, 2nd series (London: Hakluyt Society, 1940).

20 G. Mercator, *Atlas*, a2, liiir.

lesser-known, and even fragmentary, geographers were cited as authorities. The works of Ramusio, and even Thevet, for example, provide direct references not only to Plato and Aristotle, but to pre-Socratic fragmentary writers. Both Ramusio and his friend Fracastoro, for instance, cite Ionian philosophers Thales and Anaxagoras in their discourses on the flooding of the Nile, and give detailed discussions of the actual philosophical ideas of these sixth-century-BC authors. Indeed, Fracastoro devotes a whole section of his discourse to refuting ideas which originated in Thales that the flooding of the Nile derived from blockages caused by the wind at the mouth of the Nile.[21] Thevet's (or more probably Héret's) reference is less precise. He explicitly cited Thales and Pythagoras, saying that they maintained that there were uninhabitable zones, but that he knew they were wrong.[22] This statement betrays no real knowledge of Thales, and it seems it was more part of the veneer of a profound knowledge of the ancients with which Héret furnished Thevet's work. The very existence of the references, however, indicates that he thought Thales to be well enough known as a geographical and philosophical authority to be worthy of citation.

What Ramusio, Fracastoro and Thevet realised was that Thales and other pre-Socratic authors were not merely philosophers, but were also among the earliest Greek geographers. Their writings were key influences on Herodotus. Strabo, writing a specifically geographical work, cited various writers whom he considered had contributed to the science of geography, including Anaximander, Democritus and Posidonius – authors now more usually considered philosophers, although Posidonius also wrote a history. Strabo believed that geography simply could not be a closed discipline. He explicitly stated that geography could not be understood without philosophy (in its literal meaning of love of knowledge).[23] Similarly, in the renaissance, although the two disciplines had become separated, many geographical writers took the attitude of Strabo. Like their classical predecessor, they thought all forms of knowledge could contribute to the study of geography. The links with what we would now term philosophy, however, were particularly strong. The subjects were interdependent and the abstract ideas of the one were given practical application in the other.

It was philosophical reasoning combined with theology which had provided the bounds to the *oikoumene* initially. Anaximander of Miletus, for instance, drew the first known western map of the world, but this was a map depicting the theoretical shape and structure of the world in so far as it he understood it through

21 G. B. Ramusio, 'Discorso di Messer Gio. Battista Ramusio sopra il crescer del fiume Nilo, allo Eccellenitssimo Messer Ieronimo Fracastoro', *NV*, vol. 1, 400; G. Fracastoro, 'Risposta', 410.

22 A. Thevet, *Singularitez*, 35v. He also refers to Thales in A. Thevet, *Cosmographie Universelle*, Aiii.

23 Strabo, 1.1.

his natural philosophy, based on ideas of a natural balance in all aspects of the universe, which was reflected in the symmetrical relationship between the earth's continents.[24] Even when eye-witness evidence gave further shape and definition to the *oikoumene* in the classical world, a philosophical understanding continued to underpin ideas of the nature of the *oikoumene*. Herodotus, writing a century after Anaximander, wished to be an arch empiricist, but there were limits to the observational evidence he could obtain. He then wove his empirical data into his philosophical understanding of the world, and continued to propose limits of symmetry and proportionality to the *oikoumene* which were largely founded upon philosophical reasoning. In his tri-partite world there was still a proportionate relationship between continents, with Africa and Asia combined, forming the counterpart to Europe.[25]

While philosophical writings helped people to think about the shape and structure of the world and the *oikoumene*, and the natural laws which governed them, the influences of these modes of thinking can also be seen in the more overtly geographical writings of the classical world which likewise came to contribute to the geographical discourse in the sixteenth century. Certain texts are, unsurprisingly, rather prominent. No sixteenth-century geographer could overlook the works of Ptolemy, although, as I have argued elsewhere, they did not necessarily find him as useful as other ancient authorities.[26] Ptolemy did not write descriptive geography, and particularly in the written geographies which are at the heart of this study Ptolemy and the Greek mathematical geographers were not always as highly valued as other classical sources. Strabo's *Geographia*, on the other hand, with its focus on humans and descriptive geography, was extremely relevant to those interested in the type of geography which many of these humanist writers conceived.

The references to such texts were part and parcel of the classical revival, and the thirst for new geographical knowledge spurred on by the European fifteenth- and sixteenth-century voyages of exploration forged a symbiotic relationship between exploration of ancient texts and new lands. We have already seen how Thales and Anaxagoras came to be considered writers of note despite the majority of their texts having been lost to the ravages of time, but the pre-Socratic

[24] G. I. Irby, 'Mapping the World: Greek Initiatives from Homer to Eratosthenes', in *Ancient Perspectives: Maps and Their Place in Mesopotamia, Egypt, Greece, and Rome* ed. by Richard J. A. Talbert (Chicago: Chicago University Press, 2012), 89.

[25] Herodotus, 4.42. Although he thought that Europe was wider than either Asia or Africa in territorial extent, he believed in climatic zonal boundaries of the torrid and frigid zone which extended across all the continents, thereby creating symmetrically opposite limits of hot and cold to the entire *oikoumene*.

[26] M. Small, 'Warring Traditions: Ptolemy and Strabo in the Geography of Sebastian Münster', in *Ptolemy's Geography in the Renaissance*, ed. by Zur Shalev and Charles Burnett, vol. 17, Warburg Institute Colloquia (London: The Warburg Institute, 2011).

writings were certainly not the only fragmentary texts which sixteenth-century geographers used. As early as 1557 Henri Estienne, when he was launching his publishing workshop, thought it was worth publishing the Byzantine scholar Photius's epitomes from Ctesias, a fifth-century-BC Greek ethnographer, travel writer and documenter of marvels, with selections from Agatharchides, another of the Greek writers on Asia.[27] Many of the humanist geographers cited both their works even though Ctesias at any rate was almost universally derided by classical writers.[28] Megasthenes, a third-century-BC Greek historian, traveller and geographer was likewise frequently cited (though more often by scholars of the Levant than by authors discussing the newly encountered territories).[29] Sir Walter Raleigh even referred to the fragmentary nature of Megasthenes' text, speaking of 'a fragment of Megasthenes found in Eusebius, wherein this Nabonidus is called the Median'.[30] Megasthenes was certainly not the most well-known author but, as can be seen, he appeared in humanist writings ranging from those of the academic, Münster, to the scholarly adventure courtier, Walter Raleigh. Long-lost writers were witnessing a revival, and excerpts of their works were transmitted not just through Pliny but also via intermediaries such as Photius, Eusebius, Diogenes Laertius and Strabo whose works themselves were part of the new knowledge of the classical revival. They were particularly popular because they presented ideas and supposed facts about areas such as India which were being reassessed in European geographical understanding.

The beginning of the classical revival predated the voyages of discovery, of course. Byzantine and Arabic writers had kept much of the Greek geographical tradition alive, and in the Islamic tradition texts such as Ptolemy's *Geographia* had never been lost.[31] These texts then filtered into western Europe where much of the rediscovery was dependent on their survival in the Arab world. Already in

27 Ctesias, *Ex Tôn Ktêsiou, Agatharchidou, Memnonos Historikôn Eklogai, Appianou Hibêrikê Kai Annibaikê* (Geneva: Henri II Estienne, 1557).

28 On Ctesias's unreliability see Aristotle, *Hist. Animal.*, VIII, 28; Photius, *Bibl.* 27. On citations of Agatharchides by sixteenth-century geographical writers, see Acosta who uses him as an authority in smelting gold (J. de Acosta, *Historia*, 299) and J. Bodin, *De Republica Libri Sex, Latine Ab Autore Redditi, Multo Quam Antea Locupletiores. Cum Indice Copiosissimo* (Jacques du Puys: Lyons, 1586), 484). For examples of sixteenth-century geographical writers who cited Ctesias, see J. Boemus, *Omnium Gentium Mores*, 91 and A. de Torquemada, *Jardin de flores curiosas, en q̃ se tratã algunas materias de humanidad, philosophia, theologia, y geographia, con otras cosas curiosas, etc* (Salamanca: J. B. de Terranova, 1570), 52.

29 J. Boemus, *Omnium Gentium Mores*, 90; S. Münster, *Cosmographiae Universalis Lib VI* (Basle: Henric Petri, 1550), 1082, 1080. Torquemada, in a rather strange discussion on satyrs, cites him as an authority for Pliny. A. de Torquemada, *Jardin*, 23.

30 W. S. Raleigh, *The History of the World* (London: printed by William Iaggard for Walter Burre, 1617, 1st edition 1614), Book 3, cap 5, 9.

31 G. R. Tibbetts, 'The Beginnings of a Cartographic Tradition', in *History of Cartography:*

the twelfth century, the unearthing of Euclid's *Geometry* and Ptolemy's *Almagest* had altered the direction of western European thinking, and the thirteenth-century rediscovery of Aristotle in the West changed both European methods of enquiry and the ways in which they understood the world.[32] The renewed interest in the classics received more impetus in the first half of the fifteenth century when Poggio Bracciolini, perhaps the most important player of all in the rediscovery of classical knowledge, brought to light the works of Lucretius, Silius Italicus, Manilius and Ammianus Marcellinus, as well as the *Silvae* of Statius, the minor works of Tacitus, and even eight previously unknown speeches of Cicero from the library at Cluny.[33] At almost exactly the same time Jacopo d'Angelo completed a translation of the text (although not the maps) of Ptolemy's *Geographia* into Latin and introduced the work in full to Italy. [34] The history of the reception of the *Geographia* is undoubtedly the most famous example of the symbiotic relationship between the classical revival and the reformation of geographical ideas in the Age of Discovery. The thirst for new geographical knowledge fuelled the widespread revival of a classical text which in turn had a significant influence upon reconceptualising the world in the European Age of Discovery. In turn, the alterations in geographical understanding brought about by increased empirical knowledge over the course of the Age of Discovery ultimately undermined the authority of that text. Much has been written about the reintroduction of Ptolemy's *Geographia* to western Europe and the impact this work had upon people's ideas about the shape and structure of the world.[35] Although it never achieved great

Cartography in the Traditional Islamic and South Asian Societies, ed. J. B. Harley and D. Woodward, vol. 2.1 (1992).

32 M. Bowen, *Empiricism and Geographical Thought from Francis Bacon to Alexander Von Humboldt* (Cambridge: Cambridge University Press, 1981), 27.

33 L. D. Reynolds and N. G. Wilson, *Scribes and Scholars: A Guide to the Transmission of Greek and Latin Literature*, 3rd edn (Oxford: Oxford University Press, 1991), 113–14.

34 Ptolemy's book was variously known as *Geographia* and *Cosmographia* in the Renaissance. See J. L. Berggren and A. Jones, *Ptolemy's Geography: An Annotated Translation of the Theoretical Chapters* (Princeton: Princeton University Press, 2000), 51. As Patrick Dalché has shown, Ptolemy's geography was far better known in the medieval period than many have thought. See P. G. Dalché, *Ptolemée En Occident.* Nonetheless, D'Angelo's translation marked a new era. On the translation of the maps see P. G. Dalché, 'The Reception of Ptolemy's Geography (End of the Fourteenth to Beginning of the Sixteenth Century', in *Cartography in the European Renaissance: the History of Cartography*, ed. by David Woodward, vol. 3.1 (Chicago: University of Chicago Press, 2007), 292–95.

35 See for example Z. Shalev and C. Burnett, 'Ptolemy's 'Geography' in the Renaissance'; L. Rombai, 'Tolomeo e Toscanelli, fra Medioevo ed età moderna; cosmografia e cartografia nella Firenze del xv secolo', in *Il mondo di Vespucci e Verrazzano, dalla Terrasanta all'America*, ed. by Leonardo Rombai (Florence: Leo S. Olshki, 1992); M. Milanesi, *Tolemeo sostituito*, particularly chapter 1; S. Y. J. Edgerton, *Mental Matrix*,

popularity in the ancient world, it was important enough still to be circulating in the sixth century AD when Marcianus of Heraclea relied on it heavily for his *Periplus of the Outer Sea*. Thereafter it disappeared from western Europe and required a rediscovery. This was made possible by its continuing circulation in the Byzantine and Arab worlds where scholars never ceased to appreciate the Greek cosmographical tradition.[36] Around 1300 the Byzantine scholar Maximos Plaundes rediscovered the *Geographia* and thereafter it began to circulate in the Byzantine world, and was ultimately translated into Latin in Italy by Manuel Chrysoloras and Jacopo d'Angelo in 1409.[37] This translation became widely known after the Council of Constance (1414–1418) where Cardinal Guillaume Fillastre obtained a copy of the *Geographia* and took it back to France. It subsequently began to circulate outside Italy as well, and came to be in huge international demand.[38] There are more than forty extant fifteenth-century manuscripts of Ptolemy's *Geographia*, which would seem to indicate that it was copied intensively after being rediscovered, creating a wide but not unrestricted readership.[39]

The discovery of Ptolemy's work acted as a turning point in the geography of late medieval Europe, by bringing the subject of mathematical geography to greater prominence but also by providing a different context for considering the world beyond the *oikoumene*. Ptolemy's own *oikoumene* was simply a variation on the traditional model. His *oikoumene,* although extending further south than many models, was still firmly confined to the northern hemisphere. Nonetheless, what Ptolemy helped to do was provide a new means of imagining the world and the relationship of the *oikoumene* to the world. Ptolemy's work was rediscovered at a time when there was a pressing need for a new framework for geography. Arab scholars had been working on mathematical projections throughout the Middle Ages, but these had not been a feature of western European geography

10–51; W. G. L. Randles, 'Recovery of Ptolemy's Geography'; N. J. W. Thrower, 'The Rediscovery of Ptolemy and Cartography in Renaissance Europe', in *Maps and Civilization: Cartography in Culture and Society*, ed. by Norman J. W. Thrower (Chicago: University of Chicago Press, 1999), 58–75. For a recent (and contrasting) view see P. G. Dalché, *Ptolemée En Occident.*

36 There are, for instance, two ninth-century Byzantine compilations which contain references to the *Geographia*, and a further reference in Johannes Tzetzes' twelfth-century *Chiliades*. J. L. Berggren and A. Jones, *Ptolemy's Geography*, 50–51.

37 *Ibid.*, 43.

38 *Ibid.*, 52; M. D. Reeve, 'The Rediscovery of Classical Texts in the Renaissance', in *Itinerari dei testi antichi*, ed. by O. Pecere (Rome: 1991), 137; A. Codazzi, *Le edizioni quattrocentesche e cinquecentesche della 'Geografia' di Tolomeo* (Milan: La Goliardica, 1949); L. Pagani, *Claudii Ptolemaei Cosmographia* (Torriana: Stella Pollare, 1990), p. VII.

39 C. L. Joost-Gaugier, 'Ptolemy and Strabo and Their Conversation with Appelles and Protogones: Cosmography and Painting in Raphael's School of Athens', *Renaissance Quarterly*, 51 (1998), 769.

until the later Middle Ages. From the twelfth century onwards Europeans had been broadening their geographical horizons and expanding the boundaries of the *oikoumene*. A few scholars, such as Roger Bacon (1214–1294) and Sacrobosco (late eleventh or twelfth century), had begun to consider the world as a whole and to develop a mathematical geography. Sacrobosco's treaty of the sphere was the most widely read cosmographical work until the sixteenth century. In it he broke the earth into climatic subdivisions and then set the sphere of the earth into the context of the sphere of the heavens. Sacrobosco still firmly believed in the existence of inhabitable and uninhabitable territories, but used mathematics and logic to create theories about the nature of the whole earth and its cosmological setting.[40] Roger Bacon went further. He partly relied on a statement made by Aristotle in the *De Caelo* 2.14 that the earth was not of great size and that there was continuity between Asia and India.[41] He took this to indicate that it would be possible to reach the eastern borders of the *oikoumene*, by sailing westwards, arguing that the sea covered less than three quarters of the earth.[42] He also developed a co-ordinate system for plotting geographical locations which, though not widely read, did influence Peter D'Ailly, one of the foremost medieval geographers and a major influence on Christopher Columbus.[43] Bacon's coordinate system did not reach the scale of a full cartographical projection like Ptolemy's, but it shows that some Europeans were beginning to grasp the need for a mathematical geography to enable them to explore and depict the earth.[44] Ptolemy's geography therefore filled a void which men such as Sacrobosco and Bacon had begun to show existed. It enabled people to look beyond the bounds of the *oikoumene* and truly to think globally.

Ptolemy's *Geographia* is not stimulating reading, and until the Renaissance it was never widely read. The majority of the text amounts to little more than lists of places with latitudinal and longitudinal references. These have some value when accompanied by maps, but early in their history they became divorced

40 L. Thorndike, ed., *The Sphere of Sacrobosco and Its Commentators* (Chicago: University of Chicago Press, 1949), 110, 38.

41 Aristotle, *De Caelo* 2,14. W. G. L. Randles, 'The Evaluation of Columbus' India Project by Portuguese and Spanish Cosmographers in the Light of the Geographical Science of the Period', *Imago Mundi, The International Journal for the History of Cartography*, 42 (1990): 50–52.

42 A. Hiatt, *Terra Incognita*, 136.

43 The fourth book of Roger Bacon's *Opus Maius* is a treatise on geography containing many interesting ideas. See particularly 292 and 300 in R. Bacon, 'Opus Maius', ed. by John Henry Bridges (Oxford: Clarendon Press, 1897–1900). D. Woodward, 'Roger Bacon's Terrestrial Coordinate System', *Annals of the Association of American Geographers*, 80, no. 1 (1990), 109–22.

44 Dalché indeed argues that Bacon was influenced by Ptolemaic ideas transmitted via Arabic geographers. P. G. Dalché, *Ptolemée En Occident*, 135.

from cartography. Renaissance scholars seem to have believed that the maps they were familiar with were directly derived from ones which Ptolemy drew to accompany the original manuscript, when in fact the majority of these were drawn in around 1300 AD by the monk Maximos Planudes from his interpretation of Ptolemy's textual descriptions.[45] Whether Ptolemy himself ever drew maps for the geography is doubtful; if he did, they were soon separated from the text.[46] The text, devoid of maps, is dry and difficult, which perhaps helps to account for its disappearance from western European consciousness. The maps therefore make an important addition, but while making the text more readable and 'user-friendly' they also throw the focus even more onto the *oikoumene*. Although Ptolemy's *Almagest* discussed a geocentric system in which the whole sphere of the earth was pictured in relationship to the other heavenly spheres, all the maps published in Ptolemy's *Geography* did not extend much beyond the *oikoumene.*

Although the focus of Ptolemy's maps was the *oikoumene*, which he discussed as a distinct region, separate from the rest of the globe, in writing he showed the relationship of the *oikoumene* to the world (*ge*), and of the world to the universe (*kosmos*). His framework was cosmography rather than human geography. Whereas human geographers like Strabo looked at the relationship between peoples and places, between products and terrain, Ptolemy studied the relationship between the heavens and the earth, the universe and the world. Humans were not entirely incidental to Ptolemy's geography, however. Although he was uninterested in diverse ethnic groups, different creatures or human customs, he did not simply deal with global or cosmic matters, such as the exact circumference of the earth (which he estimated at about 180,000 stades – about 18% smaller than it actually is).[47] He also attempted to establish locations for cities and regions with mathematical precision, by, for example, measuring the angle of elevation to the stars at a specific location. All his calculations were based on data provided by empirical observation. Since only humans could make such observations, his data was necessarily taken from regions within the *oikoumene*, and the locations which he gave were also within the *oikoumene*.[48] The *oikoumene* therefore imposed itself upon his geography. Yet though practical necessity determined that he focussed on the *oikoumene,* it was clearly not his sole or even primary interest. He blurred distinctions between the *oikoumene* and the globe, and between the terrestrial and the celestial spheres.[49] He believed in the traditional idea of frigid

45 J. L. Berggren and A. Jones, *Ptolemy's Geography*, 46–51; A. Diller, 'The Oldest Manuscripts of Ptolemaic Maps', *Transactions and Proceedings of the American Philological Association*, 71 (1940).

46 L. Rombai, *Tolomeo e Toscanelli*, 40–41.

47 J. L. Berggren and A. Jones, *Ptolemy's Geography*, 21.

48 Ptolemy, *Geographia*, 1.2.

49 L. C. Taub, *Ptolemy's Universe: The Natural, Philosophical and Ethical Foundations of Ptolemy's Astronomy* (Chicago: Open Court Press, 1993), 4.

and torrid zones as is evident in his discussion of Ethiopians, their skin colour and location,[50] but he was much more interested in how celestial activity created these zones than in the implications of the limits. The cosmographical tradition looked beyond the bounds, focussing on the earth and the universe. This intellectual framework therefore proved useful for sixteenth-century geographers seeking to reimagine the limits of the *oikoumene*, even when much of his data proved insufficient.

After it was printed and its maps began to be mass-produced, Ptolemy's *Geographia* circulated extremely widely. By the end of the fifteenth century it was the most widely reproduced ancient scientific work, and new editions continued to be produced all over Europe throughout the sixteenth century.[51] In contrast to previous eras, many Renaissance readers had both the interest and the mathematical and theoretical tools to grasp the full complexity of Ptolemy and his method of mapping a celestial grid onto the terrestrial earth. Once the problem of language had been surmounted by translating Ptolemy into Latin, the possibilities of mathematical geography based on a cartographic projection system and coordinates of latitude and longitude were opened up to a very wide variety of Renaissance readers. The range of readership the *Geographia* acquired was astonishing. His *Geographia* became the hallmark of geographical authority. Münster, Ramusio, Boemus, and Blundeville are among the many writers who never travelled themselves, but used all the ancient writers they could find, and were particularly influenced by Ptolemy. They were not the only ones, however. Map-makers, of course, relied on his book, but so did travellers. Even those who ventured to regions never known to Ptolemy cited him. Some, André Thevet being the most notable example, used his name only to demonstrate their own familiarity with ancient authorities and their supremacy over them, but others continued to value his *Geographia*.[52] Amerigo Vespucci and Acosta, travellers to the New World at either end of the sixteenth century, both referred to him; João de Barros and Andréa Corsali, writing about eastern Asia, were also among those who used him as an authority. [53]

Fairly rapidly, however, early modern geographers and explorers saw that a Ptolemaic geography could not be followed unquestioningly but required adaptation to fit with Renaissance discoveries. Perhaps the most famous – if

[50] Ptolemy, *Geographia,* 1.9

[51] B. Weiss, *Geography in Print*, 91.

[52] For example A. Thevet, *Singularitez*, 5,37r, 42v.

[53] A. Vespucci, 'Epistola de Alberico Vesputio', in *Paesi novamente retrovati*, ed. by Fracanzano da Montalboddo (Vicenza: 1507), book 5, ch. cxv; J. de Acosta, *Historia*, 145. J. de Barros and D. de Couto, *Da Asia de João de Barros e de Diogo de Couto*, new edn (Lisbon: Na Regia Officina Typografica, 1778), passim; e.g. Decada I, vol. III, ch. I, 179, Decada I, vol. III, ch. IV, 204–5.; A. Corsali, 'Di Andrea Corsali Fiorentino allo illustrissimo signor duca Giuliano de' Medici lettera scritta in Cochin, terra dell'India, nell'anno MDXV, alli VI di gennaio', *NV*, vol. 2, 33.

also in some ways the most surprising – of these modifications are those of Columbus, who staked his life on his revisions of Ptolemaic geography. He is known to have read the 1477 Bologna edition of Ptolemy's *Geographia*, and to have been greatly interested in its contents. Even before he began exploring westwards, however, he decided that Ptolemaic geography did not agree with his image of the world. He therefore went to the somewhat extraordinary step of correcting Ptolemy's calculations before he set sail in order to create an image of the world which was more feasible for western navigation. To do this he resorted to a variety of other writers. He based much of his calculations on an estimate of the extent of the earth made by the Greek cosmographer Marinus of Tyre.[54] Marinus's book was known to the Renaissance only through Ptolemy, who despised much of his predecessor's work. It suited Columbus better, however, since while Ptolemy clung to a traditional idea of the east-west limits of the *oikoumene*, wherein Ocean covered the whole 'reverse' side of the globe, and therefore calculated that the Atlantic covered 180 degrees from Europe to Asia, Marinus thought it extended through only 155 degrees. Columbus therefore followed the contemporary scholars Martin Behaim and Paulus Toscanelli in adopting Marinus's calculations, even though Ptolemy had taken pains to show the flaws in Marinus's mathematics.[55] Columbus also turned to Aristotle, Seneca and Pliny for arguments demonstrating that the Ocean covered a sufficiently small part of the earth's surface to be navigable, and drew clinching support from a passage in the second Book of Esdras in the Apocrypha which asserts that water covered one seventh of the earth's surface (Esd 6:42).[56] In doing so he was following in the footsteps of no lesser authorities than Peter D'Ailly and Roger Bacon who had also cited the same passage.[57] All of these sources seemed to argue against a strict adherence to Ptolemy's geography and his belief that water covered half the world.[58] Here was clearly a case of pitting one ancient geographical writer against another in an effort to retain classical authority, while endorsing a changing concept of the *oikoumene*. Marinus had been debunked and corrected in the ancient world, but was now being revived,

54 G. E. Nunn, 'Marinus of Tyre's Place in the Columbus Concepts', *Imago Mundi*, 2, no. 1 (1937).

55 L. C. Jane, ed., *Introduction*, Select Documents Illustrating the Four Voyages of Columbus, 2 vols, vol. 2 (London: Hakluyt Society, 1933), 84.

56 C. Columbus, 'Raccolta di documenti e studi pubblicati dalla R. Commissione Colombiana', part I, vol. 2 (Ministero della pubblica istruzione: Rome, 1892), 184.

57 A. Hamilton and L. Garbett, *The Apocryphal Apocalypse: The Reception of the Second Book of Esdras (4 Ezra) from the Renaissance to the Enlightenment* (Oxford: Oxford University Press, 1999), 9, 28–30.

58 F. Fernández-Armesto, *Columbus and the Conquest of the Impossible* (London: Phoenix Press, 1974), 22–23.

despite his work existing only in citations in other authors, because his idea of the world best suited Columbus's.

Columbus was unusual in being an explorer who himself corrected Ptolemy. More often, it was the stay-at-home humanist scholars who were troubled by flaws in Ptolemy's calculations. Many explorers continued to use Ptolemy's *Geographia* as a reference book, but their discoveries were inconsistent with Ptolemy's theories and therefore posed problems for Ptolemaic scholars and editors. Because Ptolemy held a vision of the world, however, which had allowed for the possibility of his geography being adapted to extend beyond his known *oikoumene*, humanist geographical writers continued to use his text. He thought that some parts of the *oikoumene* had never been reached while others were inaccurately reported. He therefore believed that as humans explored further into the unknown regions of the *oikoumene,* they would provide further details which could be used to update the *Geographia.* Moreover he thought that it was not simply increased information which would require changes to be made to his cartography. He also firmly believed that the physical landscape itself changed over time and that in due course there would have to be further cartographic adjustments made to reflect this change.[59] For him, geography was not static but rather in a constant state of flux – a concept that would have appealed to the Renaissance humanists who saw their world changing almost before their eyes.

This attitude encouraged humanist-trained geographers to continue to revisit Ptolemy's texts long after the flaws in his geography had been acknowledged. In 1511, Bernardo Sylvano produced an edition of Ptolemy in which he admitted that he could no longer believe in the absolute authority of the Ptolemaic locations of sites within the *oikoumene*.[60] In the dedicatory letter at the beginning of the text, he wrote that

> Although I used to view Ptolemy as having recorded the sites and distances of places more diligently than other writers of geography, I was none the less puzzled as to why his tables agreed in very few instances with the navigations of our times: and I was all the more puzzled because I used to think that Ptolemy too had recorded what he recorded relying on navigations more than anything else. And as I frequently thought about this, and gathered numerous examples, both Greek and Latin, from everywhere in all the things that came into my hands, I found that there was a very great discrepancy in the numbers with which the intervals between places are defined.
>
> When I had noticed this, I began to examine the actual words of Ptolemy more diligently; and although in several places his numbers were in conflict

[59] Ptolemy, *Geographia*, 1.5.

[60] B. Sylvano, *Ptolemaeus Cosmographia (Venice 1511), facsimile edn with an intro. by R. A. Skelton* (Amsterdam: Theatrum Orbis Terrarum, 1969).

> with his words, I realised that almost all Ptolemy's words correspond with the navigations, but that the numbers differ very greatly. When I had seen this, I was moved by certain logic perhaps more boldly than was appropriate, and comparing Ptolemy's words with the actual navigations, I have corrected the numbers – which can be corrupted most easily, and frequently are.[61]

He decided that the problem was not with Ptolemy's data in themselves but with the transmission of the data, and proceeded to correct the numbers to agree with what the navigations of his own time indicated. For the bulk of the *Geographia*, Ptolemy provides little more than place names and numbers, so in making these corrections he was correcting the very essence of the text. He was also clearly not very happy that he was doing so. His statement that 'he was moved by certain logic perhaps more boldly than was appropriate' shows that he considered himself on dangerous ground in doing so. Yet he still relied on the newly recovered authority of Ptolemy because he offered a vision of the world and the *oikoumene* which he found useful.

For many humanist scholars, the idea of altering the text itself was unacceptable. Sebastian Münster, for instance, who, in 1540, produced one of the more influential editions of Ptolemy, was perfectly aware that Ptolemy had made mistakes in his astronomical tables from which he made his calculations about latitudes. He included the tables unaltered in his version of Ptolemy, but in the introduction to the 1545 edition he pointed out that Ptolemy had made errors. He did not, however, correct them, saying he left it to those with greater knowledge than he had.[62] Instead, employing his skills as a humanist, he tried to develop an authentic classical text. He used Pirckheimer's edition, but amended that where necessary since he felt that Pirkheimer's text, although the best available, particularly in terms of the mathematics, was not wholly reliable since he had begun with a

61 Cum Ptolemaeum inter alios geographiae scriptores diligentissime et situs et distantias locorum scripsisse conspicerem admirabar perfecto cur illius tabulae paucis admodum in rebus cum nostri temporis navigationibus consentirent: eoque magis admirabar quod Ptolemaeum quoque navigationibus comprimis innixum ea quae scripserit scripsisse arbitrabar. Cumque hac de re multa frequenter animo agitarem, diversaque et graeca et latina exemplaria undique conquirerem in omnibus quae ad manus meas pervenerunt maximam in numeris quibus locorum signantur intervalla varietatem inveni. Quod cum animadversissem coepi dilegentius ipsa Ptholemaei verba examinare: cumque plerisque in locis numeri verbis ipsis repugnarent cognovi Ptholemaei verba navigationibus prope omnia respondere: numeros maxime dis<s>entire. Quae cum vidissem audacius fortasse quam par erat ratione tunc certa permotus Ptholemaei verba cum navigationibus ipsis conferens numeros qui facillime depravari et possint et soleant correxi. 'Dedicatory Letter to Acquaviva Sylvanus', in *Ptolemaeus Cosmographia, VII.*

62 S. Münster, *Claudius Ptolemaeus: Geographia (first edition 1540)*, 4th edition (Basel: Heinrich Petri, 1552), 3v.

faulty Greek text.[63] Münster typified the problem of the humanist in approaching any classical work, which, like the *Geographia*, proved to be faulty. His solution of supplementing rather than amending the text was not wholly satisfactory for a humanist. By adding to the work and showing where it was lacking he both undermined its authority and, in a sense, contaminated the text.[64]

Münster's and Sylvano's approaches to Ptolemy show the difficulty that humanist scholars faced in maintaining the authority of the ancients in the face of new evidence. Both saw the flaws in Ptolemy's text and the need to supplement his information, yet both thought it worthwhile to present editions of his *Geographia* because they provided new ways of envisaging and understanding the *oikoumene*. Despite the clear problems with Ptolemy's view of the world, Mercator still produced a new edition of his *Geographia* as late as 1578 (although this was admittedly more for antiquarian than contemporary purposes), and as Cormack has shown, his geography came to be considered an important subject of study in the universities as late as the second half of the sixteenth century.[65] At heart, many geographical writers throughout the sixteenth century continued to believe that Ptolemy's ideas, if not his results, were worth maintaining. They were aware that even within a single ancient author there could be contradictions, but this did not mean that the ancient work needed to be jettisoned. It was merely up to the later scholars to find the kernels of truth hidden in these authorities. Ptolemy himself might have agreed. Ptolemy's texts were a crucial element in the search for new classical authorities and classical frameworks by which sixteenth-century writers could interpret the nature of an *oikoumene* that seemed to be rapidly changing in front of their eyes.

The reintroduction of Ptolemy's *Geographia* into western Europe, and the consequent relaunching of the study of mathematical geography based on astronomical observation and calculation of longitude and latitude, must be one of the most famous instances of the impact of the classical revival; yet Ptolemy provided only one facet of ancient geographical thought for Renaissance readers. Human geography, like mathematical geography, also underwent a period of unprecedented popularity with a corresponding recovery of long neglected classical texts. Human geography had developed in response to periods of rapid expansion and empire-building in the ancient world. The Homeric poems which

63 S. Münster, ed., 'Reverendo in Christo Patri Ac Domino Philippo À Gundelsheim, Episcipo Basiliensi', in Ptolemy, *Geographia* (Basel: 1540; facsimile edition Amsterdam, 1966).

64 K. Burmeister, *Sebastian Münster: Versuch Eines Biographischen Gesamtbildes*, 2nd edn (Basel: Basler Beiträge zur Geschichtswissenschaft, 1969), 117. For more on Münster's Ptolemy, see M. Small, *Warring Traditions*.

65 L. B. Cormack, 'Ptolemy at Work: The Role of the *Geography* in Geography and Mathematics Teaching in Early Modern England', in *Ptolemy's Geography in the Renaissance*.

provide some of our earliest insights into aspects of Greek geographical thought were composed in the context of Ionian colonisation. Herodotus's Histories, which contain much of our information on fifth-century Greek geographical thought, were written about the wars between the Persians and Greeks and their (broadly conceived) cause, and at the height of the Peloponnesian wars when Athens and Sparta were engaged in a bitter conflict for power and territory. The works of Polybius, Posidonius and Strabo were all written by Greeks in response to Roman imperialism.[66] Such works taught a lesson that national characteristics and human history were dependent upon their geographical environment and emphasised how the environment defined the nature of the *oikoumene*.

Herodotus, for instance, explicitly concluded his *Histories* with the words of the Persian King Cyrus, that 'people tend to become soft from soft lands; for it is not the property of the same land to produce wonderful fruit and men good in war'.[67] Polybius trying to explain the rise of Roman imperialism offered similar explanations, stating, for example, that 'the Capuans in Campania having become wealthy on account of the great fertility of their land, drifted into softness and extravagance, excelling even what is commonly said about Croton and Sybaris'.[68] The classical geographies therefore contained clues for the sixteenth-century geographers not only about how to map out the world, but also about how to approach the problems of a world where expanding margins forced people to consider the influence of the geographical environment on an imperialist agenda, and made people think about the qualities necessary to make a land part of the *oikoumene*, and whether indeed the *oikoumene* could coincide with the world.

Such clues to geographical understanding embedded in the writings of classical authors created a further reciprocal relationship between the rediscovery of classical texts and the redefining of classical knowledge. The humanist authors at the core of this study read both rediscovered classical human geographical texts and new works about recently explored lands, and fuelled further searches for and interpretations of both. The most influential of the classical human geographers in the Renaissance was the late first-century BC Stoic writer Strabo of Amasia. His lost *Histories* had received some attention in the ancient world, but his *Geographia* was almost entirely overlooked until the late fifteenth century. Josephus and Plutarch both referred to his historical work, but not to his *Geographia*. Neither Ptolemy nor Marinus of Tyre, both of whom might

66 See K. Clarke, *Between Geography and History: Hellenistic Constructions of the Roman World* (Oxford: Oxford University Press, 1999).

67 φιλέειν γὰρ ἐκ τῶν μαακῶν χώρων μαλακοὺς ἄνδρας γίνεσθαι· ο ὐ γάρ τι τῆς αὐτῆς γῆς εἶναι καρπόν τε θωμαστὸν φύειν καὶ ἄνδρας αγαθοὺς τὰ πολέμια: Herodotus, 9.122.3.

68 Καπυσίους τοὺς ἐν Καμπανίᾳ διὰ τὴν ἀρετὴν τῆς γῆς πλοῦτον περιβαλομένους ἐξοκεῖλαι εἰς τρυφὴν καὶ πολυτέλειαν, ὑπερβαλλομένους τὴν περὶ Κρότωνα καὶ Σύβαριν παραδεδομένην φήμην: Polybius, 7.1.1.

have been expected to have found much useful material in Strabo's geographies, make any reference to it.[69] Indeed, the earliest explicit reference to it is in the no longer extant second-century AD lexicon of Valerius Harpocration of Alexandria.[70] By the second century AD it was evidently circulating among Greek-speaking authors, since about the same time as Harpocration was writing, Athenaeus also referred to the *Geographia*,[71] and Dionysius Periegetes, writing in the same century, clearly knew it, and, indeed, seems to have taken an entire passage from it.[72]

Despite containing descriptive information and snippets of historical detail which might have appealed to a Latin as well as Greek audience, Strabo's *Geographia* never seems to have circulated outside the Greek-speaking world. No Roman author, not even Pliny the Elder, who made numerous references to hundreds of sources, cited the *Geographia.*[73] Strabo seems to have written primarily for an educated Greek-speaking readership.[74] Nonetheless, his work was still circulating in Late Antiquity when Procopius referred to it.[75] Then, after enjoying this period of moderate success, the *Geographia* vanished almost completely until the end of the Middle Ages. Unlike Ptolemy's *Geographia*, it was not even known to Arab geographers.[76] In 1423, however, the first Greek manuscripts of the *Geographia* were introduced into western Europe, and thereafter it became increasingly popular.[77] In the renaissance, Strabo's encyclopaedic knowledge of

69 A. Diller, *The Textual Tradition of Strabo's Geography with Appendix: The Manuscripts of Eustathius' Commentary on Dionysius Periegetes* (Amsterdam: A. M. Hakkert, 1975), 7.

70 Alexandria. Harpocration's text is no longer extant, but Keaney's fragment Λ15 cites Strabo by name and refers to material in Strabo, 8.6.22: see Harpocration, *Lexeis of the Ten Orators*, ed. by J. Keaney (Amsterdam: A. M. Hakkert, 1991). Diller states that not all people agree that the Harpocration passage is an authentic citation of Strabo since some think that they are later additions. A. Diller, *Textual Tradition of Strabo*, 10.

71 Compare Athenaeus, *Deipnosophistae*, 3.21 with Strabo, 3.4.6, and Athenaeus, 14.657 with Strabo 3.4.10.

72 A. Diller, *Textual Tradition of Strabo*, 8.

73 *Ibid.*, 7.

74 C. v. Paassen, *The Classical Tradition of Geography* (Gröningen: J. B. Wolters, 1957), 11; Strabo, 1.1.21

75 D. Dueck, *Strabo of Amasia: A Greek Man of Letters in Augustan Rome* (London and New York: Routledge, 2000), 151.

76 M. V. Anastos, 'Pletho, Strabo and Columbus,' in *Studies in Byzantine Intellectual History* (1979), 1-2.

77 On the introduction of Strabo to the west, see J. Larner, 'The Church and the Quattrocento Renaissance in Geography', *Journal of Renaissance Studies*, 12 (1998): 35; A. Diller, *Textual Tradition of Strabo*, 101–05; M. V. Anastos, *Pletho, Strabo and Columbus*, passim. On the epitomies and circulation of Strabo, see F. Sbordone, 'Exerpta ed epitomi della geografia di Strabone', *Studi Bizantini e Neollenici*, 7 (1953), 201–06; F. Sbordone, 'La tradizione umanistica della geografia di Strabone', *Bollettino*

classical geographical writings and ideas, and his interest in the relationship of humans to their physical world, made him a useful and appealing source to writers who esteemed classical knowledge above all subsequent writings, and who were interested in the role and capabilities of humans. Strabo's *Geographia* therefore became more popular than it had ever been and was a required component of any geographical library until well into the seventeenth century.[78]

Before Strabo could have a significant impact on Renaissance geographical thought, however, his work had to be translated into Latin, since knowledge of ancient Greek was still severely limited in the fifteenth and sixteenth centuries. As early as 1469, a Latin translation of Strabo by Guarino Veronensis and Gregorio Thiphernio was printed in Rome,[79] making the *Geographia* one of the earliest Greek texts to be translated and printed in Latin. By 1516 there was enough demand for a Greek version to be produced in Venice, seventeen years before the first Greek edition of Ptolemy, although the Latin versions remained consistently more popular. Greek, after all, was the language of the scholar, whereas Latin was the language of the merely educated. Even the grammar schools tended to focus solely on Latin.[80] For those not so much concerned with philological accuracy as with gleaning the general geographical knowledge and theories of the ancients, the Latin translations were adequate. There was also a sixteenth-century vernacular translation of the *Geographia* into Italian.[81]

Strabo's influence on sixteenth-century geographical thought has been understudied. Nonetheless, both geographers and explorers referred to him more often than to all other ancient geographical authors except Ptolemy and Pliny.[82]

---

del Comitato per la preparazione dell'edizione nazionale dei classici greci e latini, 4 (1961), 11–32. Pletho wrote two works, *Excerpts from Strabo*, and the *Διόρθωσις ἐνίων τῶν ούχ ορθῶς ὑπὸ Στραβωνος λεγομένων* (A correction of certain errors made by Strabo), which was published in 1439.

78 See table 7 in L. B. Cormack, *Charting an Empire*, 153.

79 Most people cite the 1472 version of the same translation produced in Venice as the first printed edition, but in fact the earliest edition was printed by C. Sweyneheym and A. Pannartz in Rome in 1469: see A. M. Biraschi et al., eds, *Strabone saggio di bibliografia 1469–1978* (Perugia: 1981), 35.

80 F. Ciccolella, 'The Greek Donatus and the Study of Greek in the Renaissance', *International Journal of the Classical Tradition*, 12 (2005), 11.

81 This was a quarto edition published in Venice and Ferrara in the 1560s. M. A. Buonaccioli, *La prima parte della geografia di Strabone, di Greco tradotta in volgare Italiano da M. Alfonso Buonaccioli* (Venice, 1562). M. A. Buonaccioli, *La seconda parte della geografia di Strabone, di Greco tradotta in volgare Italiano da M. Alfonso Buonaccioli* (Ferrara, 1565). See L. B. Cormack, 'Good Fences Make Good Neighbours: Geography in Early Modern England', *Isis* (1991), 651.

82 Although I would not claim that I had made an exhaustive analysis of classical authors cited, among the authors I have studied, Pliny seems to have been cited almost as often as Strabo and Ptolemy combined. Twenty-eight authors cite Pliny by name, compared

Moreover his readership covered a wide variety of people. Some authors more or less ignored Ptolemy, but referred to Strabo at length. The Spanish geographer Girava, for instance, opens his *Cosmography* with a reference to Strabo's description of the purpose of studying geography.[83] Girava was from Tarragona and was one of the few Spaniards to have published a whole world geography. He lived and published his work in Milan, and was heavily influenced by the classical revival which he saw around him. His cosmography was neither as influential nor as innovative as other Spanish works about individual regions, but it circulated reasonably widely, partly because it was published outside of Spain and possibly also because it followed a mode of classical allusion popular with his fellow humanists.[84] He then says that he will follow the model of writers such as Strabo, Dionysius and Pomponius Mela in describing the regions, mountains and the like.[85] Girava displayed a fair degree of familiarity with Strabo's work, but rarely referred to Ptolemy. For him Strabo provided the best method and reasoning for describing the world. Girava was not the only one who found Strabo a useful guide. João de Barros, the Portuguese humanist, also displayed real familiarity with Strabo in his *Decadas de Asia*, comparing modern places with Strabo's terminology.[86] Barros, one of the leading Portuguese humanists educated at the Royal court, certainly knew Greek, but was better trained in Latin – all his favourite authors were Roman ones, but he was clearly familiar with the Greek geographers.[87] The first two decades of *Asia*, published between 1552 and 1553 and republished in Italian by 1563, were fundamental in helping to make Asia less of a mythical place to western Europeans, but were full of classical references.[88] Barros wrote a Latin world geography, but that text was never published and disappeared, so while it seems likely from the evidence of his surviving works that it was heavily indebted Strabo, it is impossible to say for certain.[89]

Several sixteenth-century geographers, however, unequivocally relied on Strabo not only for data but also as a model. Sebastian Münster, although he produced an edition of Ptolemy and not of Strabo, was actually far more akin to Strabo in the type of material he produced.[90] Similarly, Ramusio cited Strabo in each of his own

to fifteen for Ptolemy and fourteen for Strabo. By comparison, the Roman geographers, Mela and Solinus, are cited by only seven authors each.

83 G. Girava, *Dos Libros De Cosmographia* (Milan: Christoval Caron, 1556), 1.

84 *Ibid.*

85 *Ibid.*, 2.

86 J. de Barros and D. de Couto, *Asia*, Decada I, vol. III, ch.VIII, 220, Decada III, vol. IV, ch. II, 380.

87 C. R. Boxer and J. de Barros, *João De Barros: Portuguese Humanist and Historian of Asia* (New Delhi: Concept Publishing Company, 1981), 19–26.

88 His third decade was only published in Portuguese so was far less influential. His fourth decade was not published until 1615 in Madrid. D. F. Lach, *Asia*, 505.

89 *Ibid.*, 504. G. B. Ramusio, *Alli Lettori*, 1043.

90 M. Small, *Warring Traditions*, 167–85.

discourses that he included in his *Navigazioni e viaggi.* The Greek geographer's influence stretched right across Europe. Richard Hakluyt in England, for instance, included a report by the Icelandic writer Arngrimus Ionas in which Ionas makes specific reference to Strabo's description (in book 7) of fire being kindled by water.[91] Educated travellers also referred to Strabo. The former governor Corsica who had travelled in the Levant in the late fifteenth century, Giorgio Interiano, remarked that the Circassians living in the furthest reaches of the Black Sea, still had a main diet of sturgeon, just as Strabo had described.[92] In short, Strabo's *Geographia* provided the backbone of geographical information to support the ideas of the sixteenth-century geographers.

Unlike Ptolemy and the more mathematical geographers, Strabo was almost entirely concerned with the known and knowable world. He was interested in the races, environments and physical characteristics of the *oikoumene*, and although he mentioned the possibilities of lands beyond the Ocean, and discussed Ocean itself, these were not his real area of concern. He was the archetypal human geographer, and the boundaries of the human *oikoumene* formed the limits of his *Geographia.* He was interested in global and celestial matters only in so far as they related to the *oikoumene* and in particular to human geography. For instance, he admitted that geographers had to use astronomy and geometry to explain the distinct features of particular countries, such as their shape, size and climatic conditions.[93] He also thought that if one wanted to account for the differences in climate between the countries at the extreme north and those at the extreme south of the *oikoumene*, then one had to look to the heavens, and in particular to the movement of the stars and the sun.[94] As a result those interested in geography had to observe the heavens and discover the relationship between their own individual locations and the heavens. Likewise he believed that anyone whose attention had been drawn to the heavens would also naturally consider how the *oikoumene* related to the rest of the world, because there would be an obvious gap in logic for anyone who looked at the relationship between the *oikoumene* and the heavens, and failed to look at one between the *oikoumene* and the rest of the world.[95] Even his interests in movement and stars were really aroused by concerns about their influences on the human environments at the extremes of the *oikoumene*; and the questions he set for geographers to ask about the rest

[91] A. Ionas, 'A Briefe Commentarie of Island: wherin the errors of such as have written concerning this Island, are detected and the slanders, and reproches of certaine strangers, which they have used over-boldly against the people of Island, are confuted', *PN*, vol. 4, 112.

[92] G. Interiano, 'Giorgio Interiano Genovese a Messer Aldo Manuzio Romano della vita de' Zichi, chiamati Ciarcassi', *NV*, vol. 4, 34.

[93] Strabo, 1.1.13; 2.5.

[94] Strabo, 1.1.14.

[95] Strabo, 1.1.15.

of the globe were concerned with practical matters such as why regions are not inhabited.[96] For him the *oikoumene* was not merely the area of study defined by the limits of empirical knowledge; it was the only area worthy of study. He said in so many words that 'it is not necessary for geographers to concern themselves with those things outside our inhabited world'.[97] Even for matters within the *oikoumene*, his work had a very different tone from Ptolemy's. He thought the longitude and latitude, the distances and measurements with which mathematical geographers such as Ptolemy were concerned, were tedious. He occasionally gave distances and descriptions of physical locations, but he was clearly uninterested in them – to the point that the ones which he gave were frequently inaccurate. He made his lack of interest very apparent when, while providing the distance between Heraclea and Miletus, he explicitly stated: 'regarding famous places, it is necessary to tolerate the dry part of such geography as this'.[98] He bothered with them only because they were not simply abstract mathematical matters but because they had specific implications for humans since human activity depended on geographical environment.[99]

His geography was intricately interwoven with ideas of anthropology, nature and history, for humans lay at its heart. For Ptolemy, the focus on distance and mathematical observation had a tendency to break the *oikoumene* down into mappable units where each location was related to another more than to the *oikoumene* as a whole, whereas for Strabo

> the *oikoumene* [was] not only a vast expanse made up of individual units and containing spatially separated societies, but [was] itself a spatial entity consisting of mutually related and interdependent peoples, each of whom, it is true, [had] taken root in a separate part of nature on the earth's surface, but who [were] at the same time equally bound up with the whole of *oikoumene* within which they [had] a place and a 'function'.[100]

In viewing the *oikoumene* as a whole, however, Strabo in some ways provided a closed system for his geography. Whereas Ptolemy had allowed for the possibility that his *Geographia* did not contain a portrayal of the whole *oikoumene*, Strabo by contrast, presented a geography which contrived to indicate that most of the *oikoumene* was known. It is a geography in which the Roman empire seems to have

[96] Strabo, 1.1.14.

[97] τοῖς δὲ γεωγραφοῦσιν οὔτε τῶν ἔξω τῆς καθ' ἡμᾶς οἰκουμένης φροντιστέον Strabo, 2.5.34.

[98] ἀνάγκη δ'ἐπι τῶν ἐνδόξων τόπων ὑπομένειν τὸ περισκελὲς τῆς τοιαύτης γεθγραφίας: Strabo, 14.1.9.

[99] C. v. Paassen, *Classical Tradition*, 4.

[100] *Ibid.*, 19.

reached out to conquer most of the knowable or at any rate desirable world.[101] Yet, despite this impression of a completely known world, Strabo ultimately provided a source that was more readily adaptable to sixteenth-century geography. It was a different type of use, however. Unlike Ptolemy, there was only one attempt that I am aware of to provide maps to accompany Strabo's *Geographia*, and these maps were not published with the text.[102] His work was valued as a source of miscellaneous information and as a model for how to write geography and describe the *oikoumene*, not as a means of visualising the world or indeed the *oikoumene*.

The only Latin work to rival those of Ptolemy and Strabo in its influence on sixteenth-century geography was Pliny the Elder's *Natural History*. Pliny, however, holds a different status, being more an encyclopaedist than strictly speaking a geographer. His work, written in Latin, had both popular and scholarly appeal with the result that although it had circulated widely throughout the Middle Ages, it became even more popular in the Renaissance.[103] It was in the first spate of classical texts to be printed, and one of the earlier ones to be translated into the vernacular and published therein.[104] As the surviving incunabula editions indicate, his work was initially particularly popular in Italy.[105] After 1499, however, many more regions were involved in publishing Pliny's *Natural History*. In the sixteenth century, France came to dominate with a total of twenty-seven editions, but nearly every western European country produced its own editions.[106] Even England, a country notoriously slow to print classical works, produced three.[107]

His work, then, was much more accessible than Ptolemy's or even Strabo's but it was also significantly more disjointed. At first, much of the study of Pliny in the

[101] See final three chapters in K. Clarke, *Between Geography and History*.

[102] P. Meurer, *The Strabo Illustratus Atlas: A Unique Sixteenth-Century Composite Atlas for the House of Bertelli in Venice* (Bedburg Hau: Antiquariat Gebr. Haas, 2004), 7.

[103] M. Chibnall, 'Pliny's Natural History and the Middle Ages', in *Empire and Aftermath*, ed. by T. A. Dorey (London: Routledge and Kegan Paul, 1975), 75.

[104] R. Hirsch, 'Printing and the Spread of Humanism in Germany', in *The Printed Word: Its Impact and Diffusion (Primarily in the Fifteenth and Sixteenth Centuries)*, vol. 3 (London: Ashgate, 1978), 26, 28; L. D. Reynolds, *Texts and Transmission: A Survey of the Latin Classics* (Oxford: Oxford University Press, 1983), 307–16.; L. D. Reynolds and N. G. Wilson, *Scribes and Scholars*, 101; C. Nauert, 'Humanists, Scientists, and Pliny: Changing Approaches to a Classical Author', *American Historical Review*, 83 (1979), 74.

[105] A. C. Klebs, 'Incunable Editions of Pliny's Historia Naturalis', *Isis*, 24 (1935), 120–21; E. W. Gudger, 'Pliny's 'Historia Naturalis, the Most Popular Natural History Ever Published', *Isis*, 6 (1924), 269–81. Klebs lists fifteen incunabula editions, arguing that Gudger who counts twenty-four more, grossly exaggerated the number. Both agree that all were printed in Italy.

[106] A. Labarre, 'Diffusion de l'Historia Naturalis de Pline au temps de la renaissance', in *Festschrift für Claus Nissen* (Wiesbaden: Guido Pressler, 1973), 451–571.

[107] E. W. Gudger, 'Most Popular Natural History', 274.

Renaissance focussed on creating the most accurate text, but people also became increasingly interested in the content. As a result, Renaissance writers also became more aware of the lapses and errors in the text in much the same way as they did with Ptolemy's *Geographia,* although they continued to rely on it greatly.[108] Its influence on geography is evident. Oviedo specifically stated that he used Pliny as a model for his *General and Natural History of the New World.*[109] The Jesuit, Acosta, who has himself been called the Pliny of the New World, wrote the first part of his *Natural and Moral History of the Indies* while still in Peru and far from any library, yet he was able to cite Pliny in great detail. He referred, for instance, to the voyage of Hanno cited by Pliny, and to Pliny's discussions of ancient navigations, to his descriptions of the island of Taprobana, to his discussions of eastern gold and false emeralds, among many other matters.[110] Münster similarly pillaged him for pieces of information such as his descriptions of the monsters and characteristics of Scandinavia.[111] In this way, by providing natural histories and comparanda, Pliny became vital to sixteenth-century writers trying to interpret the nature and limits of the expanding *oikoumene*, but he himself, although he discussed the *orbis terrarum,* was not particularly interested in limiting himself to the study of it.[112]

Rather, Pliny was an avid collector of all varieties of information. He did not, however, have an analytical mind. The historian of geography Edward Bunbury's damning statement, 'his voluminous treatise remains in consequence a vast compilation, bearing testimony to the unwearied diligence of its author in the collection of his facts, but showing at the same time an almost total want of critical judgment or philosophical arrangement', has some truth to it, though more recent scholarship by authors such as Mary Beagon and Trevor Murphy has shown that there were underlying themes to the work.[113] Yet even these Plinian experts have had to concede that, although Pliny arranged his work according to certain patterns, he digressed so much that at times these are barely discernible. In the sixteenth century, however, the *Natural History* was not read as a consistent thesis. Instead

108 C. Nauert, 'Humanists and Pliny'.

109 Oviedo said: 'per non incorrere io adunque in simil fallo, e non negare quello che è da Plinio, quanto all'invenzione e titolo del libro io il seguo...' G. F. de Oviedo, *Generale Istoria*, 351.

110 J. de Acosta, *Historia*, 53, 78, 89, 145.

111 For example, S. Münster, *Cosmographia*, 814, 46, 912.

112 e.g. Pliny, *NH*, 3.3

113 E. H. Bunbury, *A History of Ancient Geography among the Greeks and Romans Form the Earliest Ages Till the Fall of the Roman Empire*, 2 vols, vol. 2 (New York: Dover Publications, 1959, first printed 1879), 374. D. de Góis, *Chronica di Principe Dom Joam, Rei que foi destes regnos segundo do nome* (Lisbon: Francisco Correa, 1567). T. M. Murphy, *Pliny the Elder's Natural History: The Empire in the Encyclopedia* (Oxford University Press: Oxford, 2004); M. Beagon, *Roman Nature: The Thought of Pliny the Elder* (Oxford: Clarendon Press, 1992).

it was picked apart for information, drawn upon for an example of how to present an encyclopaedia of natural historical material, and in general held up as work containing facts, real or supposed, to be supplemented or countered. It supplied a wealth of geographical information, but unlike Strabo's and Ptolemy's works it was not considered a strictly geographical treatise.

The popularity of Strabo and Ptolemy's works can be considered the apogee of the coincidence of the Renaissance and the European expansion beyond the traditional *oikoumene,* but there were many other Greek and Roman authors who also received widespread readership and citation in this period. Homer and Herodotus, for instance, were among a vast group of classical authors who were considered geographical authorities both in classical times and in the sixteenth century. Homer was cited by numerous well-known geographers and explorers both ancient and modern, ranging from Strabo and Polybius in the ancient world to Ramusio and John Davis in the sixteenth century.[114] Ramusio used him as an example of how the ancients provided geographical knowledge and the esteem in which they held geography.[115] Herodotus had received popular acclaim when his work was first written in the 5th century BC, but had swiftly become the subject of enduring criticism in the ancient world. Ctesias, for instance, writing only a few years after Herodotus, referred to him as a word smith and teller of lies.[116] By the first century AD, criticism of Herodotus was so general that Josephus, writing towards the end of the century, went as so far as to say that he was universally attacked.[117] His writing had continued to circulate in the ancient world largely because many ancient authors applauded his style although they derided his content.[118] In the fifteenth and sixteenth centuries, however, it was his content and approach to history, geography and ethnography which gripped the readership. Style was clearly not the main focus of Herodotean readership in the Renaissance since his work circulated primarily in Latin. Guarino translated the first seventy-one chapters of Herodotus's *Histories* into Latin in 1416, effectively reintroducing him to western Europe, and in about 1452 Lorenzo Valla translated and circulated the full text, possibly at the command of Pope Nicholas V.[119] The Valla translation was then printed in 1474.[120] Herodotus provided a view of a world with a closely described *oikoumene*. His is the earliest entirely extant text to describe all the

114 In addition to Strabo, see, for example Polybius, *Histories*, 12.28.1; J. Davis, *The Worldes Hydrographical Description* (London: Thomas Dawson, 1595), 197.

115 I. Barbaro, *Viaggi Della Tana*, *NV*, vol. 3, 21.

116 Ctesias, *F.Gr.Hist.* 3C, 6888, F8, line 7.

117 Josephus, *Contra Apion*, 1.16.

118 For example, Hermogenes, *On Style*, 2.12; Dionysius of Halicarnassus, *On Literary Composition* 10; Lucian, *Herodotus*, 1.

119 B. Hemmerdinger, *Les manuscrits d'Hérodote et la critique verbale* (Genoa: Istituto di filologia classica publicaziavi, 1981), 46–48.

120 Herodotus, *Historiarum Liber Primus – Nonus*, ed. by Lorenzo Valla (Venice, 1474).

features of what became the traditional boundaries: an east-west limit defined by ocean, north-south climatic boundaries, and inner limits of wastelands and monsters. All of these had to be redefined in the course of the sixteenth-century expansion, yet his underpinning ideas of balance, proportionality and unity in nature could be reapplied in the sixteenth century to an *oikoumene* that was ceasing to have limits and was reaching out to cover the sphere. These applications of natural law, alongside his broad outlook, his widespread travel, and his fascination with the unity and variety of the world, rather than his style, made Herodotus's appeal grow in the aftermath of the American discoveries.[121]

Similarly several minor Roman writers also began to influence sixteenth-century geographical interpretation. The hunger for ancient knowledge was so great that even works which under other circumstances might simply be thought derivative and uninspired received general acclaim. As we have seen, the rather unimaginative Roman geographer Pomponius Mela, for instance, experienced a heyday in the sixteenth century. Unlike the Greek geographies, Mela's work, written about 40 AD, had not completely disappeared in the Middle Ages, but although there are a few manuscripts of Mela's *De Situ Orbis* dating to before the fourteenth century, the text had a limited circulation. Some of the most noted names in the study of medieval geography and cartography – Kimble, Wright, Skelton and Bagrow, to name but a few – have supposed that Mela was a direct influence on medieval geography, but they are mistaken, as M. Rouse, R. Rouse and C. Gormley have demonstrated.[122] Indirectly he played a part since he was a source for Pliny, but all the ideas current in the Middle Ages that are attributed to him can actually be found in Pliny's *Natural History,* and there is no need to assume any knowledge of the earlier author's work. But when Mela's *De Situ Orbis* was rediscovered in the renaissance, it became hugely important. It was first printed in Italy by 1471, and by 1509 editions had been produced in Germany, Austria, Switzerland, France and Spain.[123] Even in England it was regarded highly, as the evidence of Oxford and Cambridge wills attests.[124] F.E. Romer reports that between 1471 and the present day there were 121 recorded

121 A. Momigliano, 'The Place of Herodotus in the History of Historiography', *Historia,* 43 (1958), 10.

122 C. M. Gormley, M. A. Rouse, and R. H. Rouse, 'The Medieval Circulation of the De Chorographia of Pomponius Mela', *Medieval Studies,* 46 (1984), 266–320.

123 M. Milham, 'An Introduction to the Renaissance Tradition of Pomponius Mela' (paper given at the Acta Conventus Neo Latini Amstelodamensis: Proceedings of the Second International Congress of Neo-Latin Studies, Munich, 1979), 786–93.

124 L. B. Cormack, *Charting an Empire*, 113. In table 6 Cormack gives an indication of the number of volumes of Mela accounted for in Oxford and Cambridge wills in the late sixteenth century. She has evidence for Oxford and Cambridge scholars owning fifteen copies between 1580 and 1600.

manuscripts and 150 printed editions of his work.[125] The majority of these were made during the renaissance. The work was written in a language and intellectual tradition which made it far more easily accessible than some of the other ancient geographical writers, and it reached a relatively wide audience; it also contained explicit references to an inhabited, asymmetrical southern continent or *anticthon* that fitted reasonably well with the European idea of the southern hemisphere which was proving to be inhabitable in the sixteenth century.[126] Mela was widely referenced for a variety of reasons. Acosta, for instance, picked up on this idea of the antichthones and cited him in a general discussion of why St Augustine denied the existence of the Antipodes.[127] Sebastian Münster, on the other hand, used a very specific piece of information derived from Mela, citing him when providing evidence of Eudoxus's voyage from the Red Sea to Cadiz. He said that this voyage, combined with one mentioned by Pliny from Cadiz eastwards and with the voyages of the Portuguese, provided clear evidence that Africa was entirely surrounded by sea.[128] Interestingly he gave preference to the evidence of the ancients over the more modern sources, and clearly knew his classical sources well since there is only one reference to this journey in Mela.[129]

The encyclopaedic work of Julius Solinus was another which was read more widely at that time than at any other stage in history, and also made its mark in the geographical tradition. It was cited by many of the geographers and explorers of the sixteenth century as an authority for everything from the location of the Cape Verde Islands to a validation of the seemingly wondrous.[130] Perhaps the writer most surprising for his lack of influence of Renaissance geographical thought is Tacitus, but even he was used. His works had completely disappeared in the Middle Ages, and were only partially recovered in the fifteenth century. His *Germania* and the *Agricola*, retrieved by Poggio Bracciolini, became very influential in European geographical thought.[131] His *Annals* were occasionally

125 P. Mela, 'Description of the World', ed. by F. E. Romer (Ann Arbor: University of Michigan Press, 1998), 29.

126 Pomponius Mela, *De Situ Orbis* 3.70. See *ibid.*, 7. for discussion of the asymmetrical nature and the difference between *antipodes* and *anticthon.*

127 J. de Acosta, *Historia*, 39.

128 S. Münster, *Cosmographia*, 1115.

129 Mela, 3.9.

130 For example G. F. de Oviedo, *Generale Istoria*, 364. 'The Second Voyage to Guinea Set out by Sir George Barne, Sir John Yorke, Thomas Lok, Anthonie Hickman and Edward Casstelin, in the Yere 1554', *PN*, vol. 6. S. Münster, *Cosmographia*, 850, 1080; J. Davis, *Worldes Hydrographical*, 197; I. Barbaro, *Viaggi della Tana*, 485; S. Münster, *Cosmographia*, 850, 1080. A. Thevet, *Cosmographie Universelle*, 7r.

131 K. C. Schellhase, *Tacitus in Renaissance Political Thought* (Chicago: Chicago University Press, 1976), 1–10. D. R. Kelly, 'Tacitus Noster: The *Germania* in the Renaissance and Reformation', in *Tacitus and the Tacitean Tradition*, ed. by T. J. Luce and A. J. Woodman (Princeton: Princeton University Press, 1993), 153–4; G. Strauss,

cited by writers interested in Asia,[132] but sixteenth-century geographical writers most often turned to the *Germania* rather than the *Annals* for guidance.[133] This is largely because Tacitus's *Germania,* being one of the few strictly ethnographic tracts in classical literature, provided a model for those trying to describe other lands and peoples. Paolo Giovio, for instance, although he could extract no relevant data from Tacitus on his subject, declared that by focussing on the habits and practices of the people of Muscovy, he was going to imitate Tacitus who had separated the study of the customs of the people of Germany from his general history.[134] Less surprising perhaps is the use of Tacitus's *Germania* by those writers who travelled to Sarmatia, since Sarmatia features prominently in that work.[135] It provided a paradigm for those trying to write about the newly encountered peoples of the expanded *oikoumene.*

The influence of other writers is somewhat harder to trace, and indeed the evidence is somewhat curious. The Greek writers Cleomedes and Geminos of Rhodes, who talked about inhabitants of four different *oikoumenai*, were barely cited by sixteenth-century geographers.[136] The fifth-century AD Roman writer Martianus Capella, perhaps the last of the significant writers in the classical tradition to deal seriously with geography, and who had been hugely influential

Sixteenth-Century Germany: Its Topography and Topographers (Madison: University of Wisconsin Press, 1959), 11.

132 See for example, G. Abbot, *Brief Description*, C3v.

133 On Tacitus's interest in Asia see G. W. Bowersock, 'Tacitus and the Province of Asia', in *Tacitus and the Tacitean Tradition*, ed. by T. J. Luce and A. J. Woodman (Princeton: Princeton University Press, 1993).

134 P. Giovio, 'Paolo Iovio Istorico delle cose della Moscovia, a Monsignor Giovanni Rufo, Arcivescovo di Cosenza', *NV*, vol. 3, 672.

135 See for instance M. di Micheow, 'Trattato Il di Matteo Di Micheovo, Dottor Fisico e Canonico Cracoviense, delle due Sarmazie', *NV*, vol. 4, 639; A. Guagnino, 'La descrizione della Sarmazia Europea del Magnifico Cavalliere Alessandro Guagnino Veronese, tradotta dalla lingua Latina nel volgare Italiano dal Reverendo Messer Bartolomeo Dionigi Da Fano', *NV*, vol. 4, 327.

136 The first translation of Geminos into Latin created a complication in tracking his influence on sixteenth-century geographical writing, since Gerard of Cremona, who translated Geminos' *Introduction to the Phenomena* from the Latin, mistakenly attributed it to Ptolemy; a section of it was also misattributed in a number of manuscripts to Proclus, and went through over sixty printed editions as *The Sphere of Proklos*, so that when some sixteenth-century writers cited Proclus or Ptolemy they may in fact have meant Geminos. J. L. Berggren and J. Evans, *Geminos*, 104; R. B. Todd, 'Geminus and the Pseudo-Proclan Sphaera', in *Catalogus Translationum Et Commentariorum: Mediaeval and Renaissance Latin Translations and Commentaries; Annotated Lists and Guides*, ed. by Virginia Brown, vol. VIII (Washington DC: The Catholic University of America Press, 2003), 11–14; R. B. Todd, 'Cleomedes', in *Catalogus Translationum Et Commentariorum: Mediaeval and Renaissance Latin Translations and Commentaries; Annotated Lists and Guides*, ed. by Virginia Brown, vol. VII (Washington DC: The Catholic University of America Press, 1992), 3.

in the Middle Ages, was likewise largely overlooked in the geographical texts of the sixteenth century, even though no fewer than seven complete and five partial editions of his work were printed between 1499 and 1599.[137] Similarly, the early fifth-century Neoplatonist Macrobius, whose commentary on Cicero's *Dream of Scipio* had contained one of the most influential classical geographical texts to continue to circulate in the Middle Ages, was cited surprisingly little by name by the sixteenth-century geographers and explorers.[138] He was one of the few Roman authors to give a detailed geography of the world which included antipodal continents – indeed he went so far as to speak of four landmasses, separated from each other by an equatorial and a longitudinal Ocean. Such an image of the world might have provided useful explanatory information for sixteenth-century writers, yet it is only those writing earlier in the century who cited him much. There is no reference to him by name in either Hakluyt's or Ramusio's great works, and while Thevet refers to him in a discussion on the Antipodes – the type of discussion in which one would expect to find him cited more widely – this is in fact an aberration.[139] Acosta also cites him, but only to refute his ideas of climatic zones. Earlier in the century Richard Eden had cited him in his translations of both Martyr and Münster but these are his interpolations which are not to be found in the originals.[140] He was an author whose popularity as a geographer declined over the course of the sixteenth century, even though he held ideas which had direct relevance for the kind of theories proposed by later sixteenth-century writers.

Despite a discernible lack of enthusiasm for certain texts, some of which were, it is true, rather turgid and heavy-going, the general interest in the classical authors and their perceived authority was such that in the sixteenth century many classical works were translated into the vernacular so that even those with a lower level of education could have access to them. For instance, Pietro Andréa Mattioli's Italian version of Ptolemy's *Geographia* was printed as early as 1548. João Faras had produced a hugely influential Spanish manuscript of Mela's *De Chorographia* by about 1500,[141] and Arthur Golding translated Mela's work into English in 1585.[142] Even now, the only English translation of Solinus's

137 C. Lutz, 'Martianus Capella,' ed. by Paul Oskar Kristeller (Washington DC: The Catholic University of America Press, 1971), 398.

138 W. H. Stahl, 'Astronomy and Geography in Macrobius', *Transaction and Proceedings of the American Philological Association*, 73 (1942), 46–50.

139 A. Thevet, *Singularitez*, 111r.

140 S. Münster, *Cosmographia*, 103; P. Martyr and R. Eden, *The Decades of the Newe Worlde or West India, trans. by R. Eden* (London: in æd. G. Powell, 1555), 348.

141 J. B. de Carvalho, *La traduction Espagnole du 'De Situ Orbis' de Pomponius Mela par maître Joan Faras et les notes marginales de Duarte Pacheco Pereira* (Lisbon, 1974). See pp. 99–102 on the date of the manuscript.

142 P. Mela, 'Concerning the Situation of the World', ed. by Arthur Golding (London:

Polyhistory is that produced by Golding for Thomas Hackett in 1587.[143] It might have been expected that the classical geographical writers would have dwindled in popularity in the aftermath of the new 'discoveries', but instead the publication of old and new geographical material occurred together. The same Thomas Hackett, for whom Golding produced his translation of Solinus, made his own translation of the *Singularitez de La France Antarctique* by André Thevet.[144]

The vernacular translations and greater popularity of these non-canonical authors can, in part, be attributed to the advent of the printing press, and to the increase in lay education which provided a much larger audience for the classics. Renaissance scholars 'were not anti-Christian, but as laymen they did not subordinate the development of secular learning to its amalgamation with religious or theological doctrine'.[145] Printing houses developed all over Europe catering to this demand. Admittedly, however, some countries found a greater market for classical texts than others. Of the ninety books published by Caxton between 1477 and 1496, seventy-four were in English and the majority of these were designed to appeal strictly to the English aristocracy, focussing on matters such as chivalry.[146] Early English printing presses tended to specialise in medieval rather than classical works. In Italy, by contrast, Latin classics became 'the stock in trade of every printer'.[147] The availability of Latin and vernacular translations of Greek classics opened up vast new branches of thought. Marsilio Ficino (1433–1499) translated the Platonic corpus into Latin, bringing a new approach to ancient philosophy into a world trained in Aristotelian thought.[148] The *Histories* of Herodotus and Diodorus were both translated and printed in the fifteenth century. Poggio's Latin translation of Diodorus Siculus's *Histories* was published in Bologna in 1472, and reissued in Venice in 1476; his translation of Herodotus first appeared in the intervening two years.[149]

Such Latin and vernacular books had a wide appeal, offering the insights of classical authority to scholars seeking new methods to understand the world in

Thomas Hacket, 1585).

143 Solinus, 'The Excellent and Pleasant Work of Iulius Solinus, Polyhistor', ed. by Arthur Golding (London: Thomas Hacket).

144 A. Thevet, *The New Founde Worlde of Antarctike*, trans. by Thomas Hacket (London: Printed for Thomas Hacket, 1568).

145 P. O. Kristeller, *Renaissance Thought*, 7.

146 S. H. Steinberg, *Five Hundred Years of Printing, New Ed. Revised by J. Trevitt*, 4th ed. (London: Oak Knoll Press, 1996, first printed 1959), 74.

147 *Ibid.*, 31.

148 P. O. Kristeller, 'The First Printed Edition of Plato's Works and the Date of Its Publication (1484),' in *Studies in Renaissance Thought and Letters*, vol. 3 (Rome: 1993), 137.

149 Poggio's Latin translation of Diodorus Siculus' *Histories* was published in Bologna in 1472, and reissued in Venice in 1476. Valla's translation of Herodotus' *Histories* was first printed in 1474.

the light of the discoveries. At the same time, they offered the thrill of romance and adventure to a less scholarly public. Education had created a wide demand for such works. 'The cognoscenti and dilettanti, the gentlemen of leisure who had imbibed the taste and a little of the scholarship of the humanists, and the schoolmasters, parsons, lawyers and doctors who had passed through their university courses of *literae humaniores*, wanted books which they could carry about on their walks and travels and read at leisure in front of their fireplaces, and which would incidentally be within the financial reach of the poorer of these potential book-buyers.'[150] The printing press had made books affordable, with some books costing as little as 4 soldi, and even a complete Greek edition of Aristotle's works costing only 11 ducats.[151] Most of the ancient texts available in the Renaissance were, however, relatively inexpensive compared to the lengthy, editions of Ptolemy or Strabo (even in translation) which might explain part of the popularity of Mela and Solinus. Yet cost was clearly not the only concern. Ptolemy's maps and Pliny's and Strabo's length obviously added to their cost, but as we have seen these three works had the greatest influence on Renaissance geography. The ideas, models and even in some cases specific information which they contained did much to dictate their popularity at least among the humanist geographers with which this book is concerned.

As one model of the world failed to fit with new information, Renaissance writers were able to supplement it with other ideas gleaned from their own observations, as well as from classical writers who held differing points of view. Oddly enough, the contradictions in the ancients did not undermine their authority; instead, they allowed for flexible interpretations which enabled their authority to endure longer. The search for supporting information from classical writers outside the canon, coupled with the rediscovery of long-lost Greek and Roman texts meant that the sixteenth-century Renaissance and the Age of Reconnaissance were inextricably associated with one another. Key ideas of limit, balance and reciprocity, which pervaded Greek philosophy and were to be found, implicitly or explicitly, in nearly all classical geographical writings, became a foundation of the reinterpretation of geography in the light of the European discovery of the New World and the sea routes to Asia. Having invoked the authority of those writings, Renaissance writers then needed to refute, adapt or employ these ideas when they were creating a geography of their own world and a new framework for understanding the relationship between the *oikoumene* and the sphere.

150 S. H. Steinberg, *Five Hundred Years of Printing*, 34.

151 An Italian humanist schoolmaster would earn between 50 and 100 ducats a year. On price of books see C. B. Schmitt et al., *The Cambridge History of Renaissance Philosophy*, 30–31.

3

Defying the Limits of the World: Frigid and Torrid Zones in Sixteenth-Century Geography

The idea of the *oikoumene* created a paradigm of geography which was based not only on the knowability of the world, but also upon its usability. It relied on theories of limit and transgression. Throughout the sixteenth century the concept of the *oikoumene* persisted, in that geographers and explorers alike still wrote about the extent of the inhabitable lands, but they began to believe that all landmasses in the world were not only inhabitable but made for humans to exploit. Therefore all the different regions of the earth were made by God to interact with one another. As the chapter argues, in order for a classical geography of limit and transgression to be reimagined as one of connectedness and communication, the frigid and torrid zones which had demarcated the limits of the classical world had to be reconceptualised as inhabitable and knowable places.

The chapter first examines how and why the Greeks and Romans developed the concept of uninhabitable frigid and torrid regions denoting the limits of the *oikoumene*. In the process it demonstrates why these climatic limits became conceptual margins that hindered exploration for nearly two thousand years. It looks at the reevaluation of authority which still saw the Greeks and Romans as the arbiters of knowledge, but enabled non-canonical classical authors to become important in revising attitudes to the climatic zones which had once been thought to define the edges of the *oikoumene*. In the new geography, the arctic and equatorial regions were believed to play key roles in allowing the different parts of the world to interact with one another. Without this shift in mind-set about the frigid and torrid zones, it would have been impossible for Europeans to begin to think of the world as a single, exploitable, global unit, created by God for human dominion. The *oikoumene* could not have become coterminous with the world.

The idea of limit was fundamental to ancient geographical thought, and without exception ancient geographical writers focussed in some way on the idea of boundaries. Both Ptolemy and Strabo, as we have seen, spoke of the *oikoumene*, a term whose very existence implied a separation of one part of the world from the rest, denoting a region of human inhabitation which did not coincide with the whole world, but was restricted and limited. No classical writer thought that the *oikoumene* comprised the whole world. As we have seen, the sixteenth century was so steeped in reverence for the ancient authorities that the humanist-trained writers of the Renaissance could not simply disregard classical geographical thought with its overwhelming emphasis on boundaries to human habitation. They could find alternative classical authorities to argue against a majority view, but they could not overlook a prevalent classical theory completely. The sixteenth-century descriptive geographers had either to adapt or refute the classical boundaries explicitly. For much of the century it remained a cause for wonder that the frigid and torrid zones did not provide limits to human inhabitation. As Tomaso Giunti declared in his introduction to Ramusio's collection of travel narratives, the excitement was that 'all around, this globe of the earth is marvellously inhabited, and no part of it is empty or lacking inhabitants on account of heat or frost.[1]

Giunti's exultant declaration was challenging a widely known, clearly formulated classical geographical concept. Although to east and west the *oikoumene* came to a physical boundary in Ocean (as will be discussed in depth in the next chapter), to north and south much of its shore-line lay beyond the boundaries of the inhabitable world, so that, although it marked the end of land, it was not necessarily always reachable by humans. Instead, in these regions, the *oikoumene* was defined by climatic limits. As Strabo explicitly stated,

> geography seeks to mark off by boundaries the section (of the earth) being inhabited by us in one of the two temperate zones. To the west and to the east it is the sea which limits, but in the South and North it is the air; the air in the middle being tempered for plants and living creatures, but to either side it is intemperate with excess and with lack of heat.[2]

Those who transgressed the boundaries of the *oikoumene* into the frigid and torrid zones were bound to die. The idea that climatic zones formed an impassable boundary to the *oikoumene* became one of the biggest hurdles in the sixteenth-century reconceptualisation of geography from a world of limits to one of human communication.

1 'che d'ogni intorno questo globo della terra é meravigliosamente habitato, né vi é parte alcuna vacua: né per caldo o gielo priva d'abitatori'. T. Giunti, 'Tommaso Giunti ai Lettori', *NV*, vol. 3, 8.

2 ζητεῖ γὰρ ἡ γεωγραφία τῆς ἑτέρας τῶν εὐκράτων ἀφορίσαι τὸ οἰκούμενον ὑφ᾽ ἡμῶν τμῆμα: πρὸς δύσει μὲν οὖν καὶ ἀνατολῇ θάλαττά ἐστιν ἡ περατοῦσα, πρὸς δὲ τὰ νότια καὶ τὰ βόρεια ὁ ἀήρ, ὁ μὲν μέσος εὔκρατος ὢν καὶ φυτοῖς καὶ ζῴοις, ὁ δ᾽ ἐφ᾽ ἑκάτερα δύσκρατος ὑπερβολῇ καὶ ἐλλείψει τοῦ. Strabo 2.3.1.

In their earliest formation in the ancient world climatic zones helped to define and manage the potentially limitless space of the earth. Although the idea of zones of habitation was originally proposed by Parmenides in the fifth century BC, it received its first real geographical application in the geography of Herodotus, writing about 430 BC.[3] Herodotus had scorned the then prevalent Homeric model of a world surrounded by a perfectly circular Ocean, but, in doing so, he left no very clear idea how he envisioned the actual physical shape of the world.[4] Instead, he left his readers with the perplexing, indeed in some measure terrifying, concept of a world without limit. Such a world view was entirely alien to a society whose very philosophy was dominated by ideas of boundaries and limits and the difference between the bounded and the boundless, the limited and the unlimited.[5] By using the concept of a climatically limited *oikoumene*, however, he contrived to bring his world into a manageable framework without directly addressing the unfathomable idea of a world of indefinite extent.[6] He never explicitly used a terminology of zones, but he defined the *oikoumene* with limits formed by the climatic zones which confined human inhabitation.[7] He believed that the world was divided into three proportional parts, with Europe extending to exactly the length of Asia and Africa combined. To the north from the centre of these landmasses, the climate became increasingly colder, and to the south, correspondingly hotter. Forming the northern boundary was a territory where the air and land were said to be filled with whirling feathers, which Herodotus believed were an indication, distorted by report, of the constant snowfall in the region.[8] To the south were the scorched deserts of Libya and Egypt.[9] He described both the regions of unendurable heat and intolerable cold, as *eremoi* – wastelands or deserts – and he declared outright that the north was uninhabited on account of the cold and that the south was too hot for survival.[10] With these declarations

3 The reference to Parmenides as the originator of the zonal theory is found in Posidonius F49 (The Posidonius reference is itself fragmentary and can be found in Strabo, 2.2).

4 Herodotus, 2.23, 4.36.

5 See, for example, Parmenides and Anaximander who wrote explicitly about these subjects. Even the language and thought of the non-conformist Heraclitus were affected by ideas of boundaries and limits which were so prevalent. See DK, F94.

6 M. Small, 'Real and Theoretical Boundaries: Human Geography in Herodotus' (University of Victoria, 1998).

7 The Greeks usually used either ζώνη or κλίμα to refer to these zones. The word κλίμα placed more emphasis on the relationship with the heavens, whereas ζώνη originally merely meant 'belt' and became associated with zones, because it implied different belts of the earth. Herodotus is most explicit on the limits of the *oikoumne* in 3.106 where he discusses the southern and eastern limits. There are many other such references; see for example 3.14 on the limits of Europe.

8 Herodotus 4.7, 4.31.

9 Herodotus, 4.185.3.

10 Herodotus, 5.11, Herodotus, 2.3.1.

we see the earliest explicit statement of a doctrine that became ever more rigid: namely that no-one could venture into the equatorial and Arctic regions, for the result would be certain death. The framework for considering the *oikoumene* to be a climatically defined region was laid as early the fifth century BC.

The zonal theory was a concept that crossed divides of language, time and discipline. In the writings of Parmenides and Herodotus it was still very much in prototype form, but by the time of Aristotle, writing in the fourth century BC, it had clearly become the dominant theory. Aristotle himself talked about climatically uninhabitable regions bounding the *oikoumene*, but barely bothered to elucidate what he meant, asserting that

> It is possible to inhabit only these [temperate regions] and not beyond the tropics (for there would not be any shadow towards the North, but as it is those places become uninhabitable before the shadow fails or changes to[wards] the South) and those places under the Bear are uninhabitable on account of the cold.[11]

Many of the Latin writers who were most influential in the medieval period took over the idea of uninhabitable frigid and torrid zones from the Greeks and thereby ensured that the concept remained at the forefront of geographical throughout the Middle Ages, ensuring that it would be one of the biggest barriers to the reformulation of geographical thought in the fifteenth and sixteenth centuries. Pliny, for example, was convinced that there were regions of the world which were uninhabitable because of their climate.[12] Virgil, Ovid and Cicero likewise endorsed such an idea,[13] and as late as the fifth century Martianus Capella and Macrobius created geographies of the sphere, with widespread circulation, which were founded upon the concept.[14]

11 Ταῦτα δ’ οἰκεῖσθαι μόνα δυνατόν, καὶ οὔτ’ ἐπέκεινα τῶν τροπῶν (σκιὰ γὰρ οὐκ ἂν ἦν πρὸς ἄρκτον, νῦν δ’ ἀοίκητοι πρότερον γίγνονται οἱ τόποι πρὶν ἢ ὑπολείπειν ἢ μεταβάλλειν τὴν σκιὰν πρὸς μεσημβρίαν) τά θ’ ὑπὸ τὴν ἄρκτον ὑπὸ ψύχους ἀοίκητα: Aristotle, *Meteorologica*, 2.5 (362b).The ancients used the term τροπαί ἡλίου καί σελήνης to refer to the turnings of the sun, but the term came to mean tropics. D. L. Couprie, ‘Heaven and Earth in Ancient Greek Cosmology: From Thales to Heraclides Ponticus’ (New York: Springer, 2011), 141. Aristotle took Parmenides’ five part zonal theory and named the equator, tropics and Arctic circles as the dividers between each zone, comparing the different zones to drums. Aristotle, Meteorologica, 2.5.362a.33. See R. J. A. Talbert, *Ancient Perspectives: Maps and Their Place in Mesopotamia, Egypt, Greece, and Rome* (Chicago: Chicago University Press, 2012), 97.

12 Pliny, *NH* 2.68

13 Virgil, *Georgics,* 1.233, *Aeneid,* 7.225; Ovid, *Metamorphoses*, 1.45–50; Cicero, ‘Dream of Scipio’, *Republic*, 6.21.

14 Martianus Capella, *De Nuptiis Mercurii et Philologiae*, 6.604–5; Macrobius, *Commentarii in Somnium Scipionis*, chapters 5 and 6. A. Hiatt, ‘Blank Spaces’.

Although, as the example of Herodotus proves, the idea of a climatically defined *oikoumene* was not necessarily dependent on the concept of a spherical earth, the two became closely entwined. In the same century as Herodotus wrote about snow like whirling feathers at the northern margins, the natural philosophers Pythagoras and Thales theorised that the sphere of the heavens was divided into five zones, and Pythagoras argued that the earth had five corresponding zones.[15] Less than a century later, Aristotle provided the first clear statement of a spherical earth containing five distinct and defined climatic zones. He stated explicitly that the whole region in which we live (the temperate region of the northern hemisphere) was known, since in one direction no-one could live because of the cold, and in the other no-one could live because of the heat. He then moved on to discuss the symmetry of the earth saying that there must be a region which bears the same relationship to the South pole as ours does to the North.[16]

The reasoning used by Pythagoras and Aristotle showed how the climatic zones meant that a limited *oikoumene* could be mapped onto a spherical earth. Although not all those who believed in the idea of climatic zones accepted the sphericity of the earth, there was, from the outset, a connection between the two ideas. The zones were essentially the inscription of celestial patterns onto the earth. The heavens formed a sphere in their own right, and were divided into parallel circles defined by temperature caused by the position of the sun in the heavenly sphere. All these circles were perpendicular to the world's axis. The polar, tropical and equatorial circles of the earth corresponded exactly to the heavenly lines.[17] As the idea of a spherical world caught on, it became more and more fixed in ancient thought that there were five zones to the world, three of which were uninhabitable and two of which were temperate and could be inhabited although they were not necessarily so. The zones were defined by climatic conditions and not by physical structure. Even those writers like Strabo, who were most concerned with the human world, were working in a cosmographical context when they resorted to a zonal theory in defining the margins of the *oikoumene* mapping the heavens onto the earth.[18] The result of this, however, was that on one level the *oikoumene*, although distinct from the sphere, was inextricably associated with the sphere. In the sixteenth-century – once the idea became established that humans could enter the torrid and frigid

15 See *De Placitis Philosophorum*, 2.12 and 3.14. This dialogue traditionally forms part of the corpus of Plutarch's *Moralia*, but many now attribute it to Aetius. Plutarch, Aristotle, 'Opinions Des Philosophes,' ed. by G. Lachenaud (Paris: Belles Lettres, 1993), 15–18.

16 Aristotle, *Meteorologica*, 2.5, 362a32–362b30.

17 C. Jacob, *Géographie et ethnographie en Grèce ancienne* (Paris: Armand Colin, 1991), 106–7. See Strabo 2.3.1 for a specific statement of the correlation and harmony between celestial and terrestrial zones.

18 K. Clarke, *Between Geography and History*, 113.

zones and survive – the *oikoumene* continued to exist but was reconceptualised as coinciding with the sphere; it was a matter of readjusting the frame. Cold, temperate and hot zones remained; it was simply that humans could inhabit all of them, and therefore the *oikoumene* was not simply mapped onto the sphere but could coincide with it.

Before the *oikoumene* could be reconfigured, however, the uninhabitable nature of the torrid and frigid zones had to be challenged, and, as writer after writer accepted the validity of this concept, it is not surprising that both ancient and medieval explorers were slow to prove it wrong. With their own lives at risk, few dared test the hypothesis and venture into the torrid and frozen zones. Not only were they regions where loss of life seemed inevitable; they were places where the world ceased to function in the normal way.

Indeed the fear with which the Greeks regarded the frigid zone was so great that only one important Greek voyage to the far north is recorded – that of the fourth-century BC explorer and trader Pytheas of Massalia (Marseilles).[19] The effect of his apparently eye-witness testimony was to reinforce rather than negate the mystery of the north, and cement the fear with which southerners regarded the frigid zone. Pytheas lived about a hundred years after Herodotus, when the zonal theory was already prevalent, yet he dared to sail beyond the Pillars of Hercules to 'Ultima Thule' in the far north of Europe.[20] All that remains of Pytheas's reports are fragmentary citations in other authors, principally Pliny and Strabo, but his voyage was well known in the classical world, and his descriptions influenced all later references to the north by classical writers, so that between 300 BC and 550 AD no fewer than eighteen authors referred to Pytheas by name.[21] Pytheas was

[19] Pytheas's precise dates are hotly contested. See D. W. Roller, *Through the Pillars of Herakles: Greco-Roman Exploration of the Atlantic* (London: Routledge, 2006), 65.

[20] There is considerable debate about how far north Pytheas went, or where precisely he reached. The explorer and historian of the exploration of Scandinavia, Fridtjof Nansen, claimed that the vegetation which Pytheas described clearly indicated that he had reached Norway. Nansen's conclusions were refuted by C. F. Hawkes. Hawkes studied Pytheas in detail and decided that the extant fragments of Pytheas described Iceland, but argued that it was rapidly mistaken for the Shetlands, so that, in the writings of Tacitus and other later classical authors, their descriptions of the Shetlands became tangled up with Pytheas's description of Thule. The most recent writers on the subject, Cunliffe and Roller, come down in favour of Iceland, arguing that Norway seems too far north to be probable and the Shetlands seem a later interpretation. The only consensus is that he sailed far beyond mainland Britain into northern waters. B. Cunliffe, *The Extraordinary Voyage of Pytheas the Greek: The Man Who Discovered Britain* (London: Penguin Books, 2001). F. Nansen, *In Northern Mists, Arctic Exploration in Early Times*, 2 vols, vol. 1 (London: Heinemann, 1911), 60; Tacitus, *De Vita Agricola*, 10.6; C. F. C. Hawkes, 'Pytheas: Europe and the Greek Explorers', in *The Eighth J. L. Myres Memorial Lecture* (Oxford: Blackwell, 1975).

[21] Cunliffe is not certain that Strabo himself had read Pytheas's *On the Ocean*. He

a skilled navigator, whose data and factual material proved useful to astronomers and mathematicians.[22] Some of his tales were received with incredulity even by those who reported them. Strabo, in particular, considered him to be an outright liar.[23] Nonetheless, he cited a peculiar description which he claimed to have taken straight from Pytheas:

> the things around Thule and those places in which earth no longer existed by itself, nor indeed sea, nor air, but a compound of these things like a sea lung, in which, he says, the earth and the sea and everything else are suspended; and he says that this lung is, as it were, a fetter of everything, and is neither passable nor navigable. The thing like a lung he had seen himself but the other things he spoke of from hearsay.[24]

Nobody quite knows what Pytheas meant by a sea-lung, though most people seem to agree that he was referring to a jellyfish, but that this jellyfish was a simile for an attempted description of either the sludge ice or the thick sea mist that can form in the Arctic circle.[25] As I have argued elsewhere, the actual object of the description was of limited importance; its credibility and influence was of far more significance.[26] Pytheas's depiction was almost certainly coloured by, but also gave

---

thinks Strabo may have taken all his information on Pytheas from Polybius, Timaeus, Eratosthenes, Dicearchus, Artemidorus and Posidonius. B. Cunliffe, *Extraordinary Voyage*, 167.

22 His observations were particularly useful for mathematical geographers and astronomers. Eratosthenes used Pytheas's information to produce the northern sections of his latitude grid. Hipparchus, the second-century BC astronomer, gives us a description of the kind of information supplied by Pytheas when he says that 'no single space lies at the pole, but an empty space near which lie three stars. The spot marking the pole, aided by these, encloses a figure nearly resembling a quadrilateral – in fact, just as Pytheas the Massaliot says'. C. H. Roseman, ed., *Pytheas of Massalia: On the Ocean* (Chicago: Ares, 1994), 7–17; I. Whitaker, 'The Problem of Pytheas' Thule', *The Classical Journal*, 77, no. 2 (1981), 155, 63. The title of the book by the great stoic philosopher and geographer Posidonius (ca. 135–51 BC), Περὶ ὠκεανοῦ (*Concerning Ocean*), is derived from Pytheas's work of the same name, and he may have used him as a source for his ideas on the link between the moon and tides. I. G. Kidd, *Posidonius. Vol. 2. The Commentary: Testimonia and Fragments 1–149* (Cambridge: Cambridge University Press, 1988), 19. L. Edelstein and I. G. Kidd, *Posidonius. Vol. 1. The Fragments*, 2nd edn (Cambridge: Cambridge University Press, 1989), 19.

23 Strabo 2.4.1

24 δὲ καὶ τὰ περὶ τῆς Θούλης καὶ τῶν τόπων ἐκείνων, ἐν οἷς οὔτε γῆ καθ' αὑτὴν ὑπῆρχεν ἔτι οὔτε θάλαττα οὔτ' ἀήρ, ἀλλὰ σύγκριμά τι ἐκ τούτων πλεύμονι θαλαττίῳ ἐοικός, ἐν ᾧ φησι τὴν γῆν καὶ τὴν θάλατταν αἰωρεῖσθαι καὶ τὰ σύμπαντα, καὶ τοῦτον ὡς ἂν δεσμὸν εἶναι τῶν ὅλων, μήτε πορευτὸν μήτε πλωτὸν ὑπάρχοντα. τὸ μὲν οὖν τῷ πλεύμονι ἐοικὸς αὐτὸς ἑωρακέναι, τἆλλα δὲ λέγειν ἐξ ἀκοῆς. Strabo, 2.4.1.

25 Tacitus, 'De Vita Agricola', ed. by R. M. Ogilvie and I. Richmond (Oxford: Oxford University Press, 1967), 172–72. B. Cunliffe, *Extraordinary Voyage*, 129.

26 M. Small, 'From Jellied Seas to Open Waterways: Redefining the Northern Limit of

credence to, pre-existing Greek geographical ideas – that, at the extreme limits of the world, earth and sea, Ocean and Tartarus, meet.[27] The general trend of Greek thought, more or less substantiated by Pytheas's discoveries, was to suppose that at the furthest limits of Ocean, Ocean, and indeed land, sea or sky, all ceased to exist, and instead the world faded into primeval chaos.[28] This idea of an unnavigable indeterminate substance marking the northern limits of the world passed into first Roman and then medieval thought, so that Solinus, writing nearly five hundred years after Pytheas, described the sea beyond Thule as 'sluggish and congealed'.[29] By the eleventh century AD the idea of a 'clotted sea' was still occurring and had clearly penetrated into popular thought since it can be found in a German poem of the period. 'There is a clotted sea in the western ocean. When the strong wind drives ships upon that course,| Then the skilled seamen have no defence against it,| But they must go into the very bosom of the sea.'[30] At its northernmost margins, the world had entirely different, and dangerous, geographical and physical characteristics from the regions within the *oikoumene*.

Even when the Romans, through conquests in Germany, began to learn more of the Scandinavian peninsula and the islands north of the Germanic and Scythian lands, the northern regions beyond the limits of the *oikoumene* continued to be thought of as a region where the laws of nature ceased to prevail. By this point, the limits of the *oikoumene* had been pushed back slightly so that the beginning of the uninhabitable frigid zone was thought to coincide with the inner shores of Ocean. Tacitus, for instance, spoke of the Suiones living on the shores of Ocean, but it was inconceivable to him that anyone might live further north into Ocean than them, for nature did not prevail there.[31] As he said,

> Beyond the Suiones is another sea, sluggish and nearly motionless, with which you can trust that the inhabited world is surrounded and enclosed … If the report is true, nature extends just this far.[32]

This idea that the far northern regions of the globe lay beyond the inhabitable, and indeed beyond the natural, world was so prevalent that anyone who denied it was simply considered a liar. Strabo, for instance, said that

---

the Knowable World', *Renaissance Studies*, 21, no. 3 (2007).

27 See for example, Hesiod, *Theogony* 736–742; Plato, *Phaedo* 109–111.

28 F. Nansen, *Northern Mists*, 68.

29 'pigrum et concretum': Solinus, 22.9.

30 Old German poem *Meregarto*, cited and translated in F. Nansen, *Northern Mists*, 181.

31 Tacitus, *Germania*, 44.

32 'Trans Suionas aliud mare, pigrum ac prope immotum, quo cingi cludique terrarum orbem hinc fides, [...] Illuc usque, et fama vera, tantum natura.' Tacitus, *Germania*, 44.

Posidonius called those living in the far north περισκίους – people whose shadows go round – but made it abundantly clear that he thought Posidonius was talking nonsense, saying: 'therefore he [Posidonius] called them Periskians, although according to geography they do not exist.'[33] Although a few Greek and Roman travellers and conquerors ventured north at least as far as north Germany and Britain, and possibly as far as Scandinavia, ultimately, throughout classical times, the extreme north remained the region not only lying outside the bounds of human knowability but outside the very realms of Nature.[34] It took a major mental shift for such regions to be redefined as part of the knowable world in the sixteenth century, and required not only new empirical observation, but a rereading and reevaluation of all those writers such as Posidonius who left potential clues as to its inhabitability.[35]

Not just to the north but also to the south the limits of the *oikoumene* seemed to stretch the very limits of the laws of nature; just as at the polar regions of the world all returned to a primeval chaos, at the torrid zone there was also a blurring of the elements, where earth and sky were joined by pillars of fire. There were two possible routes for Europeans south into the torrid zone: by land into the shifting sands of the Sahara or by sea around the coast of Africa. The few Greek and Roman accounts of both routes served to reinforce the idea that the torrid zone was a liminal region, its lands desiccated and barren, its oceans a place of burning heat. The southern boundary of the *oikoumene* was better known, more explored, and, in many ways, more definite than its northern counterpart, and, if anything, posed an even greater conceptual barrier to European exploration.

The Sahara, for example, posed a visible limit to the inhabited world and though a few writers reported people crossing it, their reports served to emphasise rather than undermine the idea that to the south the land was too hot for human or indeed animal inhabitation. Herodotus, for instance, spoke of the Libyan Nasamones travelling across the Sahara until they came to a vast marshy country containing a town populated by pygmies,[36] and the Garamantes

[33] Strabo, 2.3.1; καθ' ὃ δὴ καὶ περισκίους αὐτοὺς ἐκάλεσεν, οὐδὲν ὄντας πρὸς τὴν γεωγραφίαν: Strabo, 2.5.43.

[34] Pliny uses both the terms Scatinavia (clearly the same word as Scandinavia) and Scandia (related to the Old Icelandic term Skáney) to refer to one of the northern lands. Pliny, *NH,* 4.13.96, 4.16.104. L. S. Checkin, 'Mappae Mundi and Scandinavia', *Scandinavian Studies* 65, no. 3 (1993), 490.

[35] Posidonius's opinions on the habitability of the frigid zone would have been well known to sixteenth-century humanists since, although Posidonius's own geographical text, *On Ocean,* was lost, Strabo recorded his discussion of the climatic zones in detail. Strabo 2.3.1–3.8. L. Edelstein and I. G. Kidd, *Fragments*, F49.

[36] Herodotus, 2.32.

who hunted troglodytes across the desert.[37] The descriptions of the peoples which they encountered indicated they were barely human – the pigmies were diminutive in size, unintelligible in language and fantastic in power, being a race of sorcerers; the troglodytes were sub-human reptile eaters who spoke a language like the screeching of bats.[38] Beyond their dehumanising regions, the land was even more inhospitable. Herodotus described the interior of Libya as '*eremos* and waterless and without wild animals and rainless and timberless and there is no moisture in it'.[39]

The few voyages that were attempted along the western coast also appeared to confirm the idea of a world that became oppressively hot and steadily more terrifying. In the fifth-century BC *Periplus of Hanno*, for instance, as Hanno's men approached the farthest limits of their voyage, the sailors came across 'land burning full of fragrance from which streams of fire poured down into the sea'.[40]

37 Herodotus, 4.183.

38 Herodotus, 2.32.

39 ἔρημος καὶ ἄνυδρος καὶ ἄθηρος καὶ ἄνομβρος καὶ ἄξυλος ἐστὶ ἡ χώρη, καὶ ἰκμάδος ἐστὶ ἐν αὐτῇ οὐδέν: Herodotus, 4.185.3. After Herodotus, there are virtually no other reports of travel in the Sahara. So few Greeks or Romans ventured into the desert that as late as the second century AD, Ptolemy thought it worth mentioning two Romans, Septimus Flaccus and Julius Maternus, who had journeyed with the Garamantes to the land of Agisymba 'where the rhinoceroses assemble' (Ptolemy, *Geographia*, 1.8.4). The Garamantes lived in Fazzan in southern Libya, in an area which has been extensively excavated in recent years, and had trade routes across the Sahara. Nobody knows precisely where Agisymba was but, according to Law, the evidence of the four-month journey and the rhinoceroses indicate that Maternus and Flaccus had reached southern Sudan, although there is some debate about that. R. C. C. Law, 'The Garamantes and Trans-Saharan Enterprise in Classical Times', *The Journal of African History*, 8 (1967), 181–200. N. H. Sitwell, *The World the Romans Knew* (London: Hamish Hamilton, 1984), 65. See R. Mauny, 'Trans-Saharan Contacts and the Iron Age in West Africa', in *The Cambridge History of Africa*, ed. by J. D. Fage, vol. 2 (Cambridge: Cambridge University Press, 1978), 285, for a map of how far these journeys probably penetrated into Africa. The furthest was that of the Nasamones which seems to have reached as far as the Niger. J. T. Swanson, 'The Myth of Trans-Saharan Trade During the Roman Era', *The International Journal of African Historical Studies* (1975), 582–600. While the evidence of trade with the Saharan-dwelling Garamantes indicates that the Romans did penetrate into the desert itself, and indeed had indirect trade connections with the sub-Sahara, for the most part sub-Saharan Africa remained outside their knowledge and the Sahara remained a limit to the Greco-Roman world (M. Mund-Dopchie, 'Hanno', in *Catalogus Translationum Et Commentariorum: Mediaeval and Renaissance Latin Translations and Commentaries; Annotated Lists and Guides*, ed. by Virginia Brown, vol. VIII (Washington DC: The Catholic University of America Press, 2003), 50.

40 χώραν διάπυρον θυμιαμάτων μεστ[ήν· μέγιστοι δ΄ ἀπ΄ αὐτῆς πυρώδεις ῥύακες ἐνέβαλλον εἰς τὴν θάλατταν. A. N. Oikonomides, ed., *Hanno the Carthaginian: Periplus or Circumnavigation of Africa. Greek Text with Facing English Translation, Commentary, Notes and Facsimile of Codex Palatinus Gr. 398* (Chicago: Ares Pub, 1977), lines 79-80: This is clearly a reference to a volcano, but as Paul Hair has shown

Further on they arrived by night at a land full of fire; and 'in the middle was an enormous flame greater than the others, reaching, as it seemed, as far as the stars'.[41]

Hanno's was one of the few recorded voyages down the west coast of Africa and served the same function of reinforcing the idea that the torrid region lay firmly outside the bounds of human knowledge, as Pytheas's voyage had done for the frigid zone. Unlike with Pytheas's voyages, there is no definite proof that any Greek or Roman had read Hanno's report until the time of Pliny,[42] but, thereafter, all the key Roman geographers – that is, Pliny the Elder, Pomponius Mela and Solinus – refer to him, as do Arrian and Athianus.[43] His text persisted as an authority into the sixteenth century and was deemed informative and accurate enough that Ramusio included his translation of it in his *Navigazioni e viaggi.*[44] His descriptions of columns of flame and streams of fire were further embroidered by those who cited the texts – notably Pliny, Solinus and Mela, who added mythical, semi-human creatures such as Amazons, Gorgons, Goat-pans and Satyrs which are nowhere to be found in the original text.[45] These were creatures of the peripheries rather than of the *oikoumene* proper. Transmission corrupted the text and reinforced the traditional idea that the torrid zone was uninhabitable, and even its peripheries only knowable by those who were not true humans. Beyond lay regions where earth and air, fire and water become intermixed in columns of flame and streams of fire.

Given their extreme nature, and the prospect of death involved in venturing into such regions, it is not surprising that the idea of climatic zones served to create 'walls of the mind' which were more effective deterrents to exploration than real walls would have been.[46] This was a belief that went beyond the reading public and came to be a part of generally accepted lore. Nevertheless, by the late Middle Ages some scholars and travellers were beginning to realise that to enter these zones was not tantamount to suicide. Their ideas provided a foundation for those who sought to challenge the boundaries of the *oikoumene.*

---

it is very difficult to identify this, or indeed any topographical feature in the *Periplus*. P. E. H. Hair, 'The "Periplus of Hanno" in the History and Historiography of Black Africa", *History in Africa*, 14 (1987): 56-58.

41 ἐν μέσῳ δ᾽ ἦν ἠλίβατόν τι πῦρ, τῶν ἄλλων μεῖζον, ἁπτόμενον, ὡς ἐδόκει, τῶν ἄστρων *Ibid.*, 83–85. This is possibly a reference to bush burning. P. E. H. Hair, 'Periplus', 58.

42 Pliny 2.67.3; there is a possibility that Herodotus 2.59, which refers to a Carthaginian voyage along the coast of Africa, referred back to this text. The surviving version of the text is written in Greek.

43 M. Mund-Dopchie, *Hanno*, 50.

44 hanno, 'la navigazione di annone, capitano de' Cartaginesi nelli parti dell'Africa fuori delle colonne d'Ercole', *NV*, vol. 1, 547–50.

45 M. Mund-Dopchie, *Hanno*, 50; Pliny, *NH*, 6.31.36; 5.1; Mela, 3.90-93; 3.99; Solinus, 56.12.

46 E. P. Hanson, *New Worlds Emerging* (London: Victor Gollancz Ltd, 1950), 52.

The intellectual challenge presaged the empirical one so that as early as the twelfth century Robert Grosseteste (c.1170–1253) stated that it was known from the evidence of other writers that the equatorial zone was actually a temperate region,[47] and Roger Bacon (1219–1292) referred to the torrid zone as habitable.[48] Likewise, Albertus Magnus (c.1193–1280) argued that eyewitness testimony made him refute conventional authority, specifically citing his contradiction of Avicenna and Ptolemy about the habitability of the torrid zone, although he did not think it a pleasant place to live.[49] The debate can arguably be most clearly seen in Robert Anglicus's twelfth-century commentary on Sacrobosco. Sacrobosco had denied the habitability of the equatorial zone, but Anglicus spent several pages discussing why it actually supported human life.[50] Cardinal Guillaume de Fillastre, writing in the aftermath of the Council of Constance, likewise clearly argued in favour of the habitability of the torrid zone, even bringing in the language of God's design, arguing that while God has made some areas uninhabitable merely to show his power, he would not have made the majority of the earth uninhabitable.[51] Nonetheless, the prevalent opinion remained on the side of those upholding the traditional bounds to the *oikoumene*.[52]

Such ideas were slow to filter into exploration and widespread geographical understanding, but already by the time of Marco Polo (1254–1324) the north-south divide emphasised by the idea of a torrid zone was beginning to break down.[53] Marco Polo discusses both Zanzibar and Madagascar which lie

[47] R. Grosseteste, *Die philosophischen Werke des Robert Grosseteste*, ed. by L. Baur (Münster: Beiträge zur Geschichte der Philosophie des Mittelalters, 1912), 66–67. Eastwood thinks that Grosseteste here refers to Ptolemy's *Almagest* 2.6: B. Eastwood, 'Mediaeval Empiricism: The Case of Grosseteste's Optics', *Speculum*, 43 (1968), 314.

[48] R. Bacon, 'Opus', part IV, Dist. IV, ch. 4, 136.

[49] Albertus (Magnus), Habes in hac pagina, *De natura locorum librum mira eruditione, & singulari fruge refertum, & iam primum summa diligentia revisum, in lucem aeditum, quem leges diligentius, vel si cosmographia vel physica profecisse te volueris* (Strasbourg: Atlantsee, 1515), XIII.

[50] L. Thorndike, *The Sphere of Sacrobosco and its Commentators*, Sacrobosco, 139, Anglicus, 237–39.

[51] P. G. Dalché, 'l'oeuvre géographique du cardinal Fillastre (1428). Représentation du monde et perception de la carte à l'aube des découvertes', in *Humanisme et culture géographique à l'époque du concile de Constance: Autour de Guillaume Fillastre: Actes du Colloque de l'Université de Reims*, ed. by Didier Marcotte (Turnhout: Brepols, 2002), paragraph 12, 331.

[52] I. Metzler, 'Perceptions of Hot Climate in Medieval Cosmography and Travel Literature', *Reading Medieval Studies*, 23 (1997), 72–73.

[53] Marco Polo's text was not, however, readily believed at the time. A more general acceptance came only once there was further supporting evidence from merchants and missionaries who had travelled to the east, and only gained widespread appreciation in 1557 when Tommaso Giunti published Ramusio's *Discorso* on Marco Polo in the

below the equator, in or beyond the torrid zone, and though he does not claim to have eye-witness knowledge of either of them, he does make it clear that such knowledge is obtainable.[54] Likewise, works such as those of Grosseteste and Bacon seem to have convinced the author of John Mandeville's fictional pilgrimage account that the equatorial region could, in fact, have a mild climate.[55] This was a work (written c. 1356) purporting to be a travelogue by John Mandeville, a knight who travelled from England far into Asia, both east and south. In terms of the expanding *oikoumene*, it is an extremely contradictory text. In the printed 1582 version, it states that 'at the north part of the world... that land is in some places so cold that there may no man dwell therein, and on the south side of the world it is in some places so hot that there can no man dwell'.[56] This seems a very clear affirmation of the limits of the *oikoumene.* But in an earlier version circulating in manuscript in England, while the author occasionally stated explicitly that a region was uninhabitable owing to the heat (e.g. that there are no fish in the Libyan sea, 'for the water there, because of the great heat is evermore boiling'),[57] uninhabitable regions (according to his narrative) had specific locations, and did not form a continuous uninhabitable

---

second volume of his *Navigazione e viaggi.* Therefore, although his treatise seemed to undermine support for the idea of the torrid zone, the scepticism with which his work was viewed until the sixteenth century limited the influence of his observations about the inhabitability of the region. The most popular version of his text, as J. Larner has shown, was the most scholarly version which contained the most geographical information, and manuscripts of this and other versions circulated from the time it was first written. Nevertheless, people's first reaction tended to be doubt, even if they later agreed that it must be true. J. Larner, *Marco Polo and the Discovery of the World* (New Haven: Yale University Press, 1999), 105–15.

54 M. Polo, 'Dei Viaggi di Messer Marco Polo, Gentiluomo Veneziano', *NV*, vol. 3, 286–88.

55 E. G. R. Taylor, 'The Cosmographical Ideas of Mandeville's Day', in *Mandeville's Travels; Texts and Translations*, ed. by M Letts (London: Hakluyt Society, 1953), li–liii.; A. C. Cawley, 'Down Under: A Possible New Source for Mandeville's Travels', in *Iceland and the Mediaeval World, Studies in Honour of Ian Maxwell*, ed. by G. Turville-Peter and J. Stanley Martin (Melbourne, Australia: 1974), 147.

56 J. Mandeville, 'The Voyage of John Mandevil Knight in Latin, Begun in the Raigne of Edwarde the 2. Anno. 1322. Continued for the Space of 33. Yeeres, and Ended in the Raigne of Edwarde the 3. From England to Iudea, and from Thence to India, China, Tartaria, and as Farre as 33. Degrees to the South of the Aequinoctiall', in *The Principall Navigations, Voiages and Discoveries of the English Nation*, ed. by Richard Hakluyt, David B. Quinn, and R. A. Skelton, vol. 1 (London: 1589, 1965 facsimile edition), ch. XLIII.&quot. This wording and concept is not to be found in any of the other versions which have been compiled by M. Letts in his 2 volume edition of various Mandeville texts compiled for the Hakluyt Society in 1953. (*Mandeville's Travels; Texts and Translations,* ed. by M Letts).

57 55 M. Letts, ed., *Mandeville's Travels; Texts and Translations*, 2nd vol. (London: Hakluyt Society, 1953), 103. Letts here prints the text from BL, Eg., 1982, dated

band. Instead, in this rendition of Mandeville's work, the author explicitly broke down the idea of a confined *oikoumene*, stating: 'I say sickerly that a man might go all the world about, both above and beneath, and come again to his own country, so that he had his health, good shipping and good company'.[58] He did not even mention the idea of a torrid zone; rather, he entered into regions once thought to belong to it, and indeed claimed to have travelled far into the southern hemisphere. No version that I know, of either Mandeville or Marco Polo's works, explicitly confronted the idea of an uninhabitable torrid zone – they merely overlooked it – but this was a definite step forward towards refuting the notion. It was not until the fifteenth century, however, that the ideas planted by Grosseteste, Bacon, and others took hold in any practical manner and scepticism filtered into practical exploration – although, as can be seen by the later version of Mandeville, this was not a universal challenge.[59]

In addition to sceptical attitudes to the climatic boundaries discernible in both the medieval scholarly writings and the travelogues, the classics themselves contained keys to the opening-up of the world, but the majority of these alternative explanations only came to the fore in the Renaissance. The idea of uninhabitable climatic zones was firmly entrenched in the canonical works, but among the rediscovered authors of the fifteenth-century Renaissance were some differing opinions. Ptolemy did not discuss the torrid zone much, but did state that the *oikoumene* extended sixteen and a half degrees south of the equator.[60] Geminos of Rhodes (first century BC) thought that the whole torrid region was unpleasantly but not unendurably hot.[61] Strabo, although believing in the uninhabitable realms himself, reported Posidonius's discussion of the zones under the tropics as being hot and dry, but inhabited by people with flat noses, curly hair and protruding lips;[62] and mentioned Polybius's ideas on the habitability of the tropics.[63] Aside from the

between 1410 and 1422. I have chosen to use a translation of Mandeville that was in circulation in English by the sixteenth century.

58 *Ibid.*, 129.(BL Eg.) This quotation does not appear in Hakluyt's vastly abbreviated version of the travels.

59 W. G. L. Randles, *De la Terre plate*, 33–38.

60 Ptolemy, Book 7.

61 Geminos of Rhodes, *Εἰσαγωγή εἰς τὰ Φαινόμενα* (*Elementa Astronomiae*), 16.24

62 Strabo, 2.3.1.

63 Strabo, 2.3.2. Polybius (c. 200–120 BC) wrote a treatise on the *Habitability of the Equatorial Region*, *Περὶ τῆς ὑπὸ τὸν Ισημερινὸν Οικήσεως*, which was known in the ancient world, although it did not survive along with the rest of his work, perhaps because so few ancients agreed with its argument. Geminos of Rhodes refers to this treatise, book 16.32–8. It may, in fact, have been book 34 of his *Histories*. If it is a separate treatise, nothing of it is known except its title. If it is part of book 34, only fragments remain. The two main commentators on Polybius disagree about how it fits with the rest of the corpus. See F. W. Walbank, *Polybius* (London and Berkeley:

terse Ptolemy and the little-read Geminos, all these ancient works talking of the habitability of the torrid zone survived only in fragmentary form, and so provided more scope for the imagination than actual argument. Because Strabo reported them, however, Renaissance scholars knew about these contradictory authorities and cited them in references. By the time of the Portuguese explorations, even before the advent of print, texts were circulating in manuscript form which voiced these dissenting opinions, and by the sixteenth century people were drawing attention to the contradictory opinions about the extent to which the burning zone was an uninhabitable region. Robert Recorde, for example, in a discussion about the limits of the burning and frigid zones, in which one classical authority was pitted against another, stated that:

> You maye not weighe the matter by auctoritye, for so shoulde that former doctrine continue styll, seynge I aleaged for it Parmenides, Aristotle, Polybius, Cleomedes and Proclus, & against them only Posidonius and Strabo, which maye seeme the weaker in numbre: but then considre that the firste sort bring only affirmation for their testimony, and bare autoritye: the other, confute theym by good reason and substantiall argumentes, whiche are farre to bee esteemed aboue anye autoritye.[64]

Recorde, as we have seen in chapter two, was particularly concerned with matters of reason and authority, but the searching for what constituted authority was a common feature of all the sixteenth-century descriptive geographies. Humanist authors were finding alternative explanations in the classics and then providing a rationale for assessing the merits of one classical doctrine over another. In the case of the torrid zones and frigid zones, the judgement about which classical authority was correct had to depend on reason: until the limits began to be challenged by intellectual rationale, there could be no empirical evidence to test one authority against the other, for anyone who strayed into these regions was gambling with their lives.

Reason did not just play its role in attempting to distinguish between one authority and the other; it also helped to break down belief in the climatic barriers, which had moved from the proposed theories of Herodotus into the accepted lore. The chronicler Gomes de Zurara documented six reasons for the Infante Henry's pursuing a policy of southern exploration and expansion, but he could find few sailors willing to attempt the voyages he proposed. Zurara was a librarian, keeper of the king's archives and Royal chronicler. Although he made one short journey

---

University of California Pres, 1990), 15, 123 and P. Pédech, *La méthode historique de Polybe* (Paris: Les Belles Lettres, 1964), 589–90. Pédech thinks that it formed part of book 34; Walbank is less certain.

64 R. Recorde, *Castle*, 174.

to Africa to try to understand better the people and places about which he was writing, he was not the best positioned to grasp the motivation of the sailors in refusing to cross into the torrid zone.[65] Indeed, he was slightly contemptuous of their reasoning, but his contempt in some ways gives more veracity to his reports of the mind-set of the sailors. Their fear was what Henry had to overcome. He reports that:

> up to his time, neither by writings, nor by the memory of man, was known with any certainty the nature of the land beyond that Cape.... it seemed to him that if he or some other lord did not endeavour to gain that knowledge no mariners or merchants would ever dare to attempt it.[66]

Zurara, writing c. 1455, shows the fear that Henry's command to sail past Cape Bojador held for the ordinary sailor. His speech cannot be taken to be the actual words of the Portuguese sailors, but it might legitimately be thought to portray their sentiments. He claimed that the sailors pleaded not to be sent across the Tropic of Cancer: 'How shall we, they said, pass the boundaries which our ancestors set and what profit can it bring for the Prince by the loss of our souls along with our bodies, since we shall knowingly be our own murderers?'[67] The boundaries laid down by the ancestors had become as real as if they were set in stone. One had to be very daring to defy them. In Zurara's narrative, Henry, in ultimately persuading Gil Eannes to sail past Cape Bojador and into the tropics in 1434, resorted to reason and a questioning of authority. He attempted to demonstrate the problems in authority which had led to such general credence in the idea of the torrid zone and invoked reason and the need to test by experiment.[68]

Long after the Portuguese crossed the equator and the torrid zone, people still felt the need to come to terms with the idea of a burning zone. While the Portuguese themselves recognised the significance of Gil Eanes's voyage, much of Europe seems to have remained unaware that the boundaries of the torrid zone had been tested and proven not to exist. The classical zonal paradigm

65 P. E. H. Hair, 'The Early Sources on Guinea', *History in Africa*, 21 (1994).

66 G. E. de Azurara, *Descobrimento Da Guiné*, 45–6. 'e porque elle tiinha voontade de saber a terra que hya a allem das ilhas de Canarya e huũ cabo, que se chama do Bojador, porque atea acquelle tempo, nem per scriptura, nem per memorya de nhuũs homeês, nunca foe sabudo determinadaments a callidade da terra que hya allem do dicto cabo... E porque o dicto senhor quis desto saber a verdade parecendolhe que se elle ou alguũ outro senhor se nom trabalhasse de o saber nehuũs mareantes nem mercadores nunca se delle antremeteryam porque claro sta que nunca nehuũs daquestes se trabalham de navegar...'

67 *Ibid.*, 54. Como passaremos, deziam eles, os termos que poserom nossos padres, ou que proveito pode trazer ao Iffante a perdiçom de nossas almas, juntamente com os corpos, ça conhecidamente seremos omicidas de nos mesmos?'

68 *Ibid.*, 57.

prevailed for another half-century. As late as the early 1480s, people in Italy were largely ignorant of the Portuguese achievement.[69] As awareness of the achievements of the Portuguese sailors became more widespread, however, there was a role reversal in who struggled more with the concept of an inhabitable torrid zone. Many explorers ceased to be concerned about travelling in the equatorial region, whereas the humanist scholars, accustomed to upholding the authority of the ancients, more frequently struggled with the concept.[70] Girolamo Benzoni, the Italian historian of the New World, whose work formed the fourth volume of de Bry's collection on the Americas, blamed in part the fixation on the existence of a torrid zone for people's inability to accept the truth of what Columbus had accomplished. He said that there were Portugese men

> who professing to know everything about the cosmographic art, judge that they are not ignorant of any part of it. Therefore, looking down on Columbus superciliously, they persuaded the king that he should not wish to rely on the dreams of an outsider. Rather that he was out of his mind: and that it was not possible for gold or anything of wealth to exist in the west, as they claimed to have discovered. For no-one could cross the zone of the equinoctial line, lest they be burned by the immoderate heat of the sun, since in that part the force of its seething heat is greater, where the orbit and course of the sun is constantly between the two tropics.[71]

These Portuguese scholars might be thought to have had their own agenda, since their own authority relied upon the dominance of scholarly study over experience, but not all who queried the possibility of crossing the equator were so motivated. Some truly grappled with the problems posed by a classical

69 C. H. Clough, 'The New World and the Renaissance', in *The European Outthrust and Encounter: The First Phase (c. 1400–c. 1700): essays in tribute to David Beers Quinn on his 85th Birthday,* ed. by C. H. Clough and P. E. H. Hair (Liverpool: Liverpool University Press, 1994), 291.

70 On the change in sailors' attitudes, see M. Tymowski, 'Death and Attitudes to Death at the Time of Early European Expeditions to Africa', *Cahiers d'Études Africaines*, 54, no. 215 (2014). And N. Wey Gómez, *Tropics of Empire*.

71 '[Erant tum in Lusitania homines,] qui Cosmographicam artem profitentes, nullam se eius partem ignorare arbitrarentur. Itaque superciliose Columbum despicientes, regi suadebant, nequam peregrini hominis somniis fidem adhibere vellet. Quippe amentem esse: neque in Occidente aurum, aut divitiarum quidquam existere posse, veluti exploratum affirmabant. Imo neminem subiectam Aequinoctiali lineae plagam transgredi posse, quin immodicis solis ardoribus torreretur, siquidem eo in spatio maiorem aestus esse vim, qua perpetua inter duos Tropicos solis orbita et cursus est.' G. Benzoni, 'Insignis et admirandis historia de reperta primum Occidentali India à Christophoro Columbo', in *Americae Pars Quarta*, ed. by Theodor de Bry (Frankfurt: I Feyrabend, 1594), 30.

authority that, save for the few examples of Geminos, Posidonius and Polybius – which, as mentioned above, had only recently come to people's attention – uniformly asserted the uninhabitability of the equatorial regions.

The idea of an uninhabitable torrid zone was one of the few features of classical geography which could not be adapted to fit with Renaissance discoveries, though some tried. For humanists in the period of exploration, it was important to show that humanism and discovery could be connected. In the eyes of Damiano de Góis, a sixteenth-century Portuguese chronicler, for instance, Henry the Navigator had not been undermining or ignoring the classical writings when he had compelled his men to explore, but rather was driven by the 'desire to establish the veracity of the classical sources'.[72] There is no such indication in Zurara who was Henry's contemporary. Rather, such a statement seems a desperate attempt to reconcile humanism and the discoveries.

Others simply admitted that there were flaws in the classical argument, but did not see this as a reason to reject the classics. The Jesuit Acosta, writing an eye-witness natural history of the New World more than a century after the Portuguese crossed the equator, quotes passages verbatim from Virgil, Ovid and Aristotle dealing with the five zones, and the uninhabitability of the tropics, and proceeds to elaborate upon the reason for the belief in a region too hot to sustain life.[73] He himself had clearly been swayed by the arguments of the ancients with whom he was so familiar, for he states that despite setting out on the voyage, when he knew he was reaching the equatorial region he had persuaded himself that he would not be able to tolerate the violent heat. In reality, however, he found that it was so cold that he had to go inside to stay warm. He asserts that this forced him to laugh at Aristotle, but though it proved the fallacy of Aristotle's reasoning, it did not overturn the ancient authorities for Acosta. Instead he trawled through other classical – and even medieval – works to find justification for the habitability of the region, claiming that

> Plato placed his very famous Island of Atlantis in the torrid zone, because he says that at a certain time of the year, it had the sun above it; despite that, he said that it was temperate, abundant and rich. Pliny places Taprobana or Sumatra as they now call it below the equinox, as in fact it is; and he says of it not only that it is rich, prosperous, but also very populous both with people and animals. From which one should understand that although the ancients believed that the heat of the torrid zone was unbearable, if the heat could be avoided, they could realise that it was not as uninhabitable as they presumed. The most excellent Astrologer and Cosmographer Ptolemy and the illustrious philosopher

[72] D. de Góis, *Chronica*, ch. 7, fol. 5v.

[73] See J. de Acosta, *Historia*, 46–49.

and physician Avicenna understood much better because both felt that below the equinox there was very pleasant habitation.[74]

Once again it became a matter of using reason and experience to evaluate which classical authority to prefer. The only real means of maintaining the authority of the classics was to point out those few ancient authors who denied that the equatorial region was totally uninhabitable, though unlike Recorde or the Infante, Acosta set his own experience as the means of determining which authority was correct.

Once the climatic boundary of the torrid zone had fallen by the wayside, however, the frigid zone was soon to follow. By 1527, Robert Thorne was questioning the long-accepted zonal theory, stating that:

> it is a generall opinion of all Cosmographers, that passing the seventh clime, the sea is all ice, and the colde so much that none can sunder it. And hitherto they had all the like opinion that under the line Equinoctial, for much heate the land was unhabitable.

> Yet since (by experience is proved) no land so much habitable nor more temperate. And to conclude, I thinke the same should be found under the North, if it were experimented. For as all judge, Nihil fit vacuum in rerum natura: So I judge there is no land uninhabitable nor Sea innavigable.[75]

Yet not all writers took the point of view of Thorne. There were still many who emphasised the uninhabitable nature of the frigid zone. The islands surrounding

[74] Ptolemy *Almagest*, 2.6. In the geography, Ptolemy did not go as far as to say the equinoctial was commodious, but he did discuss various places, including Pattal in India, Taprobana and Cattigara, and Garama in Libya, which he claimed lay within the tropics. 'Platón ponía su tan celebrada Isla Atlántida en parte de la Tórrida, pues dice, que en cierto tiempo del año tenía al Sol encima de sí; con todo eso dice de ella, que era templada, abundante y rica. Plinio pone á la Taprobana ó Sumatra, que ahora llaman, debajo de la equinoccial, como en efecto lo está, la cual no solo dice, que es rica y próspera, sino también muy poblada de gente y de animales. De lo cual se puede entender, que aunque los Antiguos tuvieron por intolerable el calor de la Tórrida, pero pudieron advertir, que no era tan inhabitable, como la hacían. El excelentísimo Astrólogo y Cosmógrafo Ptoloméo, y el insigne Filósofo y Médico Avicena antinaron harto mejor, pues ambos sintieron, que debajo de la equinoccial habia muy apacible habitación.' *Ibid.*, 144–45.

[75] R. Thorne, 'A Declaration of the Indies and Lands Discovered and Subdued Unto the Emperour, and the King of Portugal: And Also of Other Partes of the Indies and Rich Countries to Be Discovered Which the Worshipfull M. Robert Thorne Merchant of London (Who Dwelt Long in the Citie of Sivil) in Spaine Exhorted King Henrie the Eight to Take in Hand (Ad 1527)', *PN*, vol. 3, 178.

the pole were thought to be inhabited by pygmies or Amazons.[76] Even the sixteenth-century Swedish writer, Olaus Magnus, referred to the area north of Lithuania and east of Finland as uninhabitable because of the cold.[77]

As late as 1589, Thomas Blundeville, a Cambridge-educated humanist and mathematician, wrote that the search for northern passages was a foolish enterprise since these were places 'more meet for whales and monstrous fishes to dwell in, then for men: and specially for English men which are not able to suffer the cold winter at Wardhouse'.[78] Pliny, among other ancient authors, had written of the promontory of Tabin at the northern extreme of the *oikoumene*,[79] and various early modern geographers and cartographers had accepted Pliny's statement as portraying evidence of classical knowledge of the north. Ortelius, for instance depicted the promontory on his map of the polar regions and ascribed knowledge of it to Pliny. Blundeville, however, refused to believe those writers who tried to reinterpret the ancient authorities to say that they had first-hand knowledge of the north and of Tabin which was believed to lie at 76 degrees latitude. For he thought that no-one bred in the temperate or hot zone 'is ever able to continue the whole journey in any of those waies: no, though they were much more passable than I take them to be indeede'.[80] Blundeville had been forced to take into consideration the idea that the frigid zone contained human beings, but he still clearly thought in terms of the old zonal system and its barriers to human inhabitation. He adapted the ancient idea to make it not an absolute boundary to human survival, but nonetheless a definite limit to exploration by any from the world known to the Greeks and Romans. Blundeville was fighting a rear-guard battle, however. Already, the English navigators Stephen and William Borough, John Davis, and Martin Frobisher were among the more famous explorers making forays into the frozen zone.[81] Meanwhile, scholars such as Richard Hakluyt, John Dee, Ortelius and Mercator were actively advocating Arctic exploration, as witnessed by the

[76] Martin Behaim's map, for instance, shows a circle of islands and lands around the North Pole. On either side of the Pole are large islands, one of which has what appears to be an Amazon wearing a long dress attacking a polar bear. The pigmies are less readily identifiable, but are recorded in the texts. O. Magnus, *Historia De Gentibus Septentrionalibus* (Rome: J. M. de Viottis, 1555), 71.

[77] *Ibid.*, 98.

[78] T. Blundeville, *A Brief Description of Universal Mappes and Cardes* (London: Roger Ward for Thomas Cadman, 1589), C2.

[79] Pliny, *NH*, 6.53. Pliny actually calls it a mountain range, Tabis, but it was transformed into a promontory in the early modern period.

[80] T. Blundeville, *Brief Description*, C3a.

[81] Stephen and William had sailed for a northeast passage, reaching the White Sea in 1554; Martin Frobisher and John Davis sailed for the northwest passage in 1576–78 and 1585–87 respectively. Davis reached as far as sixty-seven degrees north.

constant searches for northeast and northwest passages, this region too having been deemed to be survivable, if not welcoming.

Ultimately, the ability to journey into the uninhabitable zones came to be seen as a triumph of autopsy, and as a reason for exploration. The classics needed to be read and tested, either to be proved or disproved. Thevet, for instance, argued that experience had shown that both the polar and the equatorial regions were inhabitable, contrary to ancient belief: 'The Ancients should not be excused, and not at all believed, since they spoke about conjecture rather than about experience.'[82] Thevet's English translator, Thomas Hackett, also pointed out that many of the classical writers were themselves inconsistent since they discussed regions supposedly in the torrid zone.[83] Although Thevet saw value in the use of classical works, it was primarily for their use as a starting point. Such an attitude did not wholly undermine the classical authorities. Ancient ideas required testing but they were accepted until disproved. The northern and southern zonal boundaries of the world had provided a mental prison until they were challenged. Once challenged, and disproved, they provided a reason for further exploration.

Even then, those people who clearly had no faith in the zonal theory sometimes simply used the classics to provide a new starting point. Olaus Magnus, for instance, while extolling the authority of the classical writers, helped turn the traditional image of the frozen north on its head.[84] He did not dispute the extreme cold of the northern lands, but rather argued that the cold in fact made the place easier for humans to know, since it opened up pathless territories to travellers, caused inns to be set up, markets to be held, and allowed for wars to be waged on water as well as land.[85] Explorers such as Sir Humphrey Gilbert and John Davis went still further. While clearly discrediting the ancient idea of an uninhabitable frozen zone, they used the classical authorities to provide reasons for the possibility of a sea passage through the Arctic waters. When advocating the existence of a northwest passage, Davis, for example, upheld the authority of experience, saying 'what needeth the repetition of authorities from writers, or wrested philosophical reasons, when playne experience maketh the matter so manifest', yet he still placed emphasis on the ancients, citing Homer, Hyginus, Pomponius Mela and Cornelius Nepos as providing evidence for the existence of such a passage. He ignored the classical authorities who dwelt on the uninhabitability, and cited those who seemed to provide an excuse for exploration. The zonal theory had

82 'Parquoy sont excusables les Anciens, et non du tout croyables, ayans parlé par coniecture, et non par expérience'. A. Thevet, *Singularitez*, 36.

83 A. Thevet, *New Founde Worlde*, 31r.

84 For more on Olaus Magnus's careful veneration of classical authority and his 'almost slavish adherence to his source material', see P. Fisher, 'Olaus Magnus and the Limits of Credence', in *Acta Conventus Neo-Latini Hafniensis* (Binghamton, NY: Medieval and Renaissance Texts and Studies, 1995), 409–19.

85 O. Magnus, *Historia*, 53.

become untenable but the classics had not entirely lost their authority, even for the explorers themselves.[86]

The challenges to the zonal theory went hand in hand with the classical revival. A theory of climatic limit to human inhabitation was overturned and the areas of more extreme climate at the margins of the world came to be seen by the humanists as the regions which contained the means of connecting the world. Sailing south into the torrid zone allowed the Portuguese and Spanish to reach Asia and the Americas. The arctic regions came to be thought of as containing the shortest sea-routes to the wonders of Cathay. These regions, once thought to be beyond the limits of the knowable world, came to be seen as key spaces for allowing global connectedness. In overturning the idea of climatic margins which determined the furthermost limits of human habitation, however, the humanist geographers and explorers had to deal with concepts of authority. The canonical classical authors could not be unthinkingly accepted since they uniformly described a limited *oikoumene* restricted to the northern temperate region. Instead, other classical authors were brought into the argument. Their authority was asserted by means of reason and experience. The classics remained fundamental to the new geography that was developing in the sixteenth century, but were used to justify the reinterpretation of the *oikoumene* from a climatically limited space, to one which embraced the whole sphere in which the torrid and frigid zones came to be seen as key regions for connecting the different parts of the world together.

86 J. Davis, *Worldes Hydrographical*, 198, 216.

4

Dispelling the Boundaries of the World: Ocean from Confine to Means of Communication

If Renaissance voyagers had to overcome the fear instilled in them by the idea of the frigid and torrid zones before they could explore beyond the northern and southern limits of the inhabitable lands, they faced a different problem in dealing with the eastern and western ones. Here the boundaries of the *oikoumene* did indeed coincide with the limits of the land which came to an end in Ocean. In ancient Greece, *Okeanos* was a body of water thought to mark the edges of the world. Just as the idea of torrid and frigid zones had filtered from Greek thought into Roman, and from there into medieval geographies, so did the idea of Ocean. While the torrid and frigid zones provided climatic limits to the *oikoumene* in the ancient world, Ocean acted as a physical boundary which also posed a limit to human habitation.

This chapter examines the way in which Ocean was redefined in the fifteenth and sixteenth centuries so that it ceased to be seen as a confine and came to be considered the primary means of communication to connect different parts of the world to each other. It begins with an overview of the three different theories of Ocean which existed in the classical world. It then looks at how the understanding of Ocean was altered in a Christian Renaissance context. Despite the overarching emphasis on Ocean as a confine throughout classical and medieval times, there was much in the enduring idea of Ocean which could enable sixteenth-century writers and travellers to reenvisage it as a means of communication. Far from limiting human inhabitation, Ocean provided the means for humans to have global interchange and exchange – the seas became the pathways of the *oikoumene*. The chapter then examines how, since Ocean surrounded all the different parts of the world, it was also thought to define the nature and extent of the landmasses. Because of their belief in the classical Ocean, humanists came to believe that the Americas must be a continent in their own right, entirely separated from Eurasia and Africa, by Ocean, but also connected with them for trade purposes by the same water that separated them. The chapter concludes by showing how the Christian and classical contexts worked together in the reenvisioning of the nature of ocean as a means of connecting all parts of the world. To do so, however, it required them not only to divest it of earlier religious connotations but also to fit it into their new conception of a world designed by God for humans to have dominion over in entirety. Rather than being the place where nature no longer worked according to law, it had to be seen as part of nature and was expressly created by God to link all parts of the earth.

The ancients had three key theories about the nature and shape of Ocean, but in each of these theories Ocean acted as a limit, and it was this idea of limit rather than the concept of Ocean that required a fundamental reassessment in the Renaissance. As we shall see, the idea of Ocean was not a static concept in the ancient world. At times it had mythical and religious connotations, and at other times it was seen an entirely physical phenomenon. But in all its manifestations it acted as single connected body of water demarcating the physical limits of the landmass or landmasses. As classical geographical knowledge expanded, different bodies of water such as the North Sea and the Atlantic Ocean came to be thought of as parts of Ocean. By the early modern period, there were several different named bodies of water at the limits of the land, but ultimately these were all thought of as part of the larger Ocean.

The idea of a circumambient Ocean is one of the oldest features of western geographical thought, and can be traced in Greek literature as far back as Homer, who described *Okeanos* as a fresh-water river surrounding a disk-like earth.[1] In this form it bears little resemblance to the great oceans crossed and re-crossed by Europeans in the sixteenth century. Yet the unchallenged belief in the existence of Ocean, which remained so fundamental to sixteenth-century thought, can ultimately be traced back to the very earliest European writings. It is possible that it predates even the Greeks who may have derived their concepts of Ocean from the Babylonians, the Mesopotamians or the Egyptians, in each of whose regions the formation of the world is exemplified every year by the emergence of the dry land above receding flood waters; but the Greeks transformed the idea and provided it with a mythology.[2] In early Greek thought, the river was the πείρατα γαίης, the boundary of all land, and marked the divide between the heavens and the earth.[3] The land formed its inner bank, while the

1 Homer, *Iliad*, 18.607–608.

2 G. S. Kirk, 'Sense and Common Sense in the Development of Greek Philosophy', *Journal of Hellenic Studies*, 81 (1961), 106; O. A. W. Dilke, *Greek and Roman Maps* (London: Thames and Hudson, 1985), 24; G. S. Kirk, J. E. Raven, and M. Schofield, *The Presocratic Philosophers: A Critical History with a Selection of Texts*, 2nd edn (Cambridge: Cambridge University Press, 1983), 11–12.

3 Homer, *Odyssey*, 11.13.

crystalline vault of the heavens signalled its outer shore.[4] Within this vault, Ocean's waters filled the whole gulf between earth and sky. Although it was not literally endless, it was of indefinite size, 'stretch[ing] out into unimaginable distance' until at some theoretical point it reached the very dome of the heavens.[5] In this earliest form, Ocean was an entirely theoretical or religious concept, belonging to the domain of myth. *Okeanos* was a god and the father of all waters. No Greek had reached the ends of the earth to discover if Ocean really existed. Belief in it was a matter of faith rather than proof. Exploration of such water would be profitless since nothing lay beyond its inner shores. If one were to sail out upon this sea, one would leave the human world behind and trespass upon the realms of the infinite and of the gods. It formed a formidable boundary, but one that was closely linked to the religious and mythical beliefs of the time. As the Greeks began to explore further afield, however, they came across large bodies of water which seemed to reinforce the idea of a πείρατα γαίης, removing Ocean from its purely mythical setting and giving it a physical presence that reinforced belief in the idea of a circumambient body of water.

By the fifth century BC, when Greek expansion and inquiry had altered the perception of the world, the idea of Ocean was one of the few features which endured from the Homeric portrayal, yet Homer himself would not have recognised the river-god Okeanos in the later descriptions of Ocean. Instead, a second theory of ocean developed. Since it was now widely accepted that the earth was a sphere, Ocean, instead of branching out on the same plane as the earth filling the void between the land and the crystalline heavenly sphere, was now thought to cover the surface of the globe, filling all the space not covered by land. By this time Ocean had a real location: Greeks had reached the Atlantic seaboard and, once there, had found a vast body of water marking the limit of the land.[6] It was not the fresh-water river which Homer had described, but in other respects it fitted with the Homeric image. Discovery and mythology

4 Homer apparently thought of the vault of heaven as a solid concave surface supported by pillars (Odyssey 1.54).

5 J. Romm, *Edges*, 22.

6 The earliest extant Greek text indicating knowledge of lands beyond the Straits of Gibraltar is a poem by Anacreon who lived between 563 and 487 BC. He wrote: 'I neither want the horn of Amaltha, nor a hundred and fifty years to rule Tartessus' (Εγώ τ' αν ούτ' αμαλθέης βουλόϊμην κέρας ούτε τα πεντήκοντά τε και εκατόν Ταρτηστού βασιλεύσα). Anacreon, 'Old Age', E. Diehl, *Anthologia Lyrica Graeca*, vol. 1 (Leipzig: Teubner, 1925), 450, no. 8. Tartessus was a region in the south of Spain round Guadalquivir, well beyond the Pillars of Hercules. Greek colonisation and exploration, particularly at this early stage, was largely conducted by sea, and any Greek who knew of the city of Tartessus must have known of the great body of water, the Atlantic, which lay just beyond. While the passage of Anacreon seems to be the first written mention of Tartessus in Greek, it was probably known to the Greeks even earlier. The Samian Kolaios reportedly fled there around 638 BC and by about 600 BC the Phocaeans had

were conflated in the interpretations of the body of water which lay beyond the Pillars of Hercules.

From the fourth century onwards, writers referred to it variously as the great sea, the outer Sea, the Atlantic, or as western Ocean, forming a deliberate connection with the Okeanos of mythology.[7] It defined the ends of the earth. In the sphere of the earth, this body of water prevented the *oikoumene* from forming a complete ring reaching round the whole earth in the temperate zone. As Aristotle stated in his *Meteorologica*, 'for the parts [of land] outside India and the Pillars of Hercules because of the sea do not seem to be consistent with the idea that the whole *oikoumene* exists continuously'.[8] Ocean, instead of filling the void between land and sky, was now thought to fill the vast space between one extreme of the *oikoumene* and the other. In this passage, Aristotle did not use the word *okeanos*, but rather employed the term *thalatta,* more usually translated as sea. Nonetheless, it is clear that he was thinking of it as a limit in the same way as others referred to Ocean. Indeed, earlier in the same passage, he stated that 'if it were not for the Ocean which prevents it, the complete circuit could be made' around the earth within the temperate region.[9] Far from acting as the means of communication which it was to become in the sixteenth century, in the classical conception it was clearly the ultimate limit to human exploration and inhabitation.

While Aristotle was not overly concerned with the specific nomenclature, other authors were much more interested in the terminology, and with connecting the waters of the global world to the Ocean of earlier geography. Strabo, who regarded Homer as the great authority on geography (in much the same way as he himself was considered an authority by the Renaissance writers), explicitly followed a Homeric model of a single *oikoumene* surrounded by Ocean. He linked the idea of the sea filling the space between India and Spain and the concept of Ocean, saying that one can learn both from the senses and experience that the inhabited world is an island:

> for wherever it has been possible for men to go to the furthest reaches of the earth, the sea, which indeed we call Ocean, has been found.[10]

---

arrived there. J. Boardman, *The Greeks Overseas: Their Early Colony and Trade*, 4th edn (London: Thames and Hudson, 1999), 213–14.

7 W. Smith, ed., *A Dictionary of Greek and Roman Geography by Various Writers*, vol. 1 (London: Walton and Maberly, 1878), 311–14.

8 τὰ δὲ τῆς Ἰνδικῆς ἔξω καὶ τῶν στηλῶν τῶν Ἡρακλείων διὰ τὴν θάλατταν οὐ φαίνεται συνείρειν, τῷ συνεχῶς εἶναι πᾶσαν οἰκουμένην: Aristotle, *Meteorologica*, 2.5 (362b).

9 εἰ μή που κωλύει θαλάττης πλῆθος, ἅπαν εἶναι πορεύσιμον: Aristotle, *Meteorologica*, 2.5 (362b).

10 Πανταχῆ γάρ, ὁπουποτοῦν ἐφικτὸν γέγονεν ἀνθρώποις ἐπὶ τὰ ἔσχατα τῆς γῆς προελθεῖν, εὑρίσκεται θάλαττα, ἣν δὴ καλοῦμεν ὠκεανόν. Strabo, 1.1.8.

Strabo continues:

> It is improbable that the Atlantic Ocean is divided into two seas, separated by narrow isthmuses which prevent circumnavigation, but rather [it is probable that] it flows together and is continuous. For those who have attempted to sail around it and turned back, say that they had to turn back, not because a continent barred and prevented them from sailing further, but because of the difficulty and isolation, since the passageway across the sea did not lessen.[11]

For Strabo, it was the vast expanse of Ocean which caused it to be a barrier to human inhabitation. There was nothing which could enable humans to overcome the sheer distance of this single continuous waterway surrounding the whole earth. Its extent rather than its conditions prevented it from becoming a means of communication. Indeed interestingly, much of Strabo's descriptions are based on the sea in a small scale being the link between areas, and enabling geography to be formed by the relationship with the sea, but this did not include Ocean.

Not all classical writers were as explicit in their terminology as Strabo. Pliny, for example, spoke of 'the northern Ocean', 'the Indian Ocean' and 'the many shores of Ocean around the Caspian', which would at first seem to undermine the idea of an interconnected water way. When discussing how all these fitted together into a single geography, he wrote of seas (*maria*) which surround the world on every side, divide it in two and allow no passage from one part to the other.[12] Clearly, even those Greek and Roman authors who accepted the sphericity of the world, still thought of the waters which surrounded the continents and divided the east from the west as having the traditional characteristic of Ocean and acting as a near inviolable boundary, but also as ultimately forming a single continuous body of water which could then be reformulated in the sixteenth century as a means of communication rather than as a confine.

While Strabo, Pliny and others emphasised Ocean as defining the limits of the single *oikoumene* on the watery globe, a third way of regarding Ocean also developed in the ancient world in the aftermath of ideas of the sphericity of the earth. According to this theory, attributed to the second-century BC Stoic mathematician Crates of Mallos, there were two distinct Oceans encircling the earth in separate directions.[13] The Atlantic formed part of one which contained a

11 Οὐκ εἰκὸς δὲ διθάλαττον εἶναι τὸ πέλαγος τὸ Ἀτλαντικόν, ἰσθμοῖς διειργόμενον οὕτω στενοῖς τοῖς κωλύουσι τὸν περίπλουν, ἀλλὰ μᾶλλον σύρρουν καὶ συνεχές. Οἵ τε γὰρ περιπλεῖν ἐγχειρήσαντες, εἶτα ἀναστρέψαντες, οὐχ ὑπὸ ἠπείρου τινὸς ἀντιπιπτούσης καὶ κωλυούσης τὸν ἐπέκεινα πλοῦν ἀνακρουσθῆναι φασίν, ἀλλὰ ὑπὸ ἀπορίας καὶ ἐρημίας, οὐδὲν ἧττον τῆς θαλάττης ἐχούσης τὸν πόρον. Strabo, 1.1.8.

12 See Pliny, *NH*, 2.47.167–168, 170.

13 Both the leading compilers of Crates's fragments, Mette and Broggiato, agree with the interpretation that he was the originator of the theory that the world was divided

north-south flowing stream, turning back on itself.[14] The other girdled the earth at the equator, forming a burning Ocean in the torrid zone.[15] All these quadripartite theorists, with the possible exception of Geminos, who specifically stated that it is erroneous to believe that there is a tropical Ocean separating the hemispheres, saw Ocean as a boundary to each of the four regions of the world.[16] As Macrobius maintained, the four landmasses were maculae or spots, divided and alienated from one another by Ocean, and breaking up its endless expanse. Crates, like other Stoics, thought that an Ocean in the torrid zone was a necessity because there the sun 'might easily keep up its warmth by the aid of vapours from the sea – for warmth was supported by moisture'.[17] Although, in this portrayal of the world, Ocean was subdivided into different sections, it continued to act as a boundary to each *oikoumene*. There might be different *oikoumenai*, but Ocean, as Crates explicitly asserted, spread over most of the earth.[18]

All these theories of Ocean could coexist, because, although on a theoretical level they provided different ideas about the shape and nature of the world, on a practical level they all proffered the same one: it was hubristic to attempt to sail beyond the boundaries of Ocean, and one need only be concerned with the *oikoumene* contained by Ocean. They all discouraged Oceanic exploration. Over

---

into four equal, separate *oikoumenai*. H. J. Mette, *Sphairopoiia: Untersuchungen zur Kosmologie des Krates von Pergamon. Mit Einem Anhang: Texte* (Munich: Beck, 1936), 253–8 where he collects together all the evidence that might refer to Crates as originator of the theory; see also his discussions 75–79 and M. Broggiato, ed., *Cratete di Mallo: I frammenti* (La Spezia: Agorà, 2001), liii, 224. It is, however, merely an interpretation, since none of the extant information is clear about his precise ideas of the form and number of the *oikoumenai*. It is possible that his back-flowing Atlantic occupied the entire second half of the globe. Indeed his emphasis on the vastness of the Ocean might lead one to suppose that he thought of the earth as divided into only two equal parts, separated from each other by the girdling equatorial Ocean. Other scholars who definitely believed in a quadripartite theory were Geminos of Rhodes c. 70 BC) and Cleomedes (c.150–200 AD), Macrobius and Martianus Capella. Geminos, Εἰσαγωγή εἰς τὰ Φαινόμενα, 16; Cleomedes, *De Motu Circulari Corporum Caelestium*, 1.2.2; Macrobius, *Commentariorum in Somnium Scipionis,* 2.9.1–6; Martianus Capella, *De Nuptiis Philologiae et Mercurii*, 6.603–605 (196G).

14 M. Broggiato, *Cratete di Mallo: I frammenti*, F5. H. J. Mette, *Sphairopoiia*, F35. The collision of the waters at the poles may have forced this back-flowing, and consequently caused the ebbing and flowing of tides. K. Clarke, *Between Geography and History*, 42. See M. Broggiato, F136 which shows that Crates did indeed discuss tides although it is by no means clear what he thought caused them.

15 See M. Broggiato, F57 (35M). = Strabo, 1.1.7 and F37 (F34M) = Geminos, Εἰσαγωγή εἰς τὰ Φαινόμενα 16.21–28 and Strabo 1.2.24; 2.3.7.

16 Geminos, 16.23, where he specifically states that it is erroneous to believe that there is a tropical Ocean separating the hemispheres.

17 F. Nansen, *Northern Mists*, 78–79.

18 M. Broggiato, F20 (33M) = Plutarch *de Fac.* 938d.

time the three theories blended together somewhat, so that by the first century AD both Seneca and Tacitus, for example, described Ocean as a real place, associated with the Atlantic in particular, but one that remained impassable and in some senses semi-mythical, just like Homer's. Just as for the earliest Greeks, the deterrents to exploration were not only its size, but also the idea that it was the domain of the gods who had designated that it should not be part of human knowledge. In both authors' descriptions, the fear of Ocean was not just an intellectual one, but had very real human implications. In practice, Ocean evoked such fear in sailors that no-one dared venture far onto its waters. The elder Seneca, for instance, concluded his first discussion in the *Suasoriae* with the supposedly factual account of Drusus Germanicus's helmsman, who gave a report of sailing on the northern Ocean (c. 15–16 AD). The helmsman cried out:

> Where are we being carried? Day itself flees and most fundamental nature encloses the remaining world with perpetual darkness. Do we seek races who are stationed beyond under another pole and another world untouched by storms? The gods call us back and forbid mortal eyes to know the end of things: why are we violating alien seas and sacred waters with our oars and disturbing the quiet homes of the gods?[19]

Similarly, Tacitus, writing about the same event, described how Germanicus was deterred from exploring further into Ocean, saying:

> In that place we have tested the Ocean itself; and rumour has it that further out still, there are columns of Hercules, so called either because Hercules went there or because we have agreed wherever something is magnificent to refer to it for his fame. Nor did Drusus Germanicus lack bravery, but Ocean prevented his inquiry concerning itself and Hercules. Soon no-one tried, and it was seen to be more reverent and wise to believe the works of the gods than to understand them.[20]

[19] 'quo ferimur? fugit ipse dies orbemque relictum/ultima perpetuis claudit natura tenebris./anne alio positas ultra sub cardine gentes/ atque alium flabris intactum quaerimus orbem?/ di revocant rerumque vetant cognoscere finem/mortales oculos: aliena quid aequora remis/ et sacras violamus aquas divumque quietas/turbamus sedes?': Seneca the Elder, 'Deliberat Alexander an Oceanum naviget.' *Suasoriae*, I, Section 15.

[20] 'ipsum quin etiam Oceanum illa temptavimus; et superesse adhuc Herculis columnas fama vulgavit, sive adiit Hercules, seu quidquid ubique magnificum est, in claritatem eius referre consensimus. Nec defuit audentia Druso Germanico, sed obstitit Oceanus in se simul atque in Herculem inquiri. Mox nemo temptavit, sanctiusque ac reverentius visum de actis deorum credere quam scire'. C. Tacitus, M. Hutton, and W. Peterson, *Agricola, Germania, Dialogus* (London: Heinemann, 1970), 185, footnote 2 – all boulders rising out of sea were attributed to Hercules who was thought to be the first builder of natural lighthouses.

Nobody meant these to be understood as the actual words of explorers. Tacitus was being ironical and Seneca rhetorical, but their writings seem to give some insight at least into how the educated thought an experienced sailor reacted to the prospect of Ocean travel. Even contemplating such a voyage was dangerous, involving a violation of divine territory. As a result of such ideas, Ocean became just as much an intellectual as a physical confine to the *oikoumene* in western European thought. It was this intellectual confine that had to be altered before the physical barrier could be overcome in the fifteenth and sixteenth centuries. While the intellectual fear remained dominant, Ocean could not be considered a God-provided means of connecting all parts of the world. Despite frequent ideas of benevolent nature, of *pronoia* in classical geographical writing, which allowed many ancient writers to interpret the geography of the world as designed for humans, I know of no classical interpretation of Ocean which thought of it as designed to be part of the natural unity of the world and connect all regions of the globe together. It was always on some level a limit in the ancient geography, and therefore required a dramatic reinterpretation to be seen as a means of communication in the sixteenth century.

Nonetheless, although all three theories, and their distorted interpretation by sailors, reinforced the idea of Ocean acting as a limit to the knowable world, there were frequent references throughout ancient literature to islands and landmasses lying beyond the inner shores of Ocean. These islands and landmasses ultimately became clues for European explorers and geographers to ideas about the opening-up of the world and the expansion of the *oikoumene*. As early as the seventh century BC, for instance, Hesiod had described the Islands of the Hesperides and the lands of the Gorgons as lying 'beyond glorious Ocean at earth's end'.[21] The Islands as Hesiod described them were merely mythical. Like the realms of the Gorgons, they were the regions of the marvellous, the monstrous, and the semi-divine, and fittingly belonged in the realms of Ocean, the territory of the divine. They eventually became identified with the Canary Islands, however, which lay at the extreme limits of the Greco-Roman world, well into Ocean. The reabsorption of these islands into European geographical knowledge in the fourteenth century (and ultimately their conquest by Europeans) played a key role in changing European attitudes to the idea of Ocean as a physical confine to the world. Elsewhere, islands such as Taprobana at the eastern end of the *oikoumene*, variously identified as Sri Lanka and Sumatra in the sixteenth century, lay beyond the boundaries of both Ocean and climate, but were seized upon by sixteenth-century writers as indicators of ancient knowledge of the world beyond the confines of Ocean.[22] Thule and Cerne were semi-mythical

[21] αἵ ναίουσι πέρην κλυτοῦ 'Ωκεανοιο ἐσχατιῇ: Hesiod, *Theogony*, 274.

[22] In the sixteenth century, most people agreed that Taprobana was Sumatra; see, for instance, J. G. de Mendoza, *Historia de las cosas mas notables, rites y costumbres del*

islands, well-known in Greek and Roman literature. Holding various different locations according to different authors, but always existing at the fringes of the *oikoumene*, making a 'fantastical boundary' to the known world, they also began to receive real locations in the sixteenth-century works.[23] Plato's Atlantis, never described as an *oikoumene* in its own right, formed an entirely distinct landmass, separated from the known world by the sea which ultimately overwhelmed it. Yet in the sixteenth century it became conflated with the Americas.[24]

For the most part, in the classical world, such islands were seen as semi-mythical, inaccessible and populated by the barely human. The case of Britain shows how an island could become incorporated into the *oikoumene* – and yet, despite its size, proximity to Europe, and connection with mainland trade, Britain remained cloaked in the mystery of Ocean for centuries. The Roman grammarian Quintilian (c. 35–c. 100 AD), teacher of the younger Pliny, believed that as late as 50 BC, no-one knew whether Britain was another land across Ocean, or a conquerable part of the *oikoumene*.[25] Quintilian's near contemporary, Tacitus, showed in his Agricola that more was known about Britain then Quintilian implied, and that it was theoretically just another region of the Roman empire, but even in his eyes it was only just incorporated into the *oikoumene*.[26] As he said, its northern shores 'have no lands confronting them,

gran reyno de la China con un itinerario del nuevo mundo (Rome: V. Accolti, 1585), 3; J. de Acosta, *Historia*, 144. Ramusio emphatically declared that because of its size it must refer to Sumatra not Sri Lanka: G. B. Ramusio, *Discorso sopra la navigazione di iambolo mercante antichissimo*, *NV*, vol.1, 906. Not everyone agreed with this assessment, however. The traveller Andrea Corsali argued just as emphatically that the details of geography indicated that it must be Sri Lanka, not Sumatra. A. Corsali, 'Lettera scritta in Cochin', *NV*, vol. 2, 53. Modern scholars now accept that the classical authors actually referred to Sri Lanka: see S. Hornblower and A. J. S. Spawforth, *The Oxford Classical Dictionary*, rev. 3rd edn (Oxford: Oxford University Press, 2003), 1473.

23 As Clarke points out, Strabo (based on Eratosthenes) locates Cerne beyond the Pillars of Hercules (1.3.2), while Pliny (based on Ephorus) says it is beyond the Persian Gulf (*NH*, 4.35). Similarly, Thule had vastly differing locations. Most people agreed that it lay in the far north. See F. Cordano, *La geografia degli antichi* (Roma: Laterza, 1992), 107. There were, however, differing ideas even about that, and by the time an anonymous 7th-century author from Vienna was writing, he located three different islands, Thile, Thyle and Thilos to the north east of Britain, in the Atlantic Ocean and in the Persian Gulf respectively. His was an aberration, however: see V. H. de P. Cassidy, 'The Voyage of an Island,' *Speculum*, 38 (1963), 598. For more on the changing location of Thule and Cerne and the role they served as boundary lands, see K. Clarke, *Between Geography and History*, 34. Solinus seemed to conflate Thule with Ireland (22.9), whereas Pytheas placed it in the Scandinavian region: see Strabo, 2.4.1.

24 Plato, *Timaeus*, 24e–25d. See below chapter 6 on Atlantis and the Americas.

25 Quintilian, 7.4.2.

26 K. Clarke, 'An Island Nation: Re-Reading Tacitus' *Agricola*', *Journal of Roman*

but are beaten by the wastes of open sea'.[27] Yet Britain became part of the Roman empire and part of the *oikoumene*. For the Renaissance humanists there was precedent for reconsidering islands once thought to be inaccessible and semi-divine as in fact part of the *oikoumene*.

Islands always formed a link between the *oikoumene* and the Ocean, but they did not turn it into part of the *oikoumene*. One of the most famous voyages on Ocean in the medieval period, for instance, was the voyage of St Brendan. On his journey St Brendan meets islands of sheep and goats, an island where Judas Iscariot is held prisoner, another where lava-throwers toss molten rock at the monks as they sail past, and of course the ubiquitous violent sea monsters. A few have tried to find direct association between the islands he encountered and the voyages he undertook, suggesting that regions as diverse as the Faros and the Canaries were the location of the islands of sheep, and that the volcanic islands of Iceland or Tenerife were the origin of the story of lava-throwing humans. Yet in a sense this is not the point.[28] The voyage still emphasises the uninhabitable nature of Ocean. For example, they were 'driven this way and that through the waters of ocean for three months and they could see nothing other than sky and sea. And they were fed every two or three days.'[29] Such tales, which were widely circulated, did nothing to entice medieval sailors to explore the Atlantic.

The reinterpretation of Ocean as a means of communication in the fifteenth and sixteenth centuries clearly involved a major shift in conceptualisation. Despite the existence of sundry dissenting voices in the classical world that showed the possibilities of Oceanic travel, the idea of a delimiting Ocean was the dominant framework throughout the Middle Ages. With very few exceptions, such as Roger Bacon, medieval authors, like their classical predecessors, did not think that the sphericity of the earth meant that one end of the inhabited lands could communicate with the other.[30] The varying classical ideas of ideas of a single *oikoumene*,

Studies, 91 (2001), 94–112.

27 'nullis contra terris, vasto atque aperto mari pulsantur'. Tacitus, *Agricola*, 10.

28 G. Ashe, *Land to the West: St. Brendan's Voyage to America* (London: Collins, 1962); G. A. Little, *Brendan the Navigator: An Interpretation* (Dublin: M. H. Gill and Son, Ltd, 1945). For more on the various interpretations see J. S. Mackley, 'The Legend of St. Brendan: A Comparative Study of the Latin and Anglo-Norman Versions', (Leiden: Brill, 2008), 53–55.

29 Igitur sanctus pater cum sua familia per equora oceani huc atque illuc agitabatur per tres menses et nihil poterant videre nisi caelum et mare. Reficienbantur autem semper per biduum aut triduum. 'Vita Sanctissimi Confessoris Christi Brendani' (Transcription du manuscrit d'Alençon à la bibliothèque municipale d'Alençon, Codex 14, fol. 1 r à 11 v., chapter xiv, 11th century' <https://www.hs-augsburg.de/~harsch/Chronologia/Lspost10/Brendanus/bre_navi.html> (accessed 30 November 2019).

30 J. Russell, *Inventing Flat Earth*; L. B. Cormack, 'Flat Earth,' 363–86. For a detailed discussion of the transition in mental approach between the idea of the separate *oikoumene* and the terrestrial sphere, see W. G. L. Randles, *De la Terre plate*.

of a quadripartite world, of an absolute boundary of Ocean and of islands both mythical and real existing within Ocean, all continued to co-exist in the medieval period just as they had done in the classical world. The one common thread, in all this eclectic barrage of conceptualisations about the world, continued to be the idea of an Ocean boundary confining humans to the known *oikoumene*. Even the Arab geographers believed that 'the green sea of Darkness' (the Atlantic) was unnavigable and that the frigid and torrid zones were uninhabitable, and although better navigational techniques were developed and people ventured somewhat further on the sea, these journeys impinged very little upon the written or indeed the graphic geography.[31]

Even those travellers who were influential in breaking down the idea of climatic limits to human inhabitation remained convinced that Ocean was a boundary rather than a conduit. Marco Polo's fear of travelling beyond Madagascar, for instance, was one instilled by Ocean.[32] He believed that the currents of the Ocean would sweep one away for ever from the inhabitable lands to the pole of the world. For all inheritors of the classical tradition, the Oceans retained their terror. Ocean or Oceans might lie between one part of the world and the other, but far from bridging the intervening gulf, on both practical and conceptual grounds, they prevented communication.

The reassessment of Ocean in the early modern period required geographers and travellers not to think of it as a 'sea of Darkness', a realm of chaos, a region beyond the laws of nature, but rather as a part of the whole natural world, subject to the same laws of nature as the lands. All three of the key classical theories were re-used in the sixteenth century to demonstrate that the ancient writers had indeed seen Ocean as an integral part of trade and travel. Perhaps the most overt such use was by Pedro de Medina who in his extremely influential work on the Art of Navigation devoted a whole section to citing narratives of ancient uses of Ocean routes.[33] The clues were there in the classics, but for the Renaissance writers it required a rereading of the Greek and Roman authors and an overturning of all the traditional theories of Ocean in order to make use of them.

The small selection of classical authors who queried the very idea of an impassable expanse of water were useful for the Renaissance humanists attempting to reconstruct Ocean as a passageway rather than a barrier. Herodotus and Ptolemy, for example, did not believe in a circumambient Ocean, but neither did they think it was impassable and unknowable. For them sea was neither a boundary to the *oikoumene* nor a means of communication; it was just a large

[31] J. H. Parry, *Reconnaissance*, 5.

[32] M. Polo, *Viaggi*, 286–88.

[33] P. de Medina, *Arte de navegar* (Francisco Fernandez de Cordoba: Valladolid, 1545), 2.5.

expanse of water difficult to navigate.[34] Ptolemy indeed believed that the lands enclosed Ocean rather than the other way round. Interestingly, however, because they believed in the navigability, if not the complete connectedness, of Ocean, they could become a useful starting point for those who sought to argue that God had designed the world for humans. The fifteenth-century cardinal and cosmologist, Guillaume Fillastre, a proponent of the idea that the east could be reached by sailing westwards from Europe, argued that Ptolemy must be mistaken in his assumption that the Ocean was not one continuous piece, since God had, as it states in Genesis, decreed that all the waters be gathered together. He then argues this is the reason that it is possible to sail from west to east, although there was little experience to prove it, and used Ptolemy's ideas about the possible inhabitability of the equatorial regions to further this argument from design.[35] Similarly Fillastre's near contemporary, cardinal Peter d'Ailly combined a cosmographical reinterpretation of Ptolemy with arguments from Aristotle and Averroes to argue that only a narrow sea separated the extreme west of Europe and east of Asia.[36] Although D' Ailly, as Evelyn Edson has pointed out was aware of contemporary discussions about the limits of the world, but was largely uninterested in the practical application of his geography, and deliberately disengaged himself from sorting out disagreements between travellers' experience and ancient 'witness', his work notoriously provided theoretical underpinnings for Columbus's first transatlantic voyage.[37]

Once this fear of crossing open Ocean had been surmounted, other classical authors who had described Ocean as the limits of the *oikoumene*, but had also shown how its location on the globe meant that it straddled the space between the landmasses, were reinterpreted to show how Ocean could become a means of communication. Although both Strabo and the elder Seneca wrote about Ocean as a boundary, they both also suggested that there was the possibility of sailing from Spain to India.[38] Aristotle, Posidonius and Eratosthenes all emphasised the impossibility of a voyage from the Pillars of Hercules to India, but they too thought

[34] Herodotus makes it clear that it was navigable, however, providing a story of Phoenicians circumnavigating Africa. 4.43

[35] P. G. Dalché, *L'oeuvre géographique*, 332, paragraph 15.

[36] P. d'Ailly, E. J. P. Buron, and Columbus, *Ymago Mundi: Texte latin et traduction française des quatre traités cosmographiques de d'Ailly et des notes marginales de Christophe Colomb. Étude sur les sources de l'auteur* (Paris: Maisonneuve frères, 1930), 2, 454–55.

[37] E. Edson, *The World Map, 1300–1492: The Persistence of Tradition and Transformation* (Baltimore Johns Hopkins University Press, 2007), 132. P. d'Ailly, et al., *Ymago Mundi.*

[38] Strabo, 1.1.8 (eastward) and 1.4.6 (westward). In the latter passage he says that there may be other lands lying in between Spain and India, however; see also Seneca, *Naturales Quaestiones*, Praefatio, 13.

that nothing save distance would prevent such a voyage.[39] Pliny the Elder, citing Cornelius Nepos, stated that some Indians had been brought ashore in Germany, clearly indicating that there was some sort of sea route between Germany and India.[40] All these authors were cited in the sixteenth century to show how Ocean could in fact be a single navigable space.[41]

Until the late Middle Ages, Ocean was perceived as a barrier unlike other bodies of water, in part because there were no navigational techniques which enabled transoceanic navigation. Its scale rather than its elemental nature defied human exploration whereas other bodies of water had always been seen as a means of transportation and communication. As Horden and Purcell point out, from at least the fifth century BC onwards, the Mediterranean had been seen as 'a network of communications'.[42] This network was, however, very much dependent on a type of seafaring which hugged the coastlines. It took the development of the Portuguese caravel and different economic pressures to turn Mediterranean interests westwards, and force them to use some of their seafaring techniques for open waters.[43] Once they had done so, the Ocean began to take shape. In the Renaissance the sea and land came to be seen as interdependent, forming together the terraqueous globe. The sea shaped the land, but the land also limited the sea. With such a conceptual change, the sea could be thought of as a water-way and not the terrifying expanse of ancient theory. Moreover, God had clearly intended it as such, for in all the great bodies of water he had scattered inhabitable places which could break up journeys and act as re-victualling stations. As Thevet said: 'they can clearly see an admirable providence of our God, who has placed in the sea, be it Ocean or the Mediterranean, a large quantity of islands, some larger, others smaller which withstand its floods and tempests.'[44] These islands allowed humans to take on victuals and water. They were part of God's design that the seas be the means

39 Aristotle, *Meteorologia*, 2.5.13; Posidonius cited in Strabo 2.3.6: Posidonius here says that if one sails west in a straight course from the Pillars of Hercules one will reach India within 70,000 stades (c. 12,950 km). This, however, was an impossibly long distance for travel – half the circumference of the world; Eratosthenes is cited in Strabo 1.4.6.

40 Pliny, *NH*, 2.58.170

41 G. B. Ramusio, 'Navigazione di Sebastiano Caboto: prefazione nella seguente navigazione', *NV*, vol. 4, 105–8.

42 P. Horden and N. Purcell, *The Corrupting Sea: A Study of Mediterranean History* (Oxford: Blackwell, 2000), 24.

43 F. Braudel and S. Reynolds, *The Mediterranean and the Mediterranean World in the Age of Philip II* (Berkeley: University of California Press, 1995), 108, 224–30.

44 En quoy pouvons voir clerement une admirable providence de nostre Dieu, lequel a mis en la mer, soit Oceane ou Mediteranée, grand quantité d'isles, les unes plus grandes, les autres plus petites, soutenans les flots et tempestes d'icelle'. A. Thevet, *Singularitez*, 480.

of connecting the whole world. God had provided the means of opening the whole earth to human eyes.

The Iberian explorations of the fifteenth century were the key factors in providing the empirical evidence for the changing the image of Ocean, just as they had destroyed the idea of frigid and torrid zones. Perhaps it is not surprising that it was countries which bordered on the Atlantic Ocean which challenged the idea of Ocean as a boundary. Not long after they discovered that one could safely sail into the burning zone, they began challenging other boundaries of the *oikoumene*, attempting to discover whether westwards exploration across the Atlantic was a practical option.

When the Italian, Toscanelli, wrote his letter promoting western exploration across the Atlantic Ocean in 1474, he argued that there was a much shorter route to the Indies by sailing westwards, continuing: 'but though I know that this could be shown by the spherical form; which is that of the world; nevertheless I have determined to show it in the way in which charts of navigation show it.'[45] As Goldstein has pointed out, Toscanelli took an entirely different view of the world from most geographers. Instead of the traditional image of a single landmass formed of the three continents of Europe and Asia and surrounded by an impenetrable Ocean, he represented the Ocean as part of a network of sea-routes, using the same techniques with which he would have portrayed the Mediterranean.[46] The tools were there for this mental shift: navigation techniques and ship-building had improved so that one could conceive of much longer journeys on open water than had heretofore been possible. In 1474 in a letter to Ferdinand Martins, canon of Lisbon Cathedral, he put into words what some (though not all) sailors were ready for, and promulgated the idea of an open-sea crossing to the east. This shows how far he and his contemporaries had come by the 1470s, for Ocean was now seen as providing a means of connecting all the land.[47] As far as he was concerned, the entire earth had

[45] H. Vignaud, *Toscanelli and Columbus. The Letter and Chart of Toscanelli on the Route to the Indies by Way of the West, Sent in 1474 to the Portuguese Fernam Martins, and Later on to Christopher Columbus. A Critical Study of the Authenticity and Value of These Documents [&C.]* (London: Sands & Co., 1902), 278 (English), 95 (Latin).

[46] T. Goldstein, 'Geography in Fifteenth-Century Florence', in *Merchants and Scholars: Essays in the History of Exploration and Trade: Collected in Memory of James Ford Bell*, ed. by John Parker (Minneapolis: University of Minnesota Press, 1965), 14.

[47] There is some controversy over whether the letter of Toscanelli is a forgery or not. On the whole I tend to side with Goldstein and take it for what it claims to be; see *ibid.*, 12–13. For the counter-argument, however, see H. Vignaud, *Toscanelli and Columbus*, 36–51, 243–52. Vignaud argues that, among other reasons for not believing in it, Toscanelli's plan is unknown in Italy and his correspondence is not mentioned by Columbus. He also thinks that the question of a route to the Indies was not yet an issue in 1474. Whether it was written in 1474 or slightly later, it is still a very early document about the crossing of the Atlantic, and the clearest evidence of the awareness

become accessible for humans.[48] But Ocean was far from fully discovered. Toscanelli's very configuration of land and sea was in part based on conjecture and an acceptance of traditional ideas of circumambient Oceans.

Columbus, a decade and a half later, combined Toscanelli's theory with the statement in the apocryphal scripture Esdras II (6.42–47) that only one seventh of the world was covered in water so that, if one were to take the scripture literally, the eastern shores of Asia could not be much further west than Toscanelli had calculated.[49] Such calculations were designed to overcome the fear of virtually boundless, uninterrupted Ocean.

These early attacks on the idea of an open Ocean acting as a boundary to the *oikoumene* and the transformation of it to a means of communication were not always aided by the process of European expansion. By 1520, Magellan's crew had circumnavigated the globe, demonstrating that there was indeed a continuous sea route around the world, but also proving that the circumference of the sphere was far greater than Columbus or even Ptolemy had estimated, and that the known coast of Asia lay an impossibly vast distance from Europe. When the French-sponsored sailor Giovanni Verrazzano returned from a westward voyage of exploration which resulted in his sailing along much of the eastern seaboard of the United States, his compatriot Fernando Carli, in a letter to his father from Lyons, praised him for how

> he displayed extremely great and noble courage in setting out on so unknown a voyage with only one ship which was scarcely – tons and with only 50 men, with the intention of, in as far as he could, discovering Cathay, taking a route through other climates than those which the Portuguese use in discovering it from the other side from Calicut, but going instead towards the northwest and north, entirely trusting that although Ptolemy and Aristotle and other cosmographers described that one would not find land towards such climates, that he would find it anyway.[50]

---

of the need to present the Atlantic as a waterway and not as the watery globe. It was certainly recorded by Las Casas in *Historia de Las Indias*. Vignaud includes a copy of this version in his book: 305–310. As Evelyn Edson observes, 'the triple evidence of the letter's existence seems too powerful to discard'. E. Edson, *World Map*, 210.

48 T. Goldstein, *Geography in Florence*, 15.

49 A. Hamilton and L. Garbett, *Apocryphal Apocalypse*, 9, 28–30.

50 in che mostrò coraggio troppo nobile e grande a metersi a tanto incognito viaggio con una sola nave che appen è una caravella di tonelli –, solo con 50 uomini, con intensione di, giusta sua possa, discoprire il Cataio tenendo camino per altri climati di quelli usano li Portoghesi in lo discoprire di verso la parte di Calicut, ma andando verso eoro e settentrione, omino tenendo, che ancor Tolomeo ed Aristotile ed altri cosmogragi descrivano verso tali climati non trovarsi terra, di trovarvene a ogni modo. L. Wroth, *The Voyages of Giovanni da Verrazzano 1524–1528* (New Haven: Yale Univeristy Press, 1970), 156.

To Carli, the bravery of a sailor willing to defy the ancient authorities such as Ptolemy and Aristotle and venture onto open Ocean, in regions where tradition held that there was only endless sea uninterrupted by any form of land where one could stop and victual, was worthy of praise and comment. As late as 1552, sixty years after Columbus, Cardinal Bembo felt constrained to explain that the place called Ocean was not 'of huge and unknown size, but was full of islands of places inhabited by people'.[51] Exploration and discovery did not immediately bring about a shift in knowledge, and in the age of the Renaissance the classical idea of Ocean still retained a strong hold. The oldest concept of Ocean as a vast, formless sea outside the realms of human inhabitation, was clearly still a dominant one.

Although the idea of boundless Ocean stretching unobstructed from Europe to Asia took a long time to die a complete death, the European encounter with the Americas gradually provoked a reassessment of the classics by a European intellectual population who continued to cling to the idea of an all-encompassing Ocean. As we have seen, classical authorities from Aristotle to Ptolemy had spoken of the sea covering the surface of the globe between Europe and Asia, unobstructed by any intervening large landmass. When the discovery of the Americas proved the inaccuracy of these authorities, people did not lose confidence in the idea of a single Ocean. The idea of water extending across the globe, uniting the whole world and making every region of it accessible, was a concept which was integral to the way in which they had come to think of the world. Just as it had become necessary to jettison the idea of climatic boundaries in order to think of the world as a single knowable unit, so it was necessary to cling to the idea of Ocean in order to show that a means of making all the world knowable existed. Ocean was clearly not exactly as many of the ancient writers described it, but there was still reason to believe in it.

The realisation that the Americas, rather than forming part of Asia, posed a significant obstacle to reaching Cathay, also played its part in how people envisaged its relationship with Ocean. Initially, various geographical writers saw the Americas simply as larger islands which were part of the means of bridging the space between Europe and Asia, or as attached either to Asia, or more rarely to Europe. As Surekha Davies has shown, as late as 1550, Sebastian Münster believed the Americas were part of an archipelago which stretched from the Canaries to the Moluccas, acting as revictualing points in the western sea-route to the Spice Islands, but by that point he was unusual.[52]

[51] '... di vana e ignave grandezza: ma piena d'Isole, e di luoghi habitati da gli huomini'. P. Bembo, *Della historia Vinitiana libri XII* (Appresso Gualtero Scotto: Venice, 1552), 73r.

[52] S. Davies, 'America and Amerindians in Sebastian Münster's Cosmographiae Universalis Libri Vi (1550)', *Renaissance Studies*, 25 (2011), 364.

In this conceptualisation the Americas were just a larger form of several intermediary locations which helped turn Ocean from a confine into a means of communication. Others marginalised the Americas even more, believing that a sea route ran between different American lands. The lands were incidental; the Ocean was the key factor. Peter Martyr, for instance, having read Sebastian Cabot's report of sailing along the coast of America as far as the Island of Cuba, and discovering a body of water flowing west, declared 'wherefore it is not onely more like to be true, but ought also of necessitie to be concluded that between both the lands hitherto unknown, there should be certaine great open places whereby the waters should thus continually passe from the east unto the west'.[53] The key word here is 'necessitie' – this open flowing connective sea was part of the essential structure of the world.

This belief in a connective Ocean dictated exploration, but exploration in turn proved the limitations of the idea of a circumambient Ocean acting as a conduit to different regions of the known *oikoumene*. Magellan's ill-fated voyage looked for a sea-route through the New World, and instead found one round the tip of South America into the vast, and largely islandless Pacific. It proved the existence of a western seaboard to the New World, but it disproved the idea of the Americas acting as part of the connectivity of the world, linking east and west. As late as 1524, more than thirty years after Columbus's first encounter with the New World, and two years after Magellan's expedition returned with reports of a vast eastern ocean, Giovanni Verrazzano set sail from France expecting to find a sea route leading more or less unhindered from France to Cathay. The idea of Ocean as a boundary had already been transformed into one of connectivity, but in between, requiring a further reconceptualisation of Ocean, lay a New World. Verazzano writes that he

> did not expect to find such an obstacle of new land as I have found; and if for some reason I did expect to find it, I estimated there would be some strait to get through to the eastern Ocean. This was the opinion of all the ancients, who certainly believed that our western Ocean was joined to the eastern Ocean of India without any land in between. Aristotle supports this theory by arguments of various analogies, but this opinion is quite contrary to that of the moderns, and has been proven false by experience.[54]

For Verazzano it was discouraging to find that Aristotle and all the ancients were wrong in their arguments from analogy. After the disappointment of his

[53] Peter Martyr, 'Another testimonie of the voyage of Sebastian Cabot to the west and northwest, taken out of the sixt [sic] chapter of the third Decade of Peter Martyr of Angleria', *PN*, vol. 7, 152.

[54] L. Wroth, *Verrazzano*, 142.

search for a northern sea route to Cathay, Verrazano became sceptical of the whole idea of Ocean. He was among the few who argued that the Americas might be joined to Russia or Norway, specifically stating that this was contrary to the ancients' belief.[55] Maximilian of Transylvania, secretary to Charles V, composing the first narrative of Magellan's voyage from interviews with survivors, likewise described a fear that the sailors had before departure that they would not find a sea route to Asia, 'not because it was thought hard, to be carried directly from the west to the east under the hemisphere, but because it was uncertain whether ingenious nature which has established nothing without the greatest foresight, distinguished the orient from the occident partly by sea and partly by land in such a way that it should be possible to reach the orient by this journey and course of navigation'.[56] Maximilian's words are particularly interesting because they show the same mindset about God and nature that made others infer that the Oceans must be connected, as indeed Magellan's voyage proved, but writing about the state of mind before the voyage, he was only willing to commit to the idea that nature did nothing without design. It is, however, the same underlying belief that made people think the world was designed so that Ocean should connect all parts of it.

Verrazzano and Maximilian aside, all the major sixteenth-century geographers and explorers accepted the classical premise of a circumambient Ocean surrounding the continents and therefore believed in Ocean bordering the shores of Canada and Russia, separating Asia from the Americas, despite a lack of empirical evidence.[57] As Paolo Giovio's letter to Pope Clement VII, written in the 1520s, shows, no western European knew of anyone, not even a Muscovite, who had reached the northern Ocean and had certain knowledge of its existence, yet sixteenth-century writings on Russia are peppered with references to it.[58] Even Ramusio and Hakluyt who normally preferred documented reports for their travel compilations, included a description of the as yet undiscovered

[55] Verrazzano, Cellere Codex in *ibid.* The Cellere Codex is the earliest account of Verrazzano's voyages. The first published account in Ramusio does not contain the same description.

[56] Non quia grave existimabatur, recti ex occidente in orientem subter hemisphaerium vehi, sed quia, incertum esset, utrum ingeniosa natura, quae nihil non summa cum providentia constituit, ita orientem ab occidente partim mari, partim terra distinxisset, ut hoc itinere & navigationis cursu ad orientem perveniri posset. M. Transylvanus, *De Moluccis Insulis* (Eucharius Cervicornus: Cologne, 1523), Aiii (v).

[57] Aside from two possible but poorly recorded journeys by Lorenzo Ferrer de Maldonada and Apostolos Valerianus (more commonly known as Juan de Fuca) in the late sixteenth century, which purported to reach the Bering Strait or the Straits of Anian as they were then known, there were no voyages along the Pacific coast of North America. R. Owens, 'The Myth of Anian', in *Journal of the History of Ideas*, 36 (1975), 136.

[58] P. Giovio, *Moscovia*, 680.

northeast passage, which relied only on hearsay, but proved 'that the Ocean sea compasseth and environeth all the east, northeast, and north parts of Asia and Europe, and by inference the Americas.[59]

This northern Ocean was an essential aspect of how the humanists thought of communication across the continents. Mercator, in a letter to Richard Hakluyt in 1580, argued that naturally there was a way through to China by the north, saying that 'the voyage to Cathay by the east is doubtless very easy and short', and that he was surprised the English had stopped exploring in such a direction 'after that more than half of your voyage was discovered'.[60] The northern ocean must not only exist, but it must be open water for it to act as such a means of communication. Although we now know that there are northeast and northwest passages, at the time there was no empirical evidence from which they could derive their arguments for sea-routes to China. Yet the idea of open polar waterways became an integral part of sixteenth-century geography.[61]

It was concerning the existence of the northern Ocean that explorers had truly to place their trust in the classics and in the humanist scholars. As Humphrey Gilbert wrote in 1576 before he set out searching for a northwest Ocean passage to Cathay:

> I am of the opinion that America by the northwest will be found favourable to this our enterprise, and am the rather imboldened to believe the same, for that I finde it not onely confirmed by Plato, Aristotle, and other ancient Phylosophers; but also by all the best moderne Geographers, as Gemma Frisius, Münsterus, Appianus, Hunterus, Gastaldus, Guccardinus, Michael Tramasinus, Franciscus Demogenitus, Bernardus Puteanus, Andréas Vavasor, Tramontanus, Petrus Martyr, and also Ortelius, who doth coast out in his general Mappe set out Annno 1569, all the countreys and Capes, on the northwest side of America, from Hochelaga to Cape de Paramantia: describing likewise the sea coastes of Cataia and Gronland, towards any part of America, making both Gronland and America, Islands disjoined by a great sea, from any part of Asia.

59 G. B. Ramusio, 'A Speciall Note Gathered by the Excellent Venetian Cosmographer M. John Baptista Ramusius out of the Arabian Geographie of Abilfada Ismael, Concerning the Trending of the Ocean Sea from China Northward Along the Coast of Tartarie and Other Unkowne Lands, and Then Running Westwards Upon the Northerne Coasts of Russia, and So Farther to the Northwest,' *PN*, vol. 3, 412. Cf. G. B. Ramusio, 'Dichiarazione d'alcuni luoghi ne' libri di messer Marco Polo, con l'istoria del reubarbaro', *NV*, vol. 3, 68.

60 G. Mercator, 'The Letter of Gerard. Mercator to Richard Hakluyt of Oxford Touching the Intended Discoverie of the Northwest Passage', *PN*, vol. 3, 278.

61 M. Small, 'Jellied Seas'.

He continued,

> What moved these learned men to affirme thus much I know not, or to what ende so many and sundry travellers of both ages have allowed the same: But I conjecture that they would never have so constantly affirmed, or notified their opinions therein to the world, if they had not had great good cause, and many probable reasons, to have led them thereunto.[62]

Gilbert's statement shows the extent to which there continued to be an interplay between classical authority, humanist scholarship, and exploration. He knew that there was no evidence as yet for the existence of Ocean. As had been the case with the torrid and frigid zones, explorers risked their lives in testing the ancient ideas. Although people repeatedly died in the attempt to find the northern location of Ocean, most geographers and explorers continued to believe in its existence.

Fortunately for those seeking classical authority for their reconceptualisation of Ocean as an open water way surrounding two entirely distinctly landmasses, there were enough differing opinions among the ancient authors to enable people to select which ones most seemed to support such an image of Ocean and the world as it was now revealed to exist. They used the classical authors to refute any idea of land bridges between the Old World and the New, and to confer continental status on the newly discovered lands.
As Humphrey Gilbert wrote

> Also Aristotle in his booke *De Mundo*, and the learned Germaine Simon Grynaeus in his annotations upon the same, saith that the whole earth (meaning thereby, as manifestly doth appeare, Asia Africk and Europe being all the countreys then known) is but one Island compassed about with the reach of the sea Atlantine: which likewise prooveth America to be an Island, and in no part adjoining to Asia, or the rest.[63]

Likewise Gilbert's countryman Richard Willes used the authority of Homer, Strabo, Aristotle, Dionysius, Pomponius Mela and Pliny to support the idea that the Ocean ran round Asia, Africa and Europe. He then equated the Americas with Atlantis and used Plato's *Timaeus* to prove that the Americas were also surrounded by water. He concluded:

> Out of Plato it is gathered that America is an Island. Homer, Strabo, Aristotle, Dionysius, Mela, Plinie, Pius II affirme the continent of Asia, Afrike, and Europe to be environed with the Ocean. I may therefore boldly say (though

[62] H. Gilbert, 'Discourse to Prove Passage', *PN*, vol. 7, 162.
[63] *Ibid.*, vol. 7, 162.

later intelligences thereof have we none at all) that Asia and the west Indies be not tied together by any Isthmos or strait of land.[64]

These humanists did not suffer the same disillusionment with classical authority that Verrazzano had done. Rather, although there was no later authority to substantiate the statements of the ancients, their word was enough for Willes to conclude that the Americas and the Old World were wholly distinct from one another, separated by vast stretches of Ocean in every direction.

The existence of an all-encompassing Ocean was never entirely proven in the sixteenth century. Nobody discovered a northern sea route to Cathay, nor did anyone know for certain that the Americas formed a landmass separate from the old *oikoumene*. Yet travellers, supported by contemporary geographers who traced their conviction back to ancient authority, built their image of the world they were investigating on an idea of Ocean. There were only a few dissonant voices. Thevet claimed to have seen with his own eyes that Ocean did not surround the land, but rather that land surrounded the Ocean, and therefore Ocean could not have the course which all the ancients attributed to it. But his claim personally to have seen that Ocean could not encircle the lands was not possible.[65] Such evidence could not be gleaned from any one person's voyage.[66] Thevet, however, needed the pretence of autopsy in order to defeat the authority of the ancients, for in the case of Ocean the ancient authorities received some confirmation from the explorers. Those who had explored Asia and Africa knew that the southern reaches of these continents were surrounded by water. Although South America was still unfolding before explorers' eyes in the sixteenth century, people had rounded its tip, and sailed along much of both its coasts.

The classics needed to be reinterpreted in order to find the means of opening up the world, but the clues existed. Within the ancient writings lay reports of Ocean navigation around Africa and arguably to the New World. They also contained tales of islands that seemed to be being rediscovered which could be used as stopping points to victual and break up the great expanse of Ocean. Even more importantly, the classics contained repeated references to circumambient Ocean which, when divested of the religious dimensions and of the fear and awe of open water, could be reimagined. In the sixteenth century, with changing technology and mindsets, people ventured out onto open Ocean, no longer fearing its vast expanse, but seeing it as a God-given means of communication by which

64 R. Willes, 'Certaine Other Reasons, or Arguments to Prove a Passage by the Northwest, Learnedly Written by M. Richard Willes Gentleman', *PN*, vol. 7, ed. by Richard Hakluyt, vol.7, 197.

65 A. Thevet, *Cosmographie Universelle*, 6v.

66 F. Lestringant, *Mapping*, 28.

all the component parts of the single unit of the whole world could trade and interact with each other.

The European Christian world view had changed completely in its ideas about the *oikoumene* and its relationship to Ocean. Instead of containing an isolated *oikoumene* distinct from the rest of the world, the whole world was the work of God's hands. In the eyes of both Catholics and Protestants, God had deliberately scattered different products throughout the world so that all the regions of the earth would be dependent on one another. As Edward VI wrote to the emperor of Cathay,

> For the God of heaven and earth greatly providing for mankinde, would not that all things should be found in one region, to the ende that one should have need of another, that by this meanes friendship might be established among all men, every one seeke to gratifie all.[67]

The world, as Scripture so frequently pointed out, had been given by God to man to rule. In providing the Ocean, God had provided the means of reaching out across the world, of uniting the nations of the world together, and of making all the resources of the world available to one another. Geographers such as Münster, Hakluyt, and Mercator felt that it was their sacred duty to describe and portray the world which God had created for mankind; similarly, explorers felt bound to discover it.[68] Echoing throughout much of the exploratory, geographical, religious and political writing was the idea that it was a Christian's duty to go to the farthest reaches of the world and spread the news of Christ. They could do so because God, having intended from the outset that the whole earth be under human sway, had provided such means as the Ocean to connect all lands. In Mercator's words,

> This, too, should be observed, how great the creator's prudence was in hollowing out those recesses, the receptacles of the waters, for he so distributed the seas throughout the whole world that all the kingdoms of the world could carry on trade among themselves and transport anywhere whatever nature or art should give them.[69]

[67] E. VI, 'Copy of Letter Sent by Edward VI to 'the Kings, Princes, and Other Potentates, Inhabiting the Northeast Partes of the Worlde, toward the Mighty Empire of Cathay, at Such Time as Sir Hugh Willoughby Knight, and Richard Chancelor, with Their Company, Attempted Their Voyage Thither in the Yeere of Christ 1553', *PN*, vol. 2, 210.

[68] F. Lestringant, *Mapping*, 29.

[69] G. Mercator, *Atlas*. At cum usus & officium marium postularet, eorum inter se communionem,, propter necessariam circumnavigationem, idem etiam postularet fluminum exoneration, quae per solam marium comminicationem, aequalieter & iuxta aequilibrium, in omnes sphaerae partes distribui potest, necessarium in primis fuit, quod dixi, terrae per omnia, aequilibrium, neque potuisset aquis illud compensari. **

Without such a change in attitude to Ocean, the great early modern empires and trade companies could never have been formed. The classical idea of Ocean remained accepted and vital to the understanding of the world, but it was transformed in the Renaissance, in such a way that it had enormous implications for world history. The Ocean of Greek and Roman mythology and theory no longer bounded the world; it connected it: it was the means, designed by God, to make every part of his created earth able to communicate and trade with every other.

5

Balance and Opposition: the Physical Structure of the World

After they had dispelled the notion of a limited *oikoumene* and had proved that the world outside the former *oikoumene* contained inhabitable and navigable places, the Renaissance geographers and explorers still had little idea about the actual shape and distribution of the lands across the earth's surface. Exploration demolished the old notions about the structure of the world, but theory preceded exploration in creating a new image of the world. In particular, European Humanist geographers and travellers believed that that the universe was organised on principles of inherent balance, symmetry and opposition, and that the structure of the world followed these universal principles. This chapter will examine how such ideas allowed Humanist geographers, proto-ethnographers, and travellers to assimilate the discoveries and organise them into a comprehensible format. It argues that ideas of balance and symmetry were used to explain everything from the relationship that human and animals had with their environments to the physical arrangement of the lands across the surface of the globe. They were particularly evident in the way in which geographers discussed the proportions and arrangement of the continents. As with the destruction of the idea of uninhabitable climatic zones and unnavigable Ocean, at the core of the ideas about the nature and proportion of the world, was a sense that this was a world made by God for human domination.

The sixteenth-century writers were elaborating upon an intellectual tradition that had existed for two millennia that the whole universe was governed by laws of balance and opposition. As early as the sixth century BC, Greek philosophers and geographers had argued that there was an inherent balance to the world, and virtually every classical writer thereafter had used ideas of balance, symmetry or opposition to explain the nature of the world (and indeed the cosmos). Aristotle, in the *Metaphysica*, a work that would have been well known to most educated sixteenth-century writers, argued that all his predecessors saw the world in terms of opposition.[1] Although one could link all preceding philosophy to ideas of opposition, opposition could function in contrasting ways, of balance and reciprocity, both of which were reflected in different aspects of classical geography. No two philosophers entirely agreed, but in the earliest stages of Greek thought, the ideas of opposition could broadly speaking be fitted into two schools. One branch focused on a tension between opposites, which kept all nature in a state of dynamic equilibrium. The sixth-century BC philosopher and geographer, Anaximander of Miletus, for instance, held that opposites were in constant war with each other, trying, as it were, to trespass on the other's territory. A cosmic balance was maintained, however, so that no one opposite could triumph. As he argued, the elements 'pay penalty to each other for their injustice according to the assessment of time.'[2] This injustice was the injustice of one of a pair of opposites overstepping its limits and encroaching upon the other, and for this a penalty was exacted proportionate to the offence. The other philosophy of opposition, exemplified by the Pythagoreans, concentrated more on ideas of harmony between all elements, including pairs of opposites such as earth and water, air and fire – a harmony which itself meant that the world was in perfect equilibrium.[3] Both such philosophies of opposition and harmony played a role in sixteenth-century geographical thought.

The contrast between the approaches was not absolute. The fifth-century BC philosopher, Heraclitus, for instance, adhered to a philosophy of tension, maintaining that there was a war between opposites which kept everything in

1 See Aristotle, *Metaphysica*, 1004b 29ff; 1075a 28; and 1087a 29f. There has been surprisingly little study of this tendency to explain the nature of the world in terms of opposites. F. M. Cornford, *From Religion to Philosophy: A Study in the Origins of Western Speculation* (New York, 1957), 60ff, and John Burnet has contributed a brief discussion which largely disagrees with Cornford in J. Burnet, *Early Greek Philosophy*, 4th edn (London: Black, 1958), 8. The only major work on the subject is G. E. R. Lloyd, *Polarity and Analogy: Two Types of Argumentation in Early Greek Thought* (Cambridge: Cambridge University Press, 1966). He focuses much more on tracing the development of the idea of opposition rather than on its implications for matters such as geography.

2 διδόναι γὰρ αὐτὰ δίκην καὶ τίσιν ἀλλήλοις τῆς ἀδικίας κατὰ τὴν τοῦ χρόνου τάξιν: DK F1 (=Simplicius, *Phys.* 24,13), translation from G. S. Kirk, J. E. Raven, and M. Schofield, *The Presocratic Philosophers: A Critical History with a Selection of Texts*, 2nd edn (Cambridge: Cambridge University Press, 1983), 118 (hereafter *KRS*). There is some controversy about whether Anaximander meant that the elements gave reparation, or whether it was the 'coming-to-be' and the 'destruction' of all things from and into the ἄπειρον (the boundless) which in fact paid reparation. I have followed Kahn's interpretation, from C. H. Kahn, *Anaximander and the Origins of Greek Cosmology* (1960), chapter 3.

3 For a Pythagorean table of opposites see Aristotle, *Metaph.* 986a 22ff; also see *Metaph.* 985b23ff. For more on the use of harmony from specifically Pythagorean and Platonist approaches, and its implications for geography, see chapter 6 below.

a state of tension and perfect balance. For example, he spoke of 'Something which is being brought together and brought apart, which is in tune and out of tune: but of all things can be made a unity, and out of a unity, all things'.[4] The balance and unity were maintained by the very fact that opposites existed in pairs which pulled apart from each other. If one opposite outweighed the other, then it would prove victorious and the 'unity and coherence of the world would cease.'[5] He also spoke, however, of *harmonia*, seeming to imply, as Kahn points out, 'a unity composed of conflicting parts'.[6] In a sense he bridged the gap between those focusing on tension and those concentrating on opposition, for he thought it was necessary to see both the connection and the separation between things – adding in a further pair of opposites.[7] In this he seems to be using *harmonia* in a similar way to the Pythagoreans who were the most famous exponents of the doctrine of harmony between opposites.[8] As Diogenes Laertius reported of Philolaus (b. 470 BC), a prominent Pythagorean, 'it seems to him that everything comes to exist from necessity and harmony'. He then quoted Philolaus (470–385 BC) as saying, 'Nature in the cosmos was fitted together from both boundless and bounding things: both the entire cosmos and everything in it.'[9] There was no question of victory for any one opposite over another, because nature was formed by a harmony rather than a tension between opposites.

When divorced from the philosophical realm and given application to physical geography, the difference of philosophical approach between those presupposing an equilibrium maintained by a dynamic tension in nature, and those adhering to a notion of a harmony and proportionality between all elements, was trivial. Both ideas of opposition involved the idea of a fundamental balance, and in terms of the physical structure of the world followers of

4 συνάψιες ὅλα καὶ οὐχ ὅλα, συμφερόμενον διαφεόμενον, συναῖδον διᾶιδον, καὶ εκ πάντων ἓν καὶ ἐξ ἑνος πάντα. DK 10 (Pseudo-Aristotle, *de Mundo*, 5, 396b7), trans. by G. S. Kirk, *Heraclitus: The Cosmic Fragments*, 1st edn (Cambridge: Cambridge University Press, 1954), 168.

5 *KRS*, 193.

6 C. H. Kahn, *The Art and Thought of Heraclitus: A New Arrangement and Translation of the Fragments with Literary and Philosophical Commentary* (Cambridge: Cambridge University Press, 1979), 200. See for example, fragments DK 10 (Pseudo-Aristotle, *de Mundo*, 5, 396b7), DK 51 (Hippolytus, *Refutatio*, IX.9.2), and DK 54 (Hippolytus, *Refutatio*, 9.9.2) for use of *harmonia*.

7 G. S. Kirk, *Heraclitus*, 176.

8 See C. A. Huffman, *Philolaus of Croton: Pythagorean and Presocratic, a Commentary on the Fragments and Testimonia with Interpretive Essays* (Cambridge: Cambridge University Press, 1993), 139.

9 Δοκεῖ δ' αὐτῷ πάντα ἀνάγκῃ καὶ ἁρμονίᾳ γίνεσθαι…ἁ φύσις δ' ἐν τῷ κόσμῳ ἁρμόχθη ἐξ ἀπείρων τεκαὶ περαινόντων καὶ ὅλος <ὁ> κόσμος καὶ τὰ ἐν αὐτῷ πάντα. Diogenes Laertius, 8.85.

both approaches thought that this equilibrium was apparent in the relationship of land to sea, and between different landmasses. The ideas of balance and opposition in the ancient world, therefore, ended up creating a notion of counterweight continents – an idea which sixteenth-century writers picked up and expounded.

In the sixth century BC, Anaximander, who was a geographer as well as a natural philosopher, made the first European map of the world, which already depicted the beginnings of a Greek obsession with counterweight continents and a north-south symmetry.[10] Anaximander, as we have seen, had been an adherent of the idea of a dynamic tension creating the balance in nature. This tension was not, however, apparent in his depiction of the inherently static nature of the physical earth in which he showed Europe and Asia, perfectly counterbalancing one another, separated from each other by water and surrounded by a ring of Ocean. It was not a map which was designed to further travel or navigation; rather it was a mental map, little more than a geometric model of the earth.[11]

At the time Anaximander was constructing his map there was certainly insufficient empirical data available for it to be anything but a theoretical construct, yet it set the pattern for future geography, ensuring that for the next two millennia ideas of balance and proportionality were reflected in the way that people thought of the physical structure of the earth. Hecataeus, a generation later, adapted Anaximander's map to a tri-continental system rather than a bi-continental system, but maintained the idea of a balance between landmasses, since in this later map Africa and Asia together formed the counterpart of equal mass to Europe. This system remained at the core of all subsequent Greek descriptions of the *oikoumene*.[12] Over time it evolved, but the same principles were at play in the later medieval depictions of the world on the medieval *mappae mundi*.

The proportionate arrangement of landmasses was fundamental to the entire way in which most Greeks and Romans thought of the *oikoumene*, the world, and the relationship of both to the universe. Anaximander seems to have thought that the world was a perfect cylinder suspended in the *apeiron* or boundlessness.[13] The earth stayed still in suspension because it was equally related to all extremes.[14] The surface effectively formed a disk which was almost entirely occupied by the two continents of Europe and Asia surrounded by the River Okeanos. These continents mirrored each other, acting as counterweights. Any inequality in mass between the continents would have been a physical

[10] *F.Gr.Hist*. Hecataeus T.12a= Agathemerus 1.1.

[11] C. Jacob, *Géographie et ethnographie*, 37.

[12] *F.Gr.Hist*. T.12a = Agathemerus 1.1.

[13] Hippol. *Haer*. 1.6.3 = DK 12a11. See Kahn, *Anaximander*, 53-56.

[14] Aristotle, *De Caelo* 295b 11–16 = DK 12A26.

impossibility for it would have disrupted the balance and the world would have fallen out of its suspension.

The balance in the world had further implications for understanding the world beyond the *oikoumene* and particularly for imputing the idea of other landmasses. Using the same kind of reasoning that made him infer the structure of the known continents, Anaximander introduced the idea of an Antipodes, saying 'we are standing on one surface of the earth, the other is set opposite'.[15] It is not entirely clear what precisely Anaximander meant when he made this statement, but he does seem to have been making an argument based on a fundamental notion of symmetry – that not only must there be counterpart continents on one surface of the earth, but the lower surface of the cylinder must also act as a counterpart to the upper.

After the switch to a spherical world rather than the cylindrical one with which Anaximander worked, there were various further models of the earth which emphasised a proportionate distribution of land and sea. Many of the classical geographical writers reimagined Anaximander's antipodes as a landmass filling the southern hemisphere and acting as a counterpoise to the northern continent.[16] Geographers seem to have disliked the idea of an imbalance between earth and water. As we have seen, in the third century BC, the geographer Eratosthenes had produced a surprisingly accurate estimate of the world which emphasised how little of the earth's surface the *oikoumene* occupied. It nestled into the northern hemisphere, filling up only about a quarter of the globe. Thereafter, an idea of counterbalancing continents became a regular feature of classical geography. It has been argued, for instance, that the idea that the world was divided into four equal counterpart landmasses – attributed to the second-century BC stoic philosopher, Crates of Mallos – was a direct product of Eratosthenes's estimates of the ratio between the size of the *oikoumene* and that of the earth.[17] Certainly all the known proponents of the theory accepted Eratosthenes's calculations, and indeed Macrobius and Martianus Capella were the key people responsible for ensuring that Eratosthenes's figure of 252,000 stades was transmitted to the medieval and Renaissance worlds.[18] Both kept the idea of an earth divided into four distinct landmasses current throughout the

15 Τῶν δὲ ἐπιπέδων ᾧ μὲν ἐπιβεβήκαμεν, ὃ δὲ ἀντιθετον ὑπάρχει. C. H. Kahn, *Anaximander and the Origins of Greek Cosmology*, F 12H, 56.

16 As Pliny says in NH 2.65, there is considerable discussion about the Antipodes in classical antiquity. On the antipodes in the ancient world, see A. Hiatt, *Terra Incognita*, chapter 2.

17 D. Woodward, *Maps as Prints in the Italian Renaissance: Makers, Distributors and Consumers* (London: The British Library, 1996).

18 A. A. T. Macrobius, 'Commentary on the Dream of Scipio', ed. by William Harris Stahl (New York: Columbia University Press, 1952), 20.

Middle Ages.[19] By 1485, no fewer than four printed editions of Macrobius's *Commentary on the Dream of Scipio* were in existence, and it seems certain that the emphasis he placed on habitable (and indeed inhabited) lands played its part in promoting the explorations of the late fifteenth century, even if it was cited remarkably little by the sixteenth-century geographers, as we have seen.[20]

Of all ancient Greek and Roman writers, Macrobius offers the clearest example of the idea of the perfect similarity of counterpart landmasses although he perhaps took the idea further than many of his contemporaries were prepared to do. As we saw in chapter four, he argued that the world was divided into four islands, each entirely without communication with one another.[21] He drew direct parallels between each quadrant of land, stating, for example, that 'just as the part which is inhabited by us is divided by many an intervening region of waste land, so it is necessary to believe that there is a similar division between deserts and cultivated lands in the other three inhabited areas'.[22] This was an argument that went beyond mere mathematical proportion. It was founded on ideas of complete symmetry which had been prevalent in classical geographical thought from the outset. Herodotus charted the course of the Nile by that of the Danube, its northern counterpart;[23] Ptolemy described two identical lakes, symmetrical counterparts to one another, as the sources of the Nile.[24] In a similar way, Macrobius argued that the very geographical features of the unexplorable lands could be presumed by analogy with those of the known *oikoumene*. It is clear that Macrobius, and indeed all the exponents of the quadripartite theory, were resorting to a speculative physical geography founded on the ideas of balance and symmetry.

Anaximander, Crates, and Macrobius came from three widely differing philosophical schools. The first was a pre-Socratic natural philosopher, the second a Stoic, the third a Neoplatonist. More than a thousand years separated the first from the last; yet all three shared the common idea of the necessity of a counterweight landmass to the northern *oikoumene*. As Alfred Hiatt has shown, many other ancient writers, among them some of those who had the greatest impact on Renaissance geography, championed the notion of a southern continent even if

19 See W. H. Stahl, 'Dominant Traditions in Early Medieval Latin Science', *Isis*, 50, no. 2 (1959), 112; A. A. T. Macrobius, 'Dream of Scipio', 41–50.

20 One was printed in Venice in 1472, one in Brescia in 1483, and a further two in Brescia in 1485. By 1600, there had been 33 editions printed, with versions having been produced in France, Italy, Switzerland, Germany and the Low Countries: see Stahl, *Commentary*, 61–62 W. H. Stahl, 'Astronomy and Geography', 252.

21 Macrobius, *Commentarii in Somnium Scipionis*, 2.9.1–6

22 Macrobius, *Commentarii in Somnium Scipionis,* 2.5.30: 'Nam sicut pars quae habitatur a nobis multa solitudinum interiectione distinguitur, credendum est in illis quoque tribus aliis habitationibus similes esse inter deserta et culta distinctiones.'

23 Herodotus, 2.33–34.

24 Ptolemy, *Geographia*, 4.9.

they did not argue for the existence of western ones. Pliny and Pomponius Mela were only two of the more influential writers who supported such an idea.[25] Ptolemy, too, argued that the Asian landmass bent westwards to enclose the Indian Ocean and join up with the southern tip of Africa.[26] The southern land still, however, counterbalanced the northern one.

Fig. 1 Macrobian Map: from a Macrobian manuscript produced in Brescia in 1485, less than a decade before Columbus's ventures altered the way people used Macrobian geography. Special Collections Department, Bryn Mawr College Library

A southern continent remained a feature of geographical thought throughout the Middle Ages (albeit a controversial one), and sixteenth-century Humanists seized upon the idea.[27] The recent and ongoing voyages provided little evidence

[25] Mela, 1.4 and 1.54; Pliny, *NH*, 6.81.
[26] Ptolemy, *Geographia*, 7.5.
[27] A. Hiatt, *Terra Incognita*.

of the existence of a southern continent, but without it Ocean seemed to stretch vast and undefined across the southern hemisphere creating a physical imbalance, implying a lack of proportion to the world that Renaissance geographical theorists found hard to accept. With the breaking down of the boundaries of the *oikoumene*, the idea of an austral landmass gained a new relevance for those eager to argue that the whole earth was designed in accordance with laws of symmetry and proportionality. The Renaissance writers drew on classical authorities to support their assertions about its existence, but at the same time they brought it into a new frame of reference. In the classical and medieval worlds the antipodes had always been an unknowable region whose existence was postulated because it acted as a counterweight to the known landmass of the *oikoumene*. Discussion of it belonged to the natural philosophers and cosmographers rather than to the human geographers. In the Renaissance, however, the voyages of discovery had proved that there was no longer a separation between human geography and global geography. The antipodes were now thought to be the desirable lands, the goal of expeditions; they created a global south which enticed explorers such as Columbus to sail west and south in search of wealth unavailable in Europe.[28]

Perhaps the most closely reasoned sixteenth-century argument for its existence is that of Mercator whose rationalisation for its existence closely followed classical ideas of balance and symmetry.[29] His thoughts echo those of Anaximander: the earth needed to be balanced; otherwise it would not remain in suspension but fly off into space. He argued that a southern continent provided just such a balance, stating:

> Therefore the opinion holds that the machine of the earth is balanced in itself... The constitution of the world and of the centre of (its) weight implies all these things which if they had been known and studied by the ancients, they would have judged almost true – concerning the size and situation of the new continent which has been found in our century, and of the southern continent which lies under the still unexplored Antarctic Pole. For since the lands known to the ancients are embraced by 180 degrees of longitude, that is to say, they occupy only half of the sphere, the same number of lands must exist in the other half. And since for the most part, Africa, Asia and Europe lie beyond the

[28] For more on this see N. Wey Gómez, *Tropics of Empire*.

[29] N. Crane, *Mercator*, 25, 41, 45. He was well trained in the classics, having studied first with the Humanist Brethren of the Common Life in Hertogensbosch and then at the University of Louvain. He did not complete his degree, but while there he spent much of his time studying Aristotle and other ancient philosophers, and it seems certain that he had picked up their ideas of balance and reciprocity. When he left the University he continued studying philosophy privately and even wrote a gloss on the book of Genesis, citing various ancient sources including Plato and Parmenides to try to and understand the workings of the 'earthly machine'.

> equinox towards the north, it is necessary that as great a continent lies under the Antarctic Pole as will counterbalance together with the southern parts of Asia and New India and America, the remaining lands.[30]

The very constitution of the world ordained that the machine of the world must be in balance and a southern continent must therefore counterbalance the northern. Few were quite as explicitly articulate about balance, but he was not alone; virtually every sixteenth-century world map depicted a *terra australis*. Cartographers particularly emphasised the existence of the continent, since the need for a counterbalancing southern landmass was instantly apparent to those dealing with the more mathematical and visual aspects of world geography. Nonetheless, some, though fewer, descriptive writers also referred to it.

The descriptive geographers were more hesitant partly because the *terra australis* had a problematic legacy, associated with discussions of its inhabitability. Although the idea of the antipodes kept recurring throughout the Middle Ages, there remained the constant problem of the inhabitability of this region. St Augustine in the City of God had firmly declared that there could not be an inhabitable antipodes and that those who argued for its existence 'fail to notice that, even should it be believed or demonstrated that the world is round or spherical in form, it does not follow that the part of the earth opposite to us is not completely covered with water, or that any conjectured dry land there should be inhabited by men'.[31] Such scepticism left its mark on the sixteenth-century geographies.[32] In the words of Gabriella Moretti, there was a 'negative tradition' for Columbus and others to overcome.[33] For while the ubiquity of presentation on the maps suggests a consensus about the existence of an austral continent, many of the human geographers, perhaps because of the legacy of concern about its inhabited nature, refrained from discussing them as much as they did every other landmass.

30 Quamobrem stat sententia machinam terrae in se ipsa aequilibrem esse... Haec omnia importat centri gravitatis et mundi constitutio, quae si a veteribus cognita et examinata fuissent, de novae continentis, quae nostro seculo inventa est, deque meridionalis continentis quae nec dum explorato polo Antarctico subiacet, situ et magnitudine prope verum iudicassent. Etenim cum terrae veteribus cognitae 180 gradibus longitudinis comprehendantur, hoc est, dimidiam tantum sphaeram occupent, necessarium erat tantumdem terrarum in altera medietate extare. Et cum Asia, Europa et Africa pro maxima parte ultra aequinoctialem versus boream sint sitae necesse erat tantam continentem sub antarctico polo existere, quae cum Asiae et novae Indiae, sive Americae partibus meridianis, reliquis terris aequiponderaret. G. Mercator, *Atlas*, 22.

31 Augustine, *City of God,* 16.9 'Nec adtendunt, etiamsi figura conglobata et rutunda mundus esse credatur siue aliqua ratione monstretur, non tamen esse consequens, ut etiam ex illa parte ab aquarum congerie nuda sit terra; deinde etiamsi nuda sit, neque hoc statim necesse esse, ut homines habeat.'

32 A. Hiatt, *Terra Incognita*, passim.

33 G. Moretti, *Gli Antipodi Avventure*, 123.

It was not just the influence of religious debates about its inhabitability that deterred discussion. Just as Strabo in the ancient world barely mentioned the lands lying beyond the *oikoumene* as he believed that geography is only concerned with studying our *oikoumene*, so certain of the early sixteenth-century geographers refused to speculate about unknown places, and particularly about the southern continent.[34] Sebastian Münster, for instance, who as we have seen used Strabo as a model for both his style and approach to geography, made little mention of the Americas, and none of the *terra australis*.[35] Indeed in his geographical summary published as the 'Introduction' to Simon Grynaeus's early collection of New World travel narratives and geographical works, the *Novus Orbis* of 1532, Münster went so far as to deny the existence of an Antarctic continent.[36] His opinion about the existence of a *terra australis* had clearly not changed by the time the first Latin edition of his *Cosmographiae Universalis Libri VI* was made. In none of his works did he use any of the words such as *terra australis, antipodes* and *antichthones*, with which Renaissance writers tended to allude to this region. Although his map shows some ambiguity about whether the land to the south of the Straits of Magellan was a continent or merely an island, his text is much clearer. He stated outright, clearly following the account of the voyage compiled by Maximilian of Transylvania, that the sailors believed that the land to their left was merely an island.[37]

Both Maximilian and Pigafetta, the two main authorities on Magellan's voyage, describe the lands on either side of the strait as snowy, rocky and generally inhospitable, but in Pigafetta's, the explorer's descriptions of the harsh southern land became conflated with his expectations of a southern continent. He was an educated Italian who had survived Magellan's circumnavigation of the world, and his account of the voyage clearly betrays his Humanist background.[38] He stated that 'the land on the left side turned to the southeast and was low'.[39] Interestingly this description is omitted from Ramusio's

34 ὁ δὲ γεωγράφος ἐπισκοπεῖ ταύτην μόνην τὴν καθ' ἡμᾶς οἰκουμένην. Strabo, 2.5.34.

35 M. Small, *Warring Traditions*. For more on the composition of Sebastian Münster's *Cosmography* and what he chose to focus on, see S. Davies, 'America and Amerindians.'

36 S. Münster, 'Introductio ad cosmographiam,' in *Novus orbis regionum ac insularum veteribus incognitarum una cum tabula cosmographica, et aliquot alii consimilis argumenti libellis, quorum omnium catalogus sequenti patebit pagina*, ed. Simon Grynaeus (Basle: Jo. Hervagium, 1532), 4v. *Novus orbis regionum ac insularum veteribus incognitarum una cum tabula cosmographica, et aliquot alii consimilis argumenti libellis, quorum omnium catalogus sequenti patebit pagina,* ed. Simon Grynaeus (Basle: Jo. Hervagium, 1532).

37 S. Münster, *Cosmographia*, 1140.

38 On Pigafetta's humanism see J. P. Rubiés, *Travel and Ethnology*, 144–45.

39 C. E. Nowell, *Magellan's Voyage around the World: Three Contemporary Accounts* (Evanston: Northwestern University Press, 1962), 118.

version of the voyage and notably differs from Richard Eden's mid-sixteenth century translation in which the land on either side of the strait is said to be mountainous and covered with snow.[40] Eden's was the only published version containing a description of the southern land in the sixteenth century, but both Eden's and the original clearly portray the southern edge of the strait as a continuous landmass. All versions agree that 'they found in this strayght at every three myles, a safe haven and excellent water to drynke: woode also and fyshe, and greate plentie of good herbes. They thynke that there is not a fayrer strayght in the worlde. [41] It is an enticing, inhabitable continent. Maximilian's rather than Pigafetta's description, however, was what informed Münster and others since it circulated widely. Pigafetta's seeming eye-witness support for the counter-balancing southern continent was less influential than it might have been.

Pigafetta was not alone, however. Other reports and stories also conjured up its exotic nature. Robert Thorne, for instance, was certain that if people travelled toward the Antarctic pole, 'without doubt they shall finde there the richest landes and Islands of the world of golde, precious stones, balmes, spices and other thinges that we here esteeme most which come out of strange countries.'[42] Such ideas contrasted with the idea of a frozen wasteland which sailors' reports of Tierra del Fuego were creating, the inconsistencies of which one can see in Eden's description.

One solution was to fragment the continent. Even in the classical world, few had suggested that the southern continent was an exact mirror image of the northern one, merely a counterweight. If this southern landmass or landmasses

[40] P. Martyr and R. Eden, *Decades*, 221r–v. The Pigafetta account of Magellan's voyage causes problems. He wrote one detailed account, and obtained a privilege for publishing this account in Venice in 1524; but for some reason the manuscript was not published until Carlo Amoretti made an edition in 1800. Ramusio seems to have been unable to obtain a copy of the full version and relied on a French summary made by Jacques Fabre. Ramusio's version was in turn translated by Purchas in his *Hakluytus posthumus*. The Eden version seems to be a different summary of the Pigafetta account. See G. B. Parks, 'The Contents and Sources of Ramusio's Navigationi', in *Terza edizione delle navigationi e viaggi, Facsimile Reprint*, ed. by R. A. Skelton, G. B. Parks, and G. B. Ramusio, vol. 1 (Amsterdam: Theatrum Orbis Terrarum, 1967–70), 14. The full account printed in 1800 lends credence to the idea that Tierra del Fuego was part of the southern land since in it, Pigafetta, contrary to the mountainous description of the summary, described the land on the left side as low and tending away towards the south-east. There is no suggestion that it might be merely an island. See C. E. Nowell, *Magellan's Voyage*, 118. Given that not even Ramusio was able to obtain this original document, however, the full version cannot be argued to have had much influence on sixteenth-century ideas of an austral continent.

[41] P. Martyr and R. Eden, *Decades*, 221r–v.

[42] R. Thorne, *Declaration of the Indies*, 163.

counterbalanced those of the north, that was sufficient. Maps and writers came to represent the antipodes not as a single landmass but as a series of them. John Laggard, writing a summary of Botero's *Relationi Universali* at the very beginning of the seventeenth century, described the antipodes as 'the sixth part of the world', saying it

> is as yet but little knowen, notwithstanding it is thought to containe many large provinces, amongst which Beach is supposed to be verie rich and abounding in gold. The chief isles are Java maior and Java minor, Timor, Banda, the Moluccas, Romeros and the Iles of Solomon.[43]

Botero was slightly less explicit than his translator, not actually claiming that the southern islands are a separate part of the world. He did, however, support the idea that they formed a distinct unit consisting of a group separate islands, though he acknowledged that the majority of cosmographers have described the land south of the Strait of Magellan as almost a continent.[44] In the sixteenth century the various navigation reports of discoveries in the southern Pacific were fitted together as evidence of a large mass of land which were either outcrops of a single southern continent or parts of a series of large islands which helped counterbalance the continents of the north. This southern territory could contain both the freezing land of Antarctica and the exotic regions of Java Major.

Those who wrote the descriptive geographies might have little to say about this region, but, apart from Münster, they did not deny its existence. A very few even asserted it. Thevet, for instance, described it in *Les Singularitez de la France Antarctique*, as a known continent, calling it the *terre australe*.[45] In the next chapter he talks about how there are antipodal lands at the north and south pole.[46] He seems to have accepted the report of Tierra del Fuego and synthesised it with classical ideas about an antipodal continent. Although successive voyages throughout the sixteenth century failed to discover it, the presumed existence of this continent remained unchallenged. It was a boundary land, and hence a region of fantastic wealth, like the fringe regions of the ancient world. Pedro de Quiros, writing in 1607, petitioned the king of Spain to sponsor a search for this territory, claiming that it extended along the length of the other continents and describing

43 G. Botero, *The Worlde or a Historicall Description of the Most Famous Kingdomes and Common-Weales Therein. Relating Their Situations, Manners, Customes, Civill Government and Other Memorable Matters, Translated into English and Enlarged by R. Johnson* (London, 1691), 13.

44 G. Botero, *Relationi universali devise in quattro parti. Novamente reviste, corrette e ampliate dall' istesse auttore* (Brescia: Compagnia Bresciana, 1598), 422.

45 A. Thevet, *Singularitez*, 110v.

46 *Ibid.*, 111.

it as an earthly paradise in terms of geography, product and climate.[47] It was also a region necessary for maintaining the balance and proportionality of the world and therefore assuring that the world accorded to the laws of nature. As José de Acosta stated it: 'For the attempt we are making, it is enough for us thus far to know in truth that there is land in this part of the south, and that it is a land as large as all of Europe and Asia, and even Africa; and that at both poles of the world there are to be found seas and land embracing one another on which the Ancients, like those who lacked experience, could cast doubt and contradict.'[48]

The sixteenth-century geographers derived the ideas of balance and proportionality in the makeup of the world from the classics, but went far beyond the classics in their application of ideas of balance in determining the physical arrangement of lands. The classical writers had never really discussed lands beyond the Arctic Ocean. Theopompus, it is true, had mentioned that Meropians had crossed 'the Ocean with myriads of thousands of men until they came to the Hyperboreans', implying that there was a land to the north beyond the Arctic Ocean, but even he had not fitted this land into a general pattern of a geography of balance.[49] Similarly, the various Greek and Roman writers who wrote about the island of Thule, which also lay beyond the inner shores of the Arctic, did not fit it into part of a picture of a world in equilibrium.[50]

To keep a sense of symmetry in the distribution of land and sea, some cartographers and geographers proposed the existence of a further northern continent, the fourth continent, which though significantly smaller than the southern continent was nonetheless its counterpart. These writers were clearly not using a pure Macrobian model of *oikoumene, anticthones, perioikes* and *antipodes*, but some seem to have been shifting the continents around to fit with a more-or-less balanced model. The Americas counterbalanced Africa and Eurasia, in one direction, while the southern continent counterbalanced all the lands of the northern hemisphere in another. The northern continent led the waters of Ocean into narrow sea-routes – the counterparts to those which lay between Africa or South America and the southern continent. Mercator's map of the polar regions provides the most famous example of this hypothetical continent (see Fig. 2). The four islands he depicted around the pole effectively formed a large continent.

47 A. Baert, *Le Paradis terrestre: un mythe espagnol en Océanie: les voyages de Mendaña et e Quirós, 1567–1606* (Paris: L'Harmattan, 1999), 193–213.

48 Para el intento que llevamos, bástanos hasta ahora saber de cierto, que hay tierra de esta parte del sur, y que es tierra tan grande como toda la Europa y Asia, y aun África; y que á ambos polos del mun- do se hallan mares y tierras abrazados entre sí, en lo cual los Antiguos, como á quienes les faltaba experiencia, pudieron poner duda, y hacer contra- dicción. J. de Acosta, *Historia*, Book 1, ch. 6, 29.

49 G. S. Shrimpton, *Theopompus the Historian* (Montreal, Buffalo: McGill-Queen's University Press, 1991), F75 C, 227.

50 See for example, Tacitus, *Agricola*, 10.6, or Solinus, 22.9.

Fig. 2 Gerard Mercator's Map of the north: 'Septentrionalium terrarum descriptio', 1620, first published 1595. National Maritime Museum, Greenwich, London

The straits dividing one from another were little more than rivers through which the Ocean was drawn in, causing the ebb and flow of tides. Strictly speaking, Mercator did not call this northern region a continent, but it was still in his eyes part of the counter-balancing structure of the earth.

Like the Antipodes, the northern continent had more of a cartographical than a textual presence in sixteenth-century geography, but similarly there were instances of it in written geographies. The English geographer and lawyer, Richard Hakluyt, uncle of Richard Hakluyt, the author of the *Principal Navigations*, for instance, assumed the existence of a northern land of Nova Zembla which lay within the Arctic Circle but extended south into the temperate zone where the coast of Asia turned south-east.[51] He was so certain that such a landmass existed that, when writing instructions in 1580 to Arthur Pet who was preparing to search for the north-east passage, he said, 'And touching the tract of land of Nova Zembla toward the east out of the circle Arcticke in the more temperate zones you are to have regard' in case the land was inhabited or was in any way suitable as a trading post.[52] His statement is a series of ifs and buts about the nature of the climate, soil, topography and building materials in the region, which proves that he had no factual knowledge that such a landmass existed. Like the ancients he was following a frame-work for the world founded on theory rather than empirical information. Nova Zembla must exist because it filled a void and balanced out the other regions of the world.

Unlike Mercator, Hakluyt the Elder seems to have thought of a single continuous continent, but others fragmented this northern land just as they had the southern *terra australis*. The unknown arctic landmass occupied less territory than the southern one, since the southern one was proportionately counterbalancing all the landmasses of the northern hemisphere. What it did, however, was to channel the Arctic Ocean into the northeast and northwest passages which separated the arctic lands from the north of Europe, and for that matter the north of America. These passages were entirely separate from the channels which drew the Ocean into the polar sea in Mercator's depiction. Rather, they formed the southernmost limit of the polar lands. These straits corresponded to those between Antarctica and Africa and South America. As Acosta explicitly stated, 'I don't know by what signs, some have affirmed that there is another strait to the north similar to that of Magellan.'[53] As far as Acosta knew there was no empirical evidence for these straits in the Arctic regions, but those who were proposing them were determining the geography because they believed that the arrangement of land and sea in the northern regions corresponded directly to those in the southern hemisphere.

The Greeks and Romans had thought particularly of a north-south balance when they wrote of counterbalancing landmasses and drew on a geography of symmetry and opposition. To a certain extent such an arrangement is unsurprising, given the concept of two poles. The idea of a physical opposition between regions of the world was first introduced into classical geography

[51] E. G. R. Taylor, *The Original Writings and Correspondence of the Two Richard Hakluyts*'. Document 25 and footnote 2, 148: 'Instruction for the north-east Passage by Richard Hakluyt Lawyer, Notes in writing, besides more privie by mouth, that were given by M. Richard Hakluyt, of Eton in the countie of Hereford, Esquire, Anno 1580: to M. Arthur Pet, and to M. Charles Jackman, sent by the merchants of the Moscovie companie for the discoveries of the northeast straight, not altogether unfit for some other enterprises of discoveries, hereafter to be taken in hand'.

[52] *Ibid.*, Document 25, 148.

[53] 'y no sé qué indicios, afirman algunos, que hay otro estrecho hacia el norte, semejante al de Magallanes'. J. de Acosta, *Historia*, 28, book 1, chapter 6.

by natural philosophers who had studied the heavens and the earth, seeking natural laws which could be universally applied to the *cosmos* and everything it contained. The relationship of the sun to the earth, for instance, had constructed the pattern of climatic zones which, as we have seen, had a symmetrical arrangement on the earth's surface. No such contrasts and counterparts could be perceived in the east-west relationships within Eurasia, and even the few classical writers who proposed the existence of a western landmass left its nature largely undiscussed.

In the early modern period, of course, east-west relationships and the western continent came to have far more significance than northern and southern ones. While the polar landmasses were undiscovered territories, whose existence was assumed because of an inherent belief that the world was in balance, the western landmass had at least a partially proven existence. As early as 1503 Antonio de Nebrija, the leading Spanish Humanist of the late sixteenth century, conferred continental status on the mainland Americas when he combined his knowledge of Macrobian geography, drawn from his reading of Martianus Capella, with the explorations in the New World, to describe those regions as a continent forming one of the four Macrobian landmasses.[54] He specifically used the term *antichthones*, employed by Capella to refer to the landmass lying south-west of the known *oikoumene*,[55] stating:

> Of the remaining part of our hemisphere and of the region opposite which the Antichthones inhabit, nothing of certainty has been reported by our ancestors, but such is the bravery of mankind in our time that in a short while they will bring us a true description of this land, either that it is made up of islands, or that it is a continuous land, of which our navigators have told about a large part of its coastline, principally the part which lies opposite the recently discovered islands (Hispania which I call Isabela and the other ones lying close by).[56]

Not much later, Vespucci in his letters to Lorenzo de Medici, published in the fifth libretto of the *Paesi novamente retrovati* in 1507, also referred to mainland Americas as a continent.[57] Nonetheless there was no consensus that this region

54 W. G. L. Randles, *Classical Models*, 56.

55 Capella, 196G (605).

56 'De reliquo huic nostro hemispherio et regione opposito, quod incolunt Antichtones, nihil certi a maioribus nostris traditum est, sed ut nostri temporis hominis audacia breui futurum est ut nobis ueram terrae illius descriptionem afferant, tum insularum tum etiam continentis, cuius magnam partem orae maritimae nautae nobis tradiderunt: illam maxime, quae ex aduerso insularum nuper inuentarum ('Histpaniam' dico 'Isabelam' reliquasque adiacentes) posita est.' A. Nebrija, *Introductorium Cosmographiae*, ed. by V.B. Sanchez (Seville: Lebrija, 2000, first published 1503), 98.

57 A. Vespucci, *Epistola*, book five, chapter cxiiii. These were first published in 1504, but gained in a new context in the compilation.

was a continent, nor is it always clear what writers meant when they used the term since it did not have a well-defined meaning in the sixteenth century. It might mean a large independent landmass, but could also be used merely to refer to a stretch of continuous land. It was only by the end of the sixteenth century that it came to have as a primary meaning what is now thought of by the term continent.[58] The earliest writers to believe it was a large landmass in its own right, all conflated it with a south-western continent, but gradually the western continent began to take shape.

It was not long after Vespucci described Brazil as an antipodean continent that it became apparent that this region was not, in fact, a separate southern landmass but was part of an immense western one.[59] Maximilian of Transylvania, for example, described the certainty of Magellan and his advisors before their departure that Regio Balcacarum, Terra Florida, and the new Terra Firma of the southern hemisphere formed a single landmass, but how they were uncertain as to whether it extended from pole to pole.[60] Such fears were assuaged when Magellan and his crew rounded the tip of Chile in 1520, thereby discovering the southern latitudinal extent of the landmass they now knew to exist in the western hemisphere. The east coast was then the next to take shape. The Spanish and Portuguese voyages had largely explored the east coast from the Gulf of Mexico to the Tierra del Fuego by the 1520s. As we have seen in the previous chapter, by 1524 Verrazzano had sailed much of the coast of America from Maine to Florida and proved that it extended a significant distance north and south, further demonstrating that it was a large, continuous landmass. John Rut in 1527 sailed from Newfoundland to the Caribbean, adding to Verazzano's information. By the 1530s, Cartier had also sailed far up the St Lawrence River, proving both that the continent was wide and that it continued northwards. By the end of the century, with the voyages of the Frobisher (1576–78) and Davis (1585–87), much of the North American east coast had been explored, although the Hudson's Bay region still remained largely unknown. There had certainly been enough exploration to provide empirical evidence that the Americas formed a single continuous landmass, a counterpart in north-south extent to Europe, Asia and Africa – though its precise east-west dimensions remained hazy, with little exploration of the western seaboard.[61] Nonetheless, while there was no convincing empirical

58 M. W. Lewis and K. E. Wigen, *The Myth of Continents, a Critique of Metageography* (Berkeley: University of California Press, 1997), 29.

59 A. Vespucci, *Epistola*, book 5, chapter cxiiii

60 M. Transylvanus, *De Moluccis*, Aiii(v).

61 Preciado, writing in the 1530s of a journey to California, was not even certain that it was the same landmass as the Spaniards had been exploring further south, and said, 'Whereupon wee began to be of divers opinions, some thinking that this coast of Santa Cruz was a firme land, and that it joyned with the continent of Nueva Espanna, others thought the contrary, and that they were nothing else but Islands, which were

evidence of a body of water cutting the Americas off from Asia, most people assumed that the Americas were a single continuous landmass separated from both Europe and Asia by sea. A vast western continent had become a reality.

In the Greek and Roman world, several classical authors had referred to western landmasses interrupting the empty expanse of water which stretched westward from Europe to Asia, whose imputed existence rested 'solely on logical deductions from the mathematical doctrine of the sphere, and on an aesthetic satisfaction with symmetry and analogy.'[62] Plato, Theopompus, and Plutarch, distanced over a space of five centuries, had all declared that a large territory lay to the west, filling out the blank void of Ocean.[63] Ptolemy, since he had miscalculated the size of the globe, by about eighteen percent, and had also shown the landmasses as potentially extending more than 180° degrees across the earth's surface, had not shown the same need to occupy some of the space which would otherwise have been entirely given over to Ocean.[64] The surviving fragments of Eratosthenes' work give no indication that he thought this vast expanse of Ocean was interrupted by any other body of land, but nor do they propose the contrary. Strabo, the chief preserver of Eratosthenes' thought, certainly argued that there were probably two other landmasses lying to the west along the parallel of Athens.[65] But the idea of the existence of such western lands had no more of an empirical foundation than had that of a southern continent. Indeed classical writers spoke of the southern continent as a reality whereas the idea of a western one was always tainted by mythology and pure conjecture.

The Renaissance Humanists, while occasionally openly admitting that ideas about the nature and structure were formed by imagination and reason, still looked for both empirical and classical authority to justify their ideas about the

to the westward.' F. Preciado, 'Ulloa's Voyage to Gulf of California', in *The Principal Navigations, Voiages, Traffiques and Discoveries of the English Nation, PN*, vol. 12, 212.

62 W. H. Tillinghast, *Geographical Knowledge of the Ancients*, 23.

63 See Plato's *Timaeus*; Plutarch, *Concerning the Face which Appears in the Orb of the Moon* 924; and Theopompus F75c in *F.Gr.Hist*. Theopompus's continent actually lies not just to the west but also to the east and extends right round Ocean as the only continuous landmass in the world. Theopompus was clearly writing about a fictionalised continent, using it as moral allegory, but the story was well known in the sixteenth century since it was recorded by Aelian, whose work was printed at least fourteen times in that century alone. It was taken as an example of classical knowledge by sixteenth-century authors. For a discussion of the moral allegory, see G. S. Shrimpton, *Theopompus*, 109–110, 143. Theopompus's idea of this continent of Merope seems quite similar to that proposed by Thevet in the sixteenth century who also argued that Ocean was itself surrounded by land.

64 J. L. Berggren and A. Jones, *Ptolemy's Geography*, 21–22.

65 Strabo, 1.4.6.

nature of the western continent.[66] Unlike the concept of a southern continent, that of a western one had failed to grasp the medieval imagination prior to the actual discovery of the Americas. Even the Viking journeys to the New World had failed to provide any enduring legacy in general European knowledge, so that it was only in the twentieth century that there was any certainty about the Norsemen actually reaching the Americas.[67] While the possibility of lands to be found in the western Ocean existed even in some of the better known texts, such as Aristotle's *De Caelo,* these lands were not a part of accepted geographical lore at the time of the voyages of discovery. Once the Atlantic Ocean was crossed, and people found unknown lands to the west, however, there were some Renaissance writers who tried to read back from the discoveries to find in the classical authors not merely signs that the ancients might have known about the western land, but also reasons for sailing south and west to rediscover such lands.

This re-reading of the classical authorities to find evidence of knowledge about the western world was a selective practice and not an entirely successful one, and could cause controversy between authors. The Italian historian of the New World, Girolamo Benzoni, for instance, argued that Spanish Humanist historian, Lopes de Gomara, diminished the importance of Columbus by saying that Columbus embarked upon his voyage, not in the spirit of pure discovery, but because he wanted to rediscover the lands which the ancients already knew; he had read tales of Atlantis in Plato's *Timaeus* and *Critias*, and of a vast and fertile south-western land discovered by the Carthaginians in pseudo-Aristotle's *De Mirabilibus*, and these had instilled in him the desire to seek not only Marco Polo's Cipangu, but also the western and antipodean lands.[68] Benzoni thought Gomara's emphasis on ancient prior knowledge did an injustice to Columbus – although it is possible that he was also defending himself, since unlike Gomara he had little classical education. Gomara was far from unique, however, in his approach. As we shall see in the next chapter, many sixteenth-century writers tried to find evidence

66 A particularly notable example of how Renaissance Humanists were aware of the use of reason and imagination in interpreting the structure of the world can be found in J. de Acosta, *Historia*, Book 1, chapter 7.

67 In 1963 Helge and Anne Ingstad discovered the site of a Norse house and various other artefacts from which archaeologists have been able to date the settlement to about 1000 AD at L'Anse aux Meadows in Newfoundland. A. S. Ingstad and E. S. Seeberg, *The Discovery of a Norse Settlement in America* (Oslo: Universitetsforlaget, 1977); H. Ingstad and A. S. Ingstad, *The Viking Discovery of America : The Excavation of a Norse Settlement in L'anse Aux Meadows, Newfoundland* (New York: Checkmark Books, 2001), 123-40.

68 G. Benzoni, *Insignis*, 24. L. de Gomara, *LaIstoria de las Indias, y conquista de Mexico* (Çaragoça: Agustín Millán, 1552), chapter CCXX. Unlike Benzoni, Gomara had a good Humanist education and access to Humanist circles. C. A. Roa-de-la-Carrera and S. Sessions, *Histories of Infamy : Francisco López De Gómara and the Ethics of Spanish Imperialism* (Boulder: University Press of Colorado, 2005), 1–2.

for classical knowledge of the New World in Plato's Atlantis story; others tried to use the same myth to explain the existence of humans in the New World.[69] They thought that before the sixteenth century no one had the technology to reach the Americas by ship, so it was possible that humans, the descendants of Adam, had arrived there by crossing over a land-bridge formed by the continent of Atlantis which had since sunk, isolating one part of the globe from the other. Some interpreted an ambiguous passage in Seneca's *Medea* as evidence for classical knowledge of the New World's existence.[70] Most, however, took Benzoni's approach and saw the existence of the New World as indicating a deficiency in classical knowledge and undermining classical authorities who had depicted a world where one could reach eastern Asia by sailing westwards should technology and other practicalities permit. The discovery of the New World, just like the crossing of the torrid zone, proved that the ancients could be deceived, and indeed could be simply wrong.

Nonetheless, the discovery of a western continent, while it had undermined the possibility of relying upon the ancients' purported facts, had strengthened the force of classical theory. In the New World, ideas of symmetry and analogy in the makeup of the world could again find expression. Different concepts of the land-sea distribution had vied for their place. Again these had been of largely academic value in the Middle Ages when humans were bound to their own *oikoumene*; but in the Renaissance they were of renewed importance. The Aristotelian notion of concentric spheres of earth, water, air and fire had supported the idea, common throughout the Middle Ages, that the surface of the globe was largely covered by water, only interrupted at occasional points where the sphere of earth broke through the sphere of water to form the landmasses of the world.[71] This notion was not entirely abandoned in the aftermath of the discoveries. Botero, writing *Observations of the Sea* near the end of the sixteenth century, more or less restated the Aristotelian doctrine, saying:

> It clearly seems that on the one hand, all the water must be greater than all the earth because the order of nature and the proportion of all the elements requires it. Therefore, just as the air gives way to water and fire to air and the sky to the fire, so it seems that the water must accede to the land because nature in all her works seeks to temper things and counterbalance one with another. Hence, because a little earth can resist much water, just as a little water can resist much air, as it has given more space and more territory to the air, counterbalancing the density of the lower elements with the amplitude of the superior ones and the

[69] See chapter 6 below.

[70] Seneca, *Medea* l. 375–9; see, for example, J. de Acosta, *Historia*, vol. 1, 54–55. For a detailed study of the *Medea* passage in Renaissance geography, see J. Romm, *Seneca in the Renaissance*, 77–113.

[71] W. G. L. Randles, *De la terre plate*, 58–59.

> amplitude of the latter with the density of the former, so it seems, the element of water should be much larger and more spacious than the earth.[72]

According to Botero, in theory the universe hung in balance between the elements. Earth took up less space than air in proportion to its greater density, but this balance was insufficient, for it created a world ill-suited to mankind. Botero therefore continued by saying that although in theory the balance between the spheres should lead to the formation of a world largely covered by water, such a design was not part of God's plan since he had made the earth for humans, and it was inconvenient for humans to have water encompassing the globe in the same manner as air does. Therefore,

> The water concedes part of her place to the land for the inhabitation and convenience of humans. It seems that it would be suitable that she should concede so much of it that the land, with this addition should come to have more than the water, particularly as the land itself only has a little of it.[73]

Even in this apparent overruling of the order of nature, however, a natural balance was maintained, for, as Botero explicitly pointed out, citing the authority of the ancient geometricians recorded by Plutarch, there was a balance between the water and the land. The depth of the sea related proportionately to the height of the mountains – 'the ordinary depth of the sea corresponds to the hills and smaller mountains; the extraordinary, to the Apennine and the Alps and other huge mountains'.[74] Nor was this the idea of just one individual; Humanists throughout western Europe held similar ideas about a balance between water and land.[75] Acosta likewise argued that 'in all parts of the world, the earth and water imbrace

72 'Ben pare, che da una parte, tutta l'acqua debba esser maggiore che tutta la terra; perche l'ordine della natura, e la proportione degli elementi cosi ricerca. Conciosia che, si come l'aria accede l'acqua, e il fuoco l'aria, e il cielo il fuoco cosi pare, che l'acqua debbe eccedere la terra perche la natura in ogni sua opera cerca di dar temperamento alle cose, e di contrapesar l'una con l'altra. Onde, perche poca terra può resistere à molta aqua, e di man in mano, poca aqua à molto aere: si come ella hà dato più luoco, e più paese all'aere, contrapesando la densità de gli elementi inferiori con l'ampiezza de i superiori, e l'ampiezza da questi con la densità di quelli, pare che dovesse far l'elemento dell'acqua molto più ampio, e spatioso di quello della terra.' G. Botero, *Relationi del Mare* (Padua: Andréa Viani, 1598), 83v.

73 '[L]'acqua cede alla terra parte del suo luogo per habitanza, e per commodità dell'huomo, pare che sia conveniente, che le ne ceda tanto, che essa terra, con questa aggiunta, ne venga ad haver più che l'acqua, massime che la terra ne ha per se stessa poco.' *Ibid.*, 84r.

74 L'ordinaria profondità del le colline, e a monti mediocre: e la straordinaria all'Apennino & alle Alpi, & all' altre montagne smisurate.' *Ibid.*, 84v–85r.

75 See e.g. J. de Acosta, *Historia*, vol. 2, ch. 7; G. Mercator, *Atlas*, 21.

one another, which truely is a thing to make vs admire and glorifie the Arte of the soveraigne Creator'.[76] Mercator referred to God's role in creating this balance and its necessity in the design of a world made for humans, saying that not only had the creator made the seas to be a means of communication (as discussed in the previous chapter), but

> secondly and most importantly, so that the landmass making a single sphere with the gathered waters, might remain in its equilibrium, for the earth has nowhere been established above the waters, but its greater weight gathered into one part has pressed down the whole mass of the land towards the centre of gravity of the world, and has made the waters lying on the opposite side to the earth raised up higher into the sky by the depression, to rain down & to occupy the nearest lands.[77]

Balance was part of God's intentional design in the world to allow all parts of the world to be in communication with one another. It was fundamental to the European interpretation of the physical geography of a world, much of which was not yet visited by Europeans. It allowed them to infer the existence and locations of lands and seas which would enable all parts of the world to trade with one another, and by inference would imply that the *oikoumene* reached across the whole world through known and unknown continents.

The whole western continent could be fitted into the ideas of balance and analogy, and of a world designed by God for humans, which governed ancient and Renaissance geographical thought. It was not easily accepted; but this was for practical rather than theoretical reasons. India was the goal of the west; it was the land of dreams and wealth. It had taken centuries for anyone to seize on to the possibility of sailing there by means of the Atlantic Ocean, but once attempted the idea of a westwards voyage was not easily surrendered. Yet, as we have seen, surprisingly rapidly, some writers started describing the lands of Columbus's and Cabot's discoveries as a single continent long before they were proved to be anything but large islands. When Mercator created his world map (1569), he determinedly demonstrated, with little more evidence than Waldseemüller,

[76] J. de Acosta, *Historia*, 1, 6. J. de Acosta, *A Naturall and Morall Historie of the East and West Indies Reprinted from Translation of Edward Grimston 1604*, ed. by C. R. Markham, 2 vols, vol. 60 (London: Hakluyt Society 1st series, 1880, first published in Latin and Spanish, Seville 1590), 16.

[77] G. Mercator, *Atlas*, 21: tum ut quod maximum est, terra cum aquis collectis unam efficiens sphaeram, in suo aequilibrio permanserit, alioqui enim terra non fuisset supra aquas firmata, sed plus ponderis in unam partem collectum, totam terrae massam versus centrum gravitatis & mundi depresisset, aquasque ex altera parte incumbentes illa terrae depressione altius in sublime sublatas defluere fecisset, & terrasproximas occupare.

that the Americas were a landmass in their own right, completely detached from Asia. Even more than the southern continent, in his image of the world, they reflected the idea of balance. The Atlantic formed the central divide; Europe, Asia and Africa formed one conglomeration of land to the east, and the Americas acted as their counterpart to the west. As Zerubavel has pointed out, by placing the Americas on the left edge of the map, cartographers helped to promote the idea of the absolute separateness of the New World from the Orient.[78] Zerubavel emphasises the idea of separation, pointing out how boundaries distort perception of space so that 'drawing a line between any two entities basically widens the gap we envision between them, thereby reinforcing their mental separateness from one another'; but the sixteenth-century maps did rather more than that. The Americas, as Gillies has said, were 'symmetrically juxtaposed with the landmass of the Old World'.[79] But this visual and descriptive arrangement went beyond the symmetrical; it reinforced the idea of the Americas as part of the equilibrium of the world. This image of east and west worked mainly on a single image of balance, but the Americas also fitted into a Macrobian scheme. When Nebrija referred to the newly discovered regions as the *antichthones*, one of the four Macrobian landmasses, he was explicitly fitting the new territories into the world of classical theory.[80]

When talking of the western continent, people began to create the implicit idea of a second, equivalent *oikoumene* that was part of a world made for humans. They used a language to refer to the New World which reflected not simply its position as a counter-balancing landmass, but its special status as a counterpart *oikoumene*. It was not a mere *anticthon*, a *terra incognita* or even, usually, a *terra occidentalis,* lying opposite our feet (*antipous*); it was not simply another continent; instead it was a *novus orbis* or a *novus mundus*. The term continent, already used in 1503 by Nebrija, gave the idea of a large body of continuous land; but it was not sufficient to indicate the complete complementarity which Europeans ascribed to the Americas.

The term *orbis* became associated with the Americas as early as 1516 when Peter Martyr, writing in the aftermath of Columbus's discoveries, gave the newly discovered lands the title of *novus orbis*. The term had significant implications for showing the ambiguous nature of the new world.[81] It had an unclear meaning throughout classical, medieval and Renaissance times, but always had physical connotations, and by implication therefore related to physical rather

[78] E. Zerubavel, *Terra Cognita: The Mental Discovery of America* (New Jersey: Rutgers University Press, 1992), 73–75.

[79] J. Gillies, *Geography of Difference*, 61.

[80] A. Nebrija, *Introductorium Cosmographiae*, 98.

[81] Martyr, *De Orbe Novo.*

than to human geography.[82] According to the *Oxford Latin Dictionary*, there are essentially thirteen different dictionary meanings for the word, of which the only constant is the emphasis on circularity.[83] In Latin, *orbis* could refer to virtually anything round, be it a sphere or a circle, and was used in both senses in Roman cosmology. It could in particular be used to refer to the earthly sphere. Pomponius Mela, for example proposed to give an account of the *orbis situs*.[84] While *orbis* could refer to the whole earthly sphere, however, it was also frequently used in conjunction with the words *terrae* or *terrarum* to denote the earth's central landmass. References to the *orbis terrarum* are common throughout the classical cannon.[85] The elder Pliny gave one of the clearest ideas of what was meant by the *orbis terrarum* when he said, 'We talk about the orb (*orbem*) of the lands with certainty and we admit that the globe (*globum*) is shut in by poles,'[86]

82 W. G. L. Randles, *Classical Models*, 5–76.

83 a) an object in the form of a plate or disc; b) wheel; c) a kind of fish; d) a circle or round plane figure; e) sphere/globe; f) ring-shaped object; g) a circuit, ring or circumference; h) orbis terrarum – the central land mass surrounded by Ocean, consisting of Africa, Asia and Europe; sometimes also the world; i) a part of the world; j) a climatic or other zone or subdivision of the earth conceived as a sphere; k) dry land, earth (as opposed to sky, heaven); l) rotation, revolution, cycle; m) the orbit of a heavenly body. *OLD*, 1263–1264. The dictionary gives three further definitions but I have omitted these as being very similar to the ones above.

84 Cicero, as Macrobius pointed out, used it in both contexts: 'For he used the term 'orb' both of a circle, as the Milky Way, and of a sphere, as in the case of your nine orbs or rather globes': 'Nam et orbem pro circulo posuit, orbem lacteum, et orbem pro sphaera, ut novem tibi orbibus vel potius globis.' Macrobius, *Commentarii in Ciceronis Somnium Scipionis*, 1.14. These nine globes are: the globe of the heavens, the seven planets, and the Earth (Cic., Somn. Scip., ch. 9). Mela, *De Chorographia*, 1.1. Mela implied, at least initially, that he was going to discuss the whole earth as the *orbis*, referring not only to the known world, but also the *antichthones*, the counter-world, which must for reasons of balance exist in the southern hemisphere. Mela was not, however, happy discussing the world beyond the realm of empirical knowledge and seems uncertain whether there is a second *oikoumene*, and if there is one, whether it is inhabited (Mela, 1.54). For a more extensive discussion of Mela's ideas about the inhabitability of the southern regions, see F. Romer, introduction to Mela's *Description of the World*, 7. Mela's book is sometimes called *De Chorographia* and sometimes called *De Situ Orbis*.

85 Cicero refers repeatedly to the orb of lands although he never clearly defines it. See for instance, Cicero, *de Lege Agraria*, speech 2, sections 33 and 37; *Epistulae ad Atticum*, book 1, letter 19, section 7; *Epistulae ad Brutum* book 1, letter 16, section 2; *Epistulae ad Familiares*, book 2, letter 16, section 4; *in Catilinam*, speech 1, section 9 and speech 4, section 11; Cicero, *Philippicae*, speech 3, section 35; *pro Murena*, section 74; *pro S. Roscio Amerino*, section 103; *pro Sestio*, sections 67 and 129. His contemporary Cornelius Nepos makes it clear that Rome was the capital and greatest city of this landmass. Cornelius Nepos, *Vitae Att*, 3.3 and chapter 20.5

86 'Orbem certe dicimus terrae, globumque verticibus includi fatemur'. Pliny, *NH*, 2.64.160.

and elsewhere: 'The whole orb (*orbis*) of the lands (*terrarum*) is divided into three parts, Europe, Asia and Africa.'[87] Many scholars have conflated the *orbis terrarum* with the Greek *oikoumene* and the Latin phrase certainly could be used to translate the Greek term – one need only look at Avienus's translation of Dionysius Periegetes' Οικουμένης περιήγησις as *Descriptio Orbis Terrarum* to see evidence of such usage, but the terms were not entirely synonymous.[88] Pliny's *orbis terrarium*, for instance, made no reference to inhabitability but rather to land. It was linked to the physical geography rather than to human inhabitability. When Roman authors really wanted to emphasise the inhabitability of the world, they used the words *habitabilis* or *colitur* in conjunction with *tellus* or *terra.*[89] Martyr's use of the term therefore related to a physical new world much more than to a new *oikoumene.*

The term *orbis terrarum* denoted the physical landmass of Europe, Asia and Africa, but the term *orbis* could be used to refer to other notional landmasses. Its emphasis was physical separation. Pliny talked about the *orbis terrarum* but he also used the term *orbis* in a different sense when he discussed Taprobana (modern-day Sumatra), saying that it is believed to be 'another orb' (*alterum orbem*) and is 'called the Anticthones' (*Anticthonum apellatione*).[90] Similarly, Ovid spoke of riches being brought back from the eastern orb.[91] In the passage of Seneca's *Suasoriae* (discussed in chapter four), the helmsman's response to being ordered to sail out on Ocean was: 'Do we seek races who have been placed beyond under another pole and another world (*orbem*)…? The gods call us back and forbid mortal eyes to know the end of things.'[92] This passage truly invokes the sense of an entirely separate space than either Pliny or Ovid's references to another orb. Seneca's orb lay beyond the abyss, beyond the limits of location that the gods had imposed upon humans. It had no counterpart in the known world, for it lay at the end of things.

In the sixteenth century, such a separation was contrary to the new image

87 'Terrarum orbis universus in tres dividitur partes, Europam Asiam Africam.' *Ibid.*, 3.1.3.

88 On the link between *orbis terrarum* and *oikoumene,* see Friedrich Gisinger: *Oikumene*, in *Paulys Realencyclopädie der classischen Altertumswissenschaft* (RE), vol. XVII.2, Stuttgart 1937, columns 2123–2174. See also J. Vogt, *Orbis* (Freiburg: Herder, 1960), 153–71.

89 Horace, *Carmina*, 4.14; Statius, *Thebais,* 11. See M. Y. Myers, 'The Frontiers of the Empire and the Edges of the World in the Augustan Poetic Imaginary' (Stanford University, 2008), 10.

90 Pliny, *NH,* 6.23–24

91. Ovid, *Fasti*, 3.4666.

92 'anne alio positas ultra sub cardine gentes/ atque alium flabris intactum quaerimus orbem?/ di revocant rerumque vetant cognoscere finem/mortales oculos'. Seneca, *Suasoriae* 1.15.26–29.

of the world. The Atlantic divided the Americas from Europe, it is true; but the *novus orbis* was part of one and the same world as the old *oikoumene*. When Peter Martyr introduced the term *novus orbis* for the western continent, he seemingly implied a separation from the rest of the world, but he recorded the size, the people and the settings of the *Novus Orbis* in relation to other parts of the globe. The arrangement of land and sea and the hidden harmonies of the world created by God for humans could only be interpreted and understood by humans because of its very unity. Patterns of balance and equilibrium, of symmetry and recompense could only be seen when the geographer or explorer stood back and looked at the world as a whole. Throughout the sixteenth and seventeenth centuries, authors of every nationality, from the German Grynaeus to the Spaniard Herrera y Tordesillas, referred to a *novus orbis*.[93] It was a term, however, which was only popular in Latin, where its links with the idea of *orbis terrarum* and the relationship that it had with the Greek term *oikoumene* were apparent. Even the near-contemporary translations of Martyr erase the emphasis on a distinct, separate entity that was so much a part of classical terminology of the word *orbis*, and draw attention to the idea of it as a whole new system instead. Eden (c. 1520–1576) translated Martyr's book into English as the *Decades of the New World*;[94] and likewise the German translator uses the term world rather than orb, entitling the book *Die new Welt der Landschaften und Insulen.*[95]

In many ways, the Latin term *mundus novus,* which was also applied to the Americas throughout the sixteenth century, was a far more apt term for the new regions than was *orbis*, for it contained much more potential for a human geographical angle.[96] The Latin term *mundus* is a strange term. In classical Latin, as we have seen, it was sometimes used as the equivalent of the Greek *kosmos*, which in its origins literally meant order.[97] *Mundus* might also be used in this sense, implying a universe that worked in accordance with laws of nature. Similarly, in medieval Latin the word tended to mean cosmos or universe, and by transference the earth and its inhabitants.[98] Before the discovery of the Americas,

93 Grynaeus, *Novus Orbis*; A. de Herrera y Tordesillas and C. Barlaeus, *Novus Orbis, Sive Descriptio Indiae Occidentalis...Metaphraste C. Barlaeo. Accesserunt & Aliorum... Descriptiones, & Navigationis...Australis Jacobi Le Maire...& Navigationum Omnium Per Fretum Magellanicum...Narratio* (Amsterdam: 'apud Michaelem Colinium', 1622).

94 Martyr, *Decades of the New Worlde.*

95 S. Grynaeus, *Die new Welt, Der Landschaften unnd Insulen, so bis hie her allen Altweltbeschrybern unbekant*, trans. Michael Herr (Strasbourg: G. Ulricher, 1534).

96 Perhaps most famously it was used in the 1504 Latin publication of Vespucci's letters after which it became an increasingly popular term.

97 H. G. Liddell, R. Scott and H. S. Jones (eds), *An Intermediate Greek Lexicon* (7th edition, Oxford, 1992), 446.

98 Cf. Cicero, *Univ.* 10; *De natura deorum* 3.11.27.

writers had already used the term to emphasise the social orders of newly discovered territories. For instance, Jordanus Catalani, a Dominican writing on the wonders of the east about 1329, had stated: 'everything in this India is marvellous, for it is in truth another world (*alter mundus*).'[99] It had a relevance in the context of a new *oikoumene*, having social and territorial aspects rather than simply special ones. Indeed the term *novus mundus*, or its vernacular equivalent, was not restricted in its application to those lands which are now called the New World. Sub-Saharan Africa was just as much a new world as the Americas, and as early as 1507 Montalboddo, in his compilation on the new 'discoveries' of the previous century, included the statement that southern Ethiopia was 'another world (*mondo*) if one could name it'.[100] Although, unlike the Americas, it was part of the same landmass as the old *oikoumene*, it lay beyond the boundaries of that enclosed world, in the region of the southern *oikoumene*, according to the Cratesian/ Macrobian model. It belonged in some sense to a different world. There were a few such examples, but geographers tended to apply the vocabulary of *mundus* or *orbis* much more rarely to the southern landmass or to the new regions of the old continents. *Terra Australis* might be a separate continent, but much of the landmass of the southern hemisphere was simply a continuation of lands of the northern hemisphere. Once people had ceased to believe in the torrid zone, there was nothing to separate southern Africa and Asia from the Africa and Asia of the northern hemisphere. The southern hemisphere was an unexplored region, but not a separate world. By contrast, even the geographical situation of the Americas separated them from the Old World. They really were the *Novus Mundus,* the new world, the complementary society where a whole new order of things existed.

The western landmass straddled all the climatic zones. It was known to have its own people, plants and animals. People saw in it a western counterweight to the known lands, it is true; but it was much more than that. It was a counterpart. Despite certain references to the Americas as a *novus orbis*, they were not a separate *oikoumene*. They were part of the one globe, and while on one level they were subject to different conditions, on another they were very much part of the one cosmos. Sixteenth-century Europeans knew that they could expect to find similar climatic and geographic conditions to their native city when on a similar latitude in the New World. There were no barriers of Ocean or torrid zone dividing

99 'Mirabilia sunt omnia in ista Yndia; est enim, vere, unus alter mundus.' J. C. de Sévérac, 'Mirabilia Descripta, Les Merveilles De L'asie', ed. by H. Cordier (Paris: Geuthner, 1925), 118.

100 'un altro mondo se poteria chiamare': 'El primo che ha trovato la navigation per l'Oceano verso mezodi', F. da Montalboddo, *Paesi novamente retrovati* (Vicenza: Henrico Vicentin, 1507), chapter 1, book 1.

the one from the other. Instead they were two systems that complemented one another, and, in Renaissance terms, were all knowable to the one humanity.

With Nova Zembla, Terra Australis, the New World and the Old, the Renaissance geographers had recreated a balanced world that worked on the same principles as the Macrobian four-continent geography. The locations bear little resemblance to those of the classical writers. They shifted round the globe to fit the known locations of the newly discovered territories, and the consequent empty spaces of Ocean. Not all geographers discussed all these landmasses. Some of the more empirically minded, such as Münster, focused only on the discovered lands, and while using the term *novus orbis* did not dwell on the New World much at all. In such writers, an idea of balance and counterpart is less apparent. The overall impression from the written geographies, however, is that the idea of a world made in proportion and balance was core to sixteenth-century geographical understanding. Theoretical and philosophical ideas of balance and opposition which resonated through ancient geography, finding expression in the ideas about the physical structure of the world, were given new life in the geography of the sixteenth century. They had a Christian context in the idea of a God-designed world purposefully created in a balanced arrangement between land and sea. This was alien to the classical texts but fitted well with the Renaissance idea of a world designed by God, in which all parts could be in communication with one another. Ideas of opposites and counterparts, of balance and proportionality, were implicit in all ancient approaches to the structure of the world, and still in the sixteenth century, they enabled cartographers and geographers to map out and discuss the shape of the world outside the regions of the discovered territories and show how it was designed to be an interconnected global space in which all regions were subject to human dominion and were accordingly part of the *oikoumene*.

6

A Parallel World: *Harmonia Mundi*, Connection and Separation in the Western Continent

Although Renaissance humanist geographical writers did not adhere strictly to the tenets of any one ancient school of philosophy, they were strongly influenced by Neoplatonism, and in particular by one Platonic dialogue: the *Timaeus*. In the first half of the sixteenth century, when the Americas were less completely absorbed into European geographical consciousness, there are fewer references to Plato, but in the latter half of the century writers increasingly made use of his work to interpret the nature and structure of the world. It is not surprising that in a society where people were scouring the classics to interpret better the geography of the New World, many humanist trained writers from all over Europe, including Ramusio, Hakluyt, Gomara, and Ortelius, came to equate the New World with Plato's Atlantis, first mentioned in the *Timaeus*. Yet the use of Plato went beyond mere reference to the Atlantis myth. Particularly when writing about the western hemisphere, many writers used Platonic ideas of *harmonia mundi* to explain the nature of the world, and in particular the relationship between the old *oikoumene* and the newly discovered continent. This chapter will examine the way in which sixteenth-century descriptive geographers reinterpreted the ideas of Atlantis, *harmonia mundi*, and the 'best possible world' to understand the New World, and its relationship to the Old.

For many of those Renaissance geographers seeking an understanding of the nature of the New World, Plato's Atlantis myth provided a source of information. Plato developed the story of Atlantis in his dialogues, the *Timaeus* and the *Critias*. The *Timaeus* was particularly influential since it was the only Platonic dialogue read throughout the Middle Ages.[1] In the Atlantic section, Plato tells the story

[1] J. Hankin, 'The Study of the Timaeus in Early Renaissance Italy', in *Natural Particulars: Nature and the Disciplines in Renaissance Europe,* ed. by Antony Grafton and Nancy Siraisi (Cambridge, MA: MIT Press, 1999), 88–89. Virtually every late Latin writer cited it. Gersh lists 119 extant manuscripts of Calcidius's Latin translation

of how, 9000 years earlier, soldiers from a large island called Atlantis – lying in the Atlantic to the west of the Pillars of Hercules – invaded Europe, Africa and Asia. A coalition of Greek armies resisted, but ultimately all drifted away, except the Athenians who were left alone to fight the invaders. The Athenians successfully drove them out of all their conquered territories and pursued them back to Atlantis, but at that time there was a sudden earthquake, in which the entire island (and everyone in it, including the pursuing Athenians) was swallowed up by the sea.

For humanist writers all over western Europe, Plato's Atlantis myth seemed to offer the best indication that the ancients had some knowledge of the Americas. The Spaniard Lopez de Gomara specifically cited the *Critias* and *Timaeus* as providing information on the history of Atlantis which he equated with the territory recently discovered and conquered by the Spanish.[2] After Gomara several others, ranging from the idiosyncratic Guillaume Postel to the hugely influential Ortelius, also referred to the Americas as Atlantis.[3] Ortelius indeed expended considerable effort in interpreting the world through the eye of history, and made special use of Plato in his interpretation.[4] Reference to Plato and the Atlantis myth was particularly common in England. John Dee frequently used Plato to further his understanding of the world, and likewise referred to the newly discovered western landmass as Atlantis. In the margins of his work *Concerning a New location of the Island of Estotitland and the Province of Drogio* (1577–78), for instance, he wrote: 'Note that Atlantis was called the New World about 200 years ago, but not by the Spanish but by the Frislandians.'[5] In the *Principal Navigations*, Richard Hakluyt the Younger, Richard Willes, and

of the *Timaeus,* forty-three of which are supplied with a commentary. S. Gersh, 'The Medieval Legacy from Ancient Platonism', in *The Platonic Tradition in the Middle Ages: A Doxographic Approach*, ed. by S. Gersh and J. F. M. Maarten (New York: de Gruyter, 2002), 9.

2 L. de Gomara, *La istoria* (Çaragoça: Agustín Millán, 1552), chapter CCXX. The *Istoria* was published in Spain in 1552 and later suppressed, but it was translated into English, French and Italian and became an international bestseller. As Roa-de-la-Carrera says, it became 'the most comprehensive and frequently cited treatment of the history and geography of the American territories colonised by Spain'. C. A. Roa-de-la-Carrera and S. Sessions, *Histories of Infamy*, 1.

3 G. Postel, *Cosmographicae Disciplinae Compendium* (Basil: Per Ioannem Oporinum, 1561), 32–33; A. Ortelius, *Abrahami Ortelli... Thesaurus Geographicus Recognitus Et Actus* (Antverpiae: ex officina Plantiniana, 1596), K4r.

4 J. Romm, *Mythes, cartes et histoire*, 104–16.

5 S. Leydi, *Sub Umbra Imperialis Aquilae: Immagini Del Potere E Consenso Politico Nella Milano Di Carlo V* (Firenze: Olschki, 1999), 3; J. Dee, 'Concerning a New Location of the Island of Estotitland and the Province of Drogio', in *The Limits of the British Empire*, ed. by Ken Macmillan and Jennifer Abeles (Westport, CT: Praefer, 2004), 38.

Humphrey Gilbert all refer to the *Timaeus* by name as a source of information on Atlantis.[6]

Several of these writers gave Plato pride of place above all the other classical authorities. Sir Humphrey Gilbert (1539–1583) ignored most of the more geographical classical works, but founded his arguments for the existence of a northwest passage on the *Timaeus*, explicitly identifying the Americas with Atlantis.[7] He drew upon Plato and upon commentators on Plato, including Philo, Proclus and Ficino, to explain why Atlantis was 'overflowen and swallowed up' so that it had become much further separated from Europe and had faded from European knowledge, saying

> that in these our dayes there can be no other mayne or Islande found or judged to bee parcell of this Atlantis, then those Westerne Islands, which beare now the name of America: countervailing thereby the name of Atlantis, in the knowledge of our age.[8]

Many of these writers were uninterested in any philosophical context within the dialogue, but focussed on the Atlantis story as a straightforward geographical treatise. Gomara, for instance, paid no attention to any philosophical argument at all. Gilbert was clearly aware of the philosophical arguments of the *Timaeus*, and cited authors who were concerned with theoretical arguments much more than with tethering the Atlantis story to reality, but he was not really interested in these theoretical debates. He simply stated that because Plato had said it, and subsequent writers had agreed, then Atlantis must be an island. As Richard Willes said, in another of the lengthy English discussions on a northern passage written in 1577, Europe, Asia and Africa together were surrounded by water, as was Atlantis.[9] If this were the case, then clearly the Americas must be surrounded by Ocean, and a northwest passage must therefore exist. In identifying the Americas with Atlantis, he was associating them with a region that was known not only to have been inhabited, but indeed to have had a great civilisation. Therefore the newly discovered regions could not be part of the Old World, but formed a second entirely disconnected *oikoumene* which counterbalanced the Old World. Gomara, Gilbert and Willes were all part of a group who believed that Plato had specific knowledge of the New World. Others did not go quite so far. Ramusio, for instance, began his third volume with a *Discourse* in which he made repeated

6 R. Hakluyt, 'Epistle Dedicatorie', *PN*, vol. 1, lxxv; H. Gilbert, *Discourse to Prove Passage*, *PN*, vol. 7, 160; R. Willes, 'Certaine Reasons', *PN*, vol. 7, 199.

7 M. Small, 'From Thought to Action: Gilbert, Davis, and Dee's Theories Behind the Search for the Northwest Passage', *The Sixteenth Century Journal*, 44, no. 4 (2013).

8 H. Gilbert, 'Discourse to Prove Passage', *PN*, vol. 7, 158.

9 R. Willes, 'Certaine Reasons', *PN*, vol. 7, 197.

reference to Plato and the *Timaeus*.[10] He did not think, however, that Plato had precise knowledge of the Americas, but rather that he inferred their existence correctly through logic and reason. This same kind of logic which made Plato postulate the existence of Atlantis, could be used to deduce the nature of the New World as a counterpart to the Old.

This idea of a counterpart drew on ideas of correspondences and hidden affinities between different parts of the globe. As we have seen in the previous chapter, there were those who believed that the world was maintained in an equilibrium derived from a dynamic tension in nature. And there were others who believed that this equilibrium was created by a harmony and proportionality between all elements. But ultimately these thinkers differed very little in their belief that there had to be an equal distribution of lands across the surface of the earth. When theorists took the idea of harmony and its implications for geography beyond the distribution of lands and seas, and looked for its application to the whole world, and in particular for hidden affinities between apparently disparate elements, the differences were greater. Geography based on ideas of *harmonia mundi* and hidden affinities became an important way of understanding the world, and in particular the relationship between the Old World and the New in the sixteenth century. The single most influential source for this idea of *harmonia mundi* was Plato's *Timaeus*.

In the *Timaeus,* Plato developed the Pythagorean idea of a *cosmos*, an ordered universe founded on harmony, making his title character and main discussant Timaeus, a Pythagorean from Locri, who expounded ideas of numerical ratios and harmonic proportions.[11] In Pythagorean thought, the order of the universe was expressed by the musical harmony of the spheres in which the harmony was produced by numerical relationships.[12] These ratios could be either evident or

[10] G. B. Ramusio, 'Discorso Sopra Il Terzo Volume', *NV*, vol. 5, 3–18.

[11] The interpreters of the *Timaeus* in the Renaissance were perhaps even more influenced by Pythagoreanism than Plato himself. Marsilio Ficino went so far as to classify it as a Pythagorean text, believing that Plato's source for the dialogue was a Neopythagorean text, Περὶ φύσιος κόσμω καὶ ψυχᾶς, purportedly written by Timaeus of Locri although in fact a first-century AD forgery. J. Hankin, 'Galileo, Ficino and Renaissance Platonism', in *Humanism and Early Modern Philosophy*, ed. by Jill Kraye and M. W. F. Stone (London: Routledge, 2000), 213. According to Ficino, Timaeus wrote a book, *De Universi Natura.* M. Ficino, 'Compendium Marsilii Ficini in Timaeum', in *Omnia Divini Platonis Opera, Tralatione M. Ficini, Emendatione et ad Graecum Codicem Collatione S. Grynæi*, ed. by Simon Grynaeus and Marsilio Ficino (Basileae: In officina Frobeniana, 1546), 672.

[12] The concept of the universe (in Latin, *mundus*) singing in harmony is alluded to by Cicero in *De natura deorum*, 3.11.27: 'nisi ... putamus .... ad harmoniam canere mundum, ut Pythagoras existimat.' The precise phrase *harmonia mundi* seems to have been coined by Francesco Giorgio F. Giorgio, *Francisci Georgii ... De Harmonia Mundi Totius Cantica Tria* (Venetiis: Bernardini de Vitalibus, 1525).

concealed, but everything in the universe was contained within these proportional relationships. Every extreme was linked in a harmonious relationship with its opposite. Plato used this idea of harmonious relationship to show that design and proportion were inherent in the world.[13]

The *Timaeus* shows how *harmonia mundi* was based on mathematical and geometrical relationships, which were known through reason (λόγος) and not through the 'unreasoning perception of the senses (αισθήσις ἀλογος)'.[14] Nevertheless, these relationships were reflected in the visible world. As Proclus, the fifth-century AD Neoplatonist points out, the dialogue is not simply concerned with mathematics, but also with physical theory. To explain the relevance of them both, 'it is necessary to tie both together and always to combine the physical with the mathematical, just as matters themselves are combined and are homogeneous and related according to the progression'.[15] The physical aspects of the dialogue cannot be overlooked and were given significant emphasis in the sixteenth century.

Harmonia mundi was an application and rationalisation of the apparent order in the world. It tied in with ideas of balance and symmetry. There was a harmony of the spheres, just as there were direct if hidden links between the primitive, pristine cities of the New World and the great city-states of Europe. Sixteenth-century geographers looked beyond simple mathematical proportionality and balance for absolute correspondences between the Old World and the New. Since they thought that hidden affinities linked the New World and the Old, they saw the Americas as the counterpart of the old *oikoumene*. Both landmasses reached vast distances across all the climatic zones, and contained an astonishing variety of peoples, animals and terrains. Certain Renaissance geographers and explorers turned to the Neoplatonic concept of *harmnonia mundi* to support the idea that the New World was a whole counter-world, which echoed the Old World not only in structure, but also in nature. *Harmonia mundi* was not, however, a term

13 A. C. Crombie, *Styles of Scientific Thinking in the European Tradition: The History of Argument and Explanation Especially in the Mathematical and Biomedical Sciences and Arts* (London: Duckworth, 1994), 117. The Platonic concept of *harmonia mundi* was so deeply indebted to Pythagoreanism that it ensured that 'the Pythagorean concept of world harmony was revived in modern civilization whenever Platonism was revived'. L. Spitzer and A. G. Hatcher, *Classical and Christian Ideas of World Harmony: Prolegomena to an Interpretation of the Word 'Stimmung'* (Baltimore: Johns Hopkins Press, 1963), 3.

14 Plato, *Timaeus,* 28a.

15 ἀλλὰ δεῖ συνάπτειν ἀμφότερα καὶ συμπλέσειν ἀεὶ τὰ φυσικὰ τοις μαθηματικοῖς, ὥσπερ καὶ αυτὰ τὰ πραγματα συμπλέχεται καὶ ἒστιν ὁμογενη καὶ ἀδελφὰ κατὰ τὴν ἀπὸ πρόοδον Proclus, *In Platonis Timaeus Commentaria*, 23.9.3 (Γ, *Tim*.31C 32A); see discussion in J. A. Coulter, *The Literary Microcosm: Theories of Interpretation of the Later Neoplatonists* (Leiden: Brill, 1976), 86–87.

widely used by sixteenth-century geographers since in the Renaissance it was frequently associated with the more pagan, esoteric aspects of Neoplatonic thought.[16] Ramusio, for example, was perfectly aware of this association with the occult, but saw the importance of harmony for understanding the world in spite of his inclination to tether all his geographical ideas to empirical proof. In the letter dedicating his first volume to Fracastoro, he wrote about how the diligence of his contemporary scholars had also brought many new things to light, saying specifically that philosophy had taught people to think of many things as occult, which they were now able to discern in many matters. One such thing is the 'natural harmony and discord which we see to be in many things'.[17] He was clearly not taking a purely Neoplatonic approach, as he discusses discord as well as harmony, but his language draws explicitly on ideas of hidden affinities and harmonies. He was aligning himself with Neoplatonic thought. In the *Discorso sopra il terzo volume*, he was even more explicit, referring directly to Plato and the *Timaeus*. He set the stage for his discussion by talking about what Plato had meant precisely when he spoke of Atlantis, observing that some thought it was simply an allegory showing the oppositions which exist in the universe, or the more specific opposition between earth and heaven, while what Plato was really writing about was the *struttura del mondo*.[18] Nor was he the only scholar to refer to the geographical importance of *harmonia mundi.* The concept is found in the works of those trained in Platonist and Neoplatonist thought, including some of the leading geographical and ethnographical authorities throughout Western Europe. John Dee has already been mentioned. Another was Louis Le Roy in France who made the first translation of the *Timaeus* into French in 1563, but also wrote a geographically oriented work, *De la vicissitude ou variété des choses*. As I have argued elsewhere, explorers such as Sir Humphrey Gilbert and John Davis based their ideas, in part, on their interpretation of Plato.[19] In short, the degree to which ideas of proportionality and *harmonia mundi* pervaded European thought at the time gave a philosophical foundation to geography even among some of the most empirical scholars.

Several modern scholars have referred to the idea of *harmonia mundi* in articles on sixteenth-century geographical works, but none of them has provided any philosophical definition of the term, which they have conflated with the less

[16] 'Car observans par icelles les merveilles és secrets du monde, au sein de nature, et mysteres de Dieu, ils ont decouvert la concorde de l'univers et convenance du ciel avec la terre... Ce qui a meu Plotin d'appeller tels Mages faisans profession de la Magie naturelle, ministres de nature. L. le Roy, *Vicissitude*, 45r.

[17] 'come è della concordia e discordia naturale, che in molte cose esser veggiamo'; G. B. Ramusio, 'All'eccellentiss. M. Ieronimo Fracastoro', *NV*, vol. 1, 4.

[18] G. B. Ramusio, 'Discorso Sopra Il Terzo Volume', *NV*, vol. 5, 6.

[19] M. Small, 'From Thought to Action'.

specific concepts of balance and reciprocity.[20] Where they have linked it with Platonic philosophical ideas, it has been to focus on one aspect, plenitude, or as Lovejoy defines it, the 'thesis that the universe is a *plenum formarum* in which the range of conceivable diversity of *kinds* of living things is exhaustively exemplified'.[21] Yet the idea that the universe contained infinite variety was not the focus of sixteenth-century geographical thought. *Harmonia mundi* was used, rather, to order and understand this infinite variety through looking for the underlying affinities and proportions.

Why the concept of *harmonia mundi* became so strongly associated with geography was that many people saw in it a key to the secrets of nature and, by implication, to understanding the world.[22] It was by no means a doctrine relegated to the fringes of society. In Renaissance Italy it provided a fundamental approach to a wide variety of subjects. Marsilio Ficino and Francesco Giorgi, for instance, saw harmony as fundamental to the structure of the universe.[23] The earth as part of the universe reflected this same order. Architects saw the order and proportion of the cosmos as a model for their own work: just as the divine harmonies were fundamental in the perfection of the edifice called the world, so harmony was to be striven for in human creations.[24] Others saw echoes of celestial harmonies in

20 J. M. Headley, 'The Sixteenth-Century Celebration of the Earth's Total Habitability: The Issue of the Fully Habitable World for Renaissance Europe', *The Journal of World History*, 8 (1997), 4–18; K. Seaver, 'Norumbega and Harmonia Mundi in Sixteenth-Century Cartography', *Imago Mundi: The International Journal for the History of Cartography*, 50 (1998), 34–58; D. Cosgrove, 'Mapping New Worlds', 65–89.; D. E. Cosgrove, *Apollo's Eye*, 114–23 and 39–75.

21 A. O. Lovejoy, *The Great Chain of Being: A Study of the History of an Idea* (Cambridge, MA: Harvard University Press, 1936), 52.

22 For a detailed study of such ideas, see F. Yates, *The Occult Philosophy in Elizabethan England* (London: Routledge and Kegan Paul, 1979).

23 'Est autem apud Platonicos interpretes divina musica duplex: alteram profecto in eterna Dei mente consistere arbitrantur, alteram vero in celorum ordine ac motibus, qua mirabilem quendam celestes globi orbesque concentum efficient.' M. Ficino, 'De Divino Furore, Lettere I', in *Epistolarum Familiarum Liber I*, ed. by S. Gentile (Florence: 1990, first published 1495), 24. For more on Ficino and musical harmony, see W. R. Bowen, 'Ficino's Analysis of Musical Harmonia', in *Ficino and Renaissance Neoplatonism*, ed. by Konrad Eisenbichler and Olga Zorzi Puglies (Ottowa: Dovehouse Editions, 1986), 18–28; F. Giorgio, *De Harmonia Mundi*, Proemio. This idea pervades the *Harmonia Mundi*, but can already be found clearly in the Proemio.

24 'Con pari ragione, dunque fù ordinato nella nostra Republica il Senato, e'l consiglio de Dieci; iquali nella Città di Vinegia (la cui Republica dissi essere misto di stato Regio, popolare, & nobile) rappresentano lo stato de nobili: & somo certi mezzi co iquali le estreme parti, ciò è lo stato popolare, il gran Consiglio, e'l Prencipe...). Cosi dice Platone nel Timeo, che gli estremi elementi, la terra, e'l fuoco, co gli elementi di mezzo si congiungono, e legano.' G. Contarini, *La Republica, Ei Magistrati Di Vinegia* (Venice: Girolamo Scotto, 1544), XXXIII. M. Tafuri, *Venice and the Renaissance*, trans. by Jessica Levine (Cambridge MA: MIT Press, 1989), 122. See also R.

human societies, so that Gasparo Contarini drew directly upon the *Timaeus* to demonstrate parallels between the government of the Venetian Republic and the harmony of the celestial spheres.[25]

At the same time, the idea of *harmonia mundi*, based on ideas of geometrical proportion, was given a direct application to the physical world. Fra Luca Paccioli, writing in 1494, for instance, took a strongly Pythagorean-Platonic view of the universe, speaking of the way in which geometry and proportion pervaded everything. Paccioli was an educated Franciscan, famous now as the populariser of double-entry book-keeping, but also known for his interest in mathematics and proportion in general, and his association with artists such as Alberti and Da Vinci. Geometrical proportion pervaded his whole world view and he singled out cosmography in particular to show its role, commenting:

> Not all the ancient authors demonstrate to us what we shall say about cosmography [But] Eratosthenes, Strabo, Marinus, Ptolemy and the other excellent cosmometers [demonstrate] how much number, measure and proportion are necessary for it [i.e. cosmography] when out of the whole universal world they have reduced provinces, cities, castles, and both maritime and inland sites, duly proportioning their co-ordinates in a small map.[26]

Paccioli's statement is fascinating, for it implies that the idea of making a scale map was dependent upon a theory of proportion in the universe. The entire development of a longitude and latitude grid system relied upon a notion of proportion. Once such a system was in place then anything within the world could be scaled according to that system. Paccioli seems to attribute a belief in this idea of proportion to ancient geographers and cosmographers. He felt that numbers and arithmetic were usable tools to provide an accurate representation of the world in microcosm. He uses a strange terminology, 'cosmometers', which underscores this idea of measurement and proportion. He calls *cosmometers*

Wittkower and I. Warburg, *Architectural Principles in the Age of Humanism* (1949), 89–135.

25 M. Tafuri, *Venice*, 108. Contarin, *La Republica*, XXXII.

26 'Che diremo della cosmographia non ci demonstran tutti li antichi Erathostene, Strabone, Marino, Ptholomeo e gli altri excellenti cosmometri, quanto lisia necessario el numero la misura e la proportione quando de tutto lo universo mondo debitamente proportionando lor gradi in una piccola carta provincie, cita, castelli e siti marittimi, e mediterranei hanno redatto.' L. a. Pacioli, *SūMa De Arithmetica, Geometria, Proportioni & Proportionalita* (Venice: Paganino de Paganini, 1494), Introduction.** Alternatively the passage might read 'What should we say about cosmography? Do not all the ancients demonstrate how necessary are number, measure and proportion, when out of the whole world…?' Either way, the passage emphasises the importance of number and proportionality. Interestingly, the passage was omitted from the 2nd edition of the text, produced in 1523.

those geographers who were as uninterested in the *kosmos* even as Strabo was, for the way in which they divided and measured the world. The apportioning of the world, in Paccioli's eyes, was not separable from the grand scale of proportion in the universe.

Harmonia mundi went far beyond general ideas of order and geometrical proportion to have strong implications for geography. Many Neoplatonists thought that the universe, and the physical world within it, consisted of 'a web of hidden affinities originating in the world soul and other cosmic souls'.[27] Girolamo Fracastoro, the Veronese physician and leading humanist with whom Ramusio discussed geography, and to whom he dedicated the *Navigazioni e viaggi* in his dialogue *On Sympathy and Antipathy*, asserted that the whole universe was governed by a *latens rerum consensus*, a hidden harmony of things which, working within the laws of nature, governed the universe and ensured that it stayed in equilibrium.[28] In the universe as a whole, this hidden harmony ensured a balance between all contradictory forces which kept everything in its proper place. Fracastoro was not writing in a specifically geographical context, but was studying the idea of hidden harmonies as they worked in all nature. Nonetheless, he saw place as a part of the harmony. He devoted a large section of the dialogue to showing how different, and indeed opposing, elements all had their place in the universe, and argued that

> just as the parts of an animal must have no small harmony and relationship among themselves, and must demand certain positions within the animal, so indeed in the universe... the parts demand in turn their own compatible locations, otherwise the whole universe would not be duly ordered.[29]

As Fracastoro states, harmony was the means of understanding the world order. Yet it was not merely in terms of location, but also in the very character of the different regions of the world that Renaissance geographers saw the implications of *harmonia mundi*. Its hidden affinities affected all aspects of the relationship between the Old World and the New.

Europeans who knew little about the newly discovered lands could apply the ancient theory to understand them. As Kirstin Seaver has said, 'accordingly,

27 P. O. Kristeller, *Renaissance Thought*, 52. Kristeller was talking specifically about Plotinus, but Plotinus himself influenced many later scholars, and this idea of hidden affinities is not restricted to his work.

28 G. Fracastoro, 'De Sympathia et Antipathia Liber I', in *Opera Omnia* (Venetiis: Giunti Press, 1555), 79r. See S. Pearce, 'Nature and Supernature in the Dialogues of Girolamo Fracastoro', *The Sixteenth Century Journal*, 27, no. 1 (1996), 114–16.

29 'Sicut enim in animali partes inter se consensus et relationem non parva habent, et in eo certos exposcunt situs, ita et in universo... partes eius situs invicem consentientes expostulant, alioquin universum ipsum debite constitum non erit.' G. Fracastoro, *De Sympathia*, 80v.

when Europeans heard about the New World, they anticipated that it would contain at least an embryonic counterpart of features of the geography and human behaviours of the Old World'.[30] Old World Venetians became fascinated with Tenochtitlan, seeing in it the New World complement of their own city in a more primitive stage of development.[31] Certainly, as Cosgrove has pointed out, they placed a disproportionate emphasis upon the Aztec city in their cartography of the New World.[32] In a letter to Alvise Cornaro, Fracastoro made an explicit comparison between 'Temestitan' and Venice.[33] Nor were Italians the only ones to draw comparisons between the two cities. The Spanish explorer Ribera, according to Peter Martyr, had reported that Tenochtitlan was situated on a rock in the middle of salt water, just as we read of Venice. The marginal notes in the translation of Martyr by the Englishman Michael Lok, an educated London mercer and one of the leading experts in New World geography at the time, make an even clearer comparison, referring explicitly to 'Temistitan situated as Venice'.[34] A certain similarity between Venice and Tenochtitlan was obvious given that both were built on water, but Martyr thought the affinities went far deeper, existing also in the man-made features of the cities. He noted that Ribera's maps showed that:

> To the North there are mountains distant from one another with very fruitful valleys separating them in turn through whose gorges the force of the North winds is strong in that plain. And so that side of that city of Tenochtitlan which faces towards the North is fortified with ramparts of fixed beams and great rocks, by which means the city is saved from the force of the whirlwind. I saw the same invented in Venice to withstand the raging of the Adriatic sea, lest it should shake the houses: The Venetians call that construction the shore, in common parlance, the lido.[35]

30 Seaver, 'Norumbega', 34.

31 For a discussion of the relative architectural structures of Tenochtitlan and Venice, see M. Tafuri, *Venice*, 152–53.

32 Cosgrove, 'Mapping', 83.

33 *Lettere di Girolamo Fracastoro sulle Lagune di Venezia*, cited in Tafuri, 152.

34 P. Martyr, *De Nouo Orbe, or the Historie of the West Indies ... Comprised in Eight Decades ... Three Formerly Translated into English, by R. Eden ... The Other Fiue by M. Lok*, trans. Michael Lok (London: Thomas Adams, 1612), Decade 5, 227r.

35 'A Borrea sunt montes alicubi distantes invicem vallibus uberrimis illos dirimentibus, per quarum fauces vis ventorum septentrionalium est in ea planicie valida, propterea latus illud civitatis Tenustitanae, quod tendit ad Boream, et fixarum trabium de saxorum ingentium latis aggeribus munitum, quo ab impetu turbinis civitas tuta sit. Id idem vidi ego Venetiis ad furorem Adriatici maris sustinendum, ne domos quasset, inventum: littus dicunt eam struem Veneti, vulgo el lido.' Martyr, *De Orbe*, 84v.

Venice was not the only region of the Old World for which people sought to find exact counterparts in the New. As Kirsten Seaver has explored in her article, 'Norumbega and "Harmonia Mundi" in Sixteenth-Century Cartography', the use of the term Norumbega for a region on the east coast of North America was directly linked to European geographers' ideas of *harmonia mundi.*[36] The corruption of the name of a North American region from Oranbega to Norombega was linked to geographical notions of *harmonia mundi*. Oranbega as a term first appeared in the cartographic rather than the geographic record. Giovanni Verrazzano, the first European to see the region, had simply called the territory Refugio, but by 1529, his brother Girolamo had recorded the name Oranbega for posterity, marking it on his map of the New World.[37] This name seemed to be derived from a native language.[38] Slowly, however, the name became corrupted, and by 1548 the Venetian cartographer Giacomo Gastaldi had explicitly labelled the region as 'Tierra de Nuremberg' on his map of the New World.[39] Just as Tenochtitlan in the New World corresponded to Venice, one of the great cities of the Old World, so Nuremburg, another important European city, might be expected to have its own New World counterpart.[40] Nuremburg had topographical features similar to some of those described by Verrazzano, thus reinforcing the parallels. People knew little about Norombega. Verrazzano had first recorded the place, but its location was not firmly set. It roved around the latitudes in the

36 K. Seaver, 'Norumbega'.

37 The name Refugio is in fact only given in the *Cèllere Codex*, a manuscript version of the text produced in 1524. See Wroth, *Voyages of Verrazzano*, 140. The main published edition produced by Ramusio mentions the same site, but leaves it nameless; see G. Verazzano, 'Al Cristianissimo re di Francia Francesco primo, relazione di Giovanni da Verrazzano Fiorentino della terra per lui scoperta in nome di sua Maestà, scritta in Dieppa, Adì 8 Luglio MDXXIIII', *NV*, vol. 6, 902. The first English version of the voyage, like the Ramusio version, gives no name within the body of the text. The text itself, however, carries the heading 'the discoverie of Morum Bega', clearly yet another corruption of the term Oranbega. This text was first translated into English in R. Hakluyt, 'Divers Voyages Touching the Discoverie of Ameria, and the Ilands Adiacent Unto the Same, Made First of All by Our Englishmen, and Afterward by the Frenchmen and Britons: And Certaine Nots of Adverstisements for Observations, Necessarie for Such as Shall Heereafter Make the Like Attempt, with Two Mappes Annexed Heereunto for the Plainer Understanding of the Whole Matter,' ed. by David B. Quinn (Amsterdam: Theatrum Orbis Terrarum, 1967; first published London: Thomas Woodcocke, 1582). Seaver, 'Norumbega', 35.

38 The map historian Ganong argues that it was an indigenous term for the region between Narragansett Bay and New York. W. F. Ganong, *Crucial Maps in the Early Cartography and Place-Nomenclature of the Atlantic Coast of Canada* (Toronto: University of Toronto Press, 1964), 132.

39 G. Gastaldi, 'Tierra Nueva', in *La Geografia Di Ptolemeo* (Venice: G. B. Pedrezano, 1548).

40 K. Seaver, 'Norumbega', 42–43.

travellers' reports, until eventually it became fixed at roughly the same latitude as Nuremberg, thus reemphasising the idea of *harmonia mundi*.[41]

By the end of the sixteenth century there ceased to be such avid searches for absolute parallels. Tenochtitlan was no longer the new Venice, and people ceased to find a complete counterpart for Nuremburg in the New World. The name became corrupted yet again. On Michael Lok's map of 1582, for example, the name Noranbega, clearly an amalgam of Oranbega and Nuremberg, was applied to an island lying off the east coast of America immediately south of Newfoundland. The complete equation of New World locations with Old World ones was one of the most evident of uses of *harmonia mundi.* It allowed the old to be mapped onto the new and allowed a means for Europeans to interpret the newly encountered but unexplored territories. Yet as the territories themselves became better known to Europeans, they ceased to seek direct correspondences as this proved an inadequate approach. They continued, however, to turn to Plato for geographical understanding. Many of those interested in the geographical implications of the *Timaeus* also picked up on the idea of the best possible world, which became an even more influential aspect of geographical discourse than the idea of hidden correspondences.

In particular, in the sixteenth century, writers placed emphasis on the idea of a divinely created world, made by God in accordance with rules of number and harmony, because this made the best possible world. In the *Timaeus*, Plato had described a divine creator (a *demiurgos*) and the sixteenth-century writers Christianised Plato's divine being, equating him with the Christian God. The two were in fact very different. Christians took the idea of a divine creator as a starting point, and worked forward from God to the ordered creation. Plato, in the *Timaeus*, by contrast worked backwards from the perceived order of creation to postulate the existence of a single divine being, a *demiurgos* who imposed his own order on the physical world.[42] This *demiurgos* was 'supremely good' and consequently his work could not be anything other than the best, since 'the creator's motive for creation is to make things as good as possible'.[43] Both Platonists and Christians, however, believed that the creator ensured that his creation was rational and not arbitrary. Plato argued that 'desiring, then, that all things should be good and, so far as might be, nothing imperfect, the *demiurgos* took over all that is visible – not at rest but in discordant and

41 *Ibid.*, 42 and D. B. Quinn, 'The Early Cartography of Maine in the Setting of Early European Exploration of New England and the Maritimes', in *American Beginnings: Exploration, Culture and Cartography in the Land of Norumbega*, ed. by Emerson W. Baker (Lincoln, NE: University of Nebraska Press, 1995), 45–49.

42 A. C. Crombie, *Styles of Scientific Thinking*, 119.

43 H. D. P. Lee, 'Introduction', in *Timaeus, trans. and intro. by H. D. P. Lee* (Harmondsworth: Penguin Books, 1965), 11.

unordered motion – and brought it from disorder into order, since he judged that order was in every way better (30a)'. Order therefore could be discerned in all aspects of creation.

Plato thought that for things to be as good as possible, there had to be an order to the world, for order is the closest copy of perfection.[44] The order, balance and proportionality of the universe and everything in it were seen by Pythagoreans and many Neoplatonists as the effect of number, for 'numbers contain and manifest the laws of God's universe'.[45] Neoplatonists writing in a Christian context frequently equated Plato's *demiurgos* with the Christian God. They found theological support for their belief in a world in which numbers were a key component in the book, the 'Wisdom of Solomon' (in the *Apocrypha*), where the writer declares that God created everything in 'number, weight and measure'.[46] Renaissance natural philosophers, theologians and those interested in geography, used this passage to demonstrate the importance of mathematics for understanding the natural world.[47] Nicholas Cusanus, for instance, wrote about the mathematical properties which God had given to the earth, saying:

> For he united it through arithmetic, he shaped it through geometry so that from this it would follow firm and stable movement according to its place, through music he proportioned it so that there would not be more earth in earth than water in water and air in air and fire in fire, so that none of the elements could

44 Plato of course used the Greek word *kosmos* rather than the Latin *mundus*, but the two meant essentially the same thing. See Plato, *Timaeus*, 29a: 'εἰ μὲν δὴ καλός ἐστιν ὅδε ὁ κόσμος ὅ τε δημιουργὸς ἀγαθός, δῆλον ὡς πρὸς τὸ ἀίδιον ἔβλεπεν· εἰ δὲ ὃ μηδ' εἰπειν τινὶ θέμις, πρὸς τὸ γεγονός.' 'If on the one hand the *kosmos* is beautiful and the *demiourgos* is good, clearly he [the demiourgos] looked towards the eternal. If not, which is impious for anyone to say, he looked towards the becoming. It is therefore entirely clear that he looked towards the eternal; for it is the most perfect of those things which have come to be and he is the best of causes.'

45 R. Waterfield, 'Introduction', in *The Theology of Arithmetic: On the Mystical, Mathematical and Cosmological Symbolism of the First Ten Numbers by Iamblichus* (Grand Rapids, MI: Phanes Press, 1988), 26.

46 'Wisdom of Solomon', 11.20. A. G. Debus, 'Introduction', in *Mathematicall Praeface to the Elements of Geometrie of Euclid of Megara (1570),* ed. by John Dee (New York: Science History Publications, 1975), 9. In equating the *demiurgos* with the Christian God, they were changing the text hugely, since for Plato he was simply a divine creator and not an object of worship.

47 St Augustine had first used this passage to link the Platonic concept of number found in the *Timaeus* with biblical ideas (*City of God*, 12.19), and thereafter a great many natural philosophers throughout the Middle Ages and early modern period continued to equate the two. C. B. Kaiser, *Towards a Theology of Scientific Endeavour: The Descent of Science* (Aldershot: Ashgate, 2007), 157.

be entirely dissolved into another. On account of which the machine of the world was not able to perish.[48]

The whole shape and makeup of the physical earth, the subject of geography, was formed by harmony and number. As John Dee wrote a century later, 'All thinges which from the very first originall being of thinges, have bene framed and made, do appeare to be formed by the reason of Numbers. For this was the principall example or pattern in the minde of the Creator.'[49]

Such thinkers were influenced by Renaissance and antiquarian Neoplatonism, which took the ideas of Plato's dialogues far beyond their original context.[50] In this case, they removed discussions of number and proportion from the theoretical level and made them relevant to the earth and therefore to the study of geography. There might be apparent disunity on the earth, but underneath there was mathematical order. In this sense, using number, balance and proportion, people could understand the physical structure of the whole world, even those parts of the world for which they had no eye-witness information. The influential fifth-century Neoplatonist exponents of the theory of four *oikoumenai*, Macrobius and Martianus Capella, developed physical geographies founded on ideas of number and proportionality, which clearly show the evidence of Platonic thought. They, like virtually all Neoplatonists, drew on Plato's *Timaeus* to place a strong emphasis on the importance of number and numerology for understanding the underlying order in the world, and the means by which the creator had made

[48] 'Per arithmeticam enim ipsa coadunavit; per geometriam figuravit, ut ex hoc consequerentur firmitatem et stabilitatem atque mobilitatem secundum contiones suas, per musicam proportionavit taliter, ut non plus terrae sit in terra quam aquae in aqua et aëris in aëre et ignis in igne, ut nullum elementorum in aliud sit penitus resolubile. Ex quo evenit mundi machinam perire non posse.' Nicholas of Cusa, *De Docta Ignorantia*, 2.13.

[49] J. Dee, *Mathematicall Praeface to the Elements of Geometrie of Euclid of Megara (1570),* ed. by Allen G. Debus (New York: Science History Publications, 1975), 1.

[50] Neoplatonism is a notoriously difficult subject to define. It is a term coined only in modern times, and has at least two applications. Some scholars use it to refer to later classical authors who considered themselves Platonists in that they believed that they did not uphold any philosophy which could not be discerned in Plato. The three most important of these classical Neoplatonists were Plotinus (b. 204 AD), his student Porphyry (b. 233 AD) and Proclus (b. 410 AD). Students of the Renaissance, on the other hand, often use the term to refer to the European-wide fifteenth- and sixteenth-century revival of Platonic thought exemplified by the Italian Renaissance philosophers Marsilio Ficino and Pico della Mirandola. Unlike the late classical Neoplatonists, the Renaissance ones used Plato as a starting point, but they did not believe that everything could be traced back to Plato. Therefore the two systems of thought are fundamentally separate. That said, the Renaissance philosophers were strongly influenced by the classical Neoplatonists. While Plato's own writings had all but disappeared in the Middle Ages, much Neoplatonic literature had continued to circulate.

order out of chaos.[51] In Capella's mind, the Platonic interconnection between number and geography is even more evident, for Capella's long excursus on geography forms the major part of his sixth book, the section ostensibly devoted to geometry. The emphasis of both writers on symmetry and proportionality in geography – shown in their belief in four equal and proportionately arranged landmasses with hidden affinities between them – is indicative of their debt to classical Platonism, and the ideas in the *Timaeus*. They demonstrate that the practice of seeking a reflection of Plato's world of ideas in the structure of the physical, tangible earth was not a new feature of the sixteenth century. It received a much wider application in that period, however, partly because Plato was undergoing his own revival, but also because the dialogue provided a means of supplementing empirical evidence with theoretical understanding. While the dialogue provided elements of apparent geographical detail, such as the Atlantis myth, it was particularly useful for providing theories about the nature and order of the universe which could be used to explain the physical structure of the world and the relationship between its component parts.[52] In particular, while explorations of the early modern period were providing repeated evidence for variety in the world, people trained in Platonic philosophy believed that 'a study of the multiple being in nature [could] only lead to our belief in an underlying unity pervading all things'.[53]

People had looked for similarities between Plato and Christianity for centuries, and continued to do so in the early modern period. Marsilio Ficino, who had done more than any other to make the Platonic corpus widely available in the Renaissance by translating it into Latin, was himself an ordained priest who stressed the essential harmony between Platonic philosophy and Christianity.[54] The idea of the best possible world took on a specific meaning for many sixteenth-century writers. According to the Bible, God created the world for humans.[55] While philosophers delved into the implications of the difference between things which come to be and things which are, thus complicating Plato's ideas, geographers were not interested in such matters. By Plato's definition, *the things which*

51 Stahl, one of the few scholars to write on Macrobius's geography, has pointed out that scholars now think that Macrobius may have derived much of his cosmographical information from the Neoplatonist Porphyry's lost *Commentary on the Timaeus*, written in the first half of the third century AD. See W. H. Stahl, 'Dominant Traditions', 112. He did not, however, necessarily derive the doctrine of four *oikoumenai* from such a clearly Neoplatonist source; it was present in non-Neoplatonic writings such as those of Geminos.

52 Plato, *Timaeus*, 29c–d.

53 A. G. Debus, *Introduction*, 9.

54 P. O. Kristeller, 'Ficino and His Work', in *Marsilio Ficino e il retorno di Platone*, ed. by G. C. Garfagnini (1986), 23.

55 Genesis, 1.26.

are belonged only to the world of ideas, while *the things which are becoming* were physical things.[56] Therefore geographers only needed to be concerned with the latter. They seized on the idea that the divine craftsman, or in their terms God the creator, had wanted to make things as good as possible and had fashioned 'a world that is as good and as beautiful as the character of the materials out of which it is made will allow.'[57] In Christian terms, this also involved the idea that the best and most beautiful world must be entirely accessible to humans.

Ramusio therefore argued that Plato had been deliberately drawing on ideas of a world designed for humans in the *Timaeus*, and that such an idea lay behind the Atlantis story. He thought that God the creator had deliberately designed the world so that humans, looking at it, should see the handiwork of their Lord and admiring it, praise him. In this, he was already stretching the understanding of Plato, for as the philosopher Lee has pointed out in his introduction to the *Timaeus*, the divine craftsman was never in Plato's work an object of worship.[58] Ramusio's distortion of Plato was extended by his ensuing statement, that Plato thought it defied all logic that people should assume that God would have created the world so that

> two parts should be inhabited and the rest devoid of people: and the sun and the stars in their splendour should make half their course in vain and without fruit, lighting nothing except the sea and the places deserted and devoid of animals. And therefore understanding that he [Solon] got this story from the priests of Egypt in which mention was made of another part of the world beyond Asia, Europe and Africa, he [Plato] admired it greatly and wished to set it at the beginning of his aforementioned dialogue like a sacred thing and conforming to his thought.[59]

Further on in the same *Discorso*, Ramusio continued,

> And so it is not reasonable to believe that the maker of such beautiful and perfect works as the sky, the sun and the moon, should have wished, after it

56 D. J. Zeyl, 'Introduction', in *Plato, Timaeus* (Cambridge: Cambridge University Press, 2000), xxxix.

57 *Ibid.*, xxxvii.

58 H. D. P. Lee, *Introduction*, 8.

59 'due parti d'esso fossero abitate e l'altre prive d'uomini: e 'l sole e le stelle con loro splendore facessero la metà del corso indarno e senza frutto, non lucendo se non al mare e a' luoghi deserti e privi d'animali. E però, intesa che egli ebbe questa istoria de' sacerdoti d'Egitto, nella quale si faceva menzione d'un'altra parte del mondo oltra l'Asia e l'Europa e l'Africa, l'ammirò grandemente e, come cosa sacra e conforme a' suoi pensieri, la volse porre nel principio del predetto Dialogo.' G. B. Ramusio, 'Discorso Sopra Il Terzo Volume', *NV*, vol. 5, 6. Solon was an Athenian lawmaker of the sixth century BC.

> had been made with such stupendous and marvellous order, that the sun should not illuminate one particle of this globe which we call the earth and that the rest of its course should be in vain over sea, snow and ice; but he has covered it in each region with diverse animals and above all others with man for whom it [the earth] had been made as patron and master of all, having endowed him with that divine and celestial part which is the soul: and he has set at hand and distributed in each location, more or less, the gifts necessary for life according to his divine providence and pleasure.[60]

Ramusio was clearly reaching beyond the implications of the purely theoretical ideas of Plato and anchoring them firmly in the tangible and physical world, but he was still using Platonic ideas to provide insight into the nature of this world. Ramusio moved from a Platonic context to a more strictly Christian one but in its implications of the understanding of the earth's perfection, he demonstrates well the trend of geographers and historians to use Plato's *Timaeus* to interpret the nature and structure of the physical world and to blend that with Scriptural authority. Genesis stated that God had given the entire world to humans and had given them dominion over all living things;[61] Plato argued that the world was the best possible world. The philosophy provided an argument for the shape and structure of the lands, and reasons for their accessibility; theology argued that all these lands were to be under human jurisdiction. Between the two they provided a new Renaissance image of the world, which rather than setting bounds to human inhabitation, saw the whole world as a single *oikoumene* set at humankind's feet.

[60] 'Perciochè ragionevolmente non è da credere che il fattore di cosí bella e perfetta fabrica come sono i cieli, il sole e la luna, non abbia voluto che, essendo ella fatta con tanto stupendo e maraviglioso ordine, il sole non illumini se non una particella di questo globo che chiamano terra, e il resto del suo corso sia in vano sopra mari, nevi e ghiacci; ma l'ha coperta in ciascuna sua parte di diversi animali, e sopra gli altri dell'uomo, come padrone e signor di tutti, per cagion del quale ella era stata fabricata, avendolo dotato di quella divina e celeste parte che è l'anima: e appresso ha disposti e in ciascun luogo compartiti i doni necessarii al vivere, piú e meno, secondo che alla divina sua providenza è piaciuto'; *ibid.*, 8.

[61] Genesis 1. 28–30.

7

Moving Boundaries: The Monstrous and the Marvellous

When Tommaso Giunti declared in his introduction to Ramusio's *Navigazioni e viaggi* that all parts of the globe were 'marvellously inhabited', he demonstrated the true amazement of Europeans at this idea.[1] Although by the end of the sixteenth century, most people came to believe that humans were the lords of an entire world created by God for mankind, this was a change that occurred over the course of the century. Throughout much of this period the continuing emphasis on monstrous races dwelling in liminal regions of the world undermined the idea of a coinciding globe and *oikoumene*. Europeans knew neither the topography of all the lands that were now believed to be part of the *oikoumene*, nor the characteristics of their inhabitants. While there were a few European forays into the interior of Africa, Asia and the Americas, these were insufficient to provide much information about these lands and their inhabitants. As a result, European cultural geography of the period was still marked by a sense of centre and periphery which had been one of the marked features of the classical geography of the *oikoumene*. There was a European norm at the heart of the world, and zones of inhabitation surrounding this, so that in the regions furthest from Europe all was monstrous and abnormal.

These realms of marvels and monstrous peoples acted as a means of managing space, of distinguishing the known from the unknown, enabling Europeans to maintain mental control of their expanding geographical horizons. One of the functions of the monstrous was to repel the traveller's wonder and curiosity.[2] As Frank Lestringant has pointed out, the monstrous peoples lived in areas of unverifiable knowledge where nothing marked the distinctions between true and false, between possible and impossible.[3] These monstrous and exotic regions,

1 'che d'ogni intorno questo globo della terra é meravigliosamente habitato, né vi é parte alcuna vacua: né per caldo o gielo priva d'abitatori'. T. Giunti, 'Ai lettori', *NV*, vol. 3, 8.

2 J. J. Cohen, *Monster Theory: Reading Culture* (Minneapolis: University of Minnesota Press, 1996), 12.

3 F. Lestringant, 'Fiction de l'espace brésilien à la Renaissance; l'exemple de Guanabara',

on some level undermined the concept of an *oikoumene* which coincided with the sphere, effectively continuing to separate the habitable regions from the world beyond. It was only when the monstrous races began largely to disappear from the mainstream geographical space that the *oikoumene* and the world could truly be thought to coincide, and the whole world be accepted as made for human domination.

These monsters were, however, more than simply a contrived system of creating a horizon to the known world. Classical, medieval and early modern lore held that some monsters were a product of the extreme environments of liminal regions. As a result, at the outset of the sixteenth century, geographers and explorers alike were certain that such lands with their monstrous inhabitants must exist.[4] This chapter contends that not only did the humoral and ethnographic thought prevalent at the time create an expectation of the existence of monstrous peoples in regions of more extreme climate and environment, but throughout the expansion of the fifteenth and early sixteenth centuries, Europeans needed to believe in the monstrous as a means of defining geographical space. Moreover, their widespread reading of classical literature provided further support for such belief. The classics not only proffered a generic framework in terms of concepts such as Ocean, balance and *harmonia mundi* for the sixteenth-century geographers and explorers, but – in terms of the monstrous and the marvellous – were even thought to provide specific information. Stories of monstrous peoples gained new authority because of the coincidence of the Age of Reconnaissance with the Renaissance. As with the boundaries of climatic zones and Ocean, these realms of monstrosities had to be overturned before the *oikoumene* really could coincide with the sphere, but this was a slow process.

In the 1520s, Maximilian of Transylvania denied outright the existence of all the monstrous humanoid peoples such as the one-eyed monoscoli, the umbrella-footed sciapods, the air-eaters, and even the pygmies, arguing that the survivors of Magellan's circumnavigation had refuted and repudiated all such

---

in *Arts et légendes d'espaces: figures du voyage et rhétoriques du monde, ed. by Christian Jacob et Frank Lestringant* (Paris: Presses de l'Ecole normale supérieure, 1981), 208.

4 On humoral theory and the relationship of monstrous and human characteristics to environment, see C. J. Glacken, *Traces on the Rhodian Shore: Nature and Culture in Western Thought from Ancient Times to the End of the Eighteenth Century* (Berkeley: University of California Press, 1967); S. Davies, *Renaissance Ethnography*. See also *ibid.*, chapter 1. Rebecca Earl has drawn particular attention to humoral theory and the concerns it raised for conquistadors, both in their beliefs about how it shaped the nature of the New World Indians, and how different diet and climate might affect their own characteristics. R. Earle, *The Body of the Conquistador: Food, Race, and the Colonial Experience in Spanish America, 1492–1700* (Cambridge: Cambridge University Press, 2012).

stores.[5] Yet half a century later, the very peoples such as sciapods and monoscoli, whose existence Maximilian had denied, are still to be found within the pages of Thevet's work written in the 1570s. They also gained new life by receiving pictorial reinforcement in the engravings published to accompany new editions of earlier travel narratives in the *Great Voyages* of the de Bry father and sons in the 1590s and early 1600s. Their heyday, however, was in the period immediately following the discoveries, when their function in providing a mental shape to the world was at its most important. By the end of the century they were past their prime. There were still mentions of monstrous peoples but they were no longer thought to control whole regions of the earth.

Perhaps one of the clearest examples of this transition is Samuel Purchas's edition of the narratives of Ulrich Schmidel (whom he called Hulderike Schnirdel). Schmidel had travelled to South America in the 1530s, and had included a description of Amazons living there who were very much in the mould of the description of semi-monstrous Amazon warrior women by Ancient Greeks and Romans.[6] Schmidel clearly believed in these warrior women, but Purchas's marginal note betrays how attitudes to classical monstrous peoples had changed by the time he was publishing Schmidel's narratives nearly a century later, in 1627. He declared, 'the Amazons are still further off: I doubt beyond the region of Truth; if the title be properly meant of such as are here described. For warlike wives living in societies with men, are many.'[7] Purchas was writing in a new century and in a new era of geographical knowledge. Until the very end of the sixteenth century, however, they remained an integral part of European geographical thought. Even then they did not disappear entirely from explorers' and geographers' texts, but their function changed. They ceased to mark the periphery of the world, but instead were aberrations within an *oikoumene* which truly coincided with the world.

In recent years, monsters and wonder have become hugely popular topics, yet the role that monsters have played in creating a flexible boundary to the inhabitable world in the early modern period has been surprisingly understudied.[8]

5 M. Transylvanus, *De Moluccis*, A2. S. Davies, *Renaissance Ethnography*, 158–62.

6 S. Purchas, *Hakluytus Posthumus or, Purchas His Pilgrimes*, vol. 17 (Glasgow: Maclehose, 1905), 33–34.

7 *Ibid.*, 35.

8 Surekha Davies has shown in detail how monstrous peoples related to growing ethnographic interest and cartographic depiction. S. Davies, *Renaissance Ethnography*, passim. She has also discussed other aspects of the role of monstrous races while looking at their liminal locations. S. Davies, 'Unlucky, Bad and Ugly: Categories of Monstrosity from the Renaissance to the Enlightenment', in *The Ashgate Research Companion to Monsters and the Monstrous*, ed. by Asa Simon Mittman (Farnham, Surrey, England: Ashgate, 2012). James Romm has written the most detailed account of how monsters helped to create the edges of the knowable world in classical times,

There is a large body of literature on monsters as prodigies and portents, and on monsters as anomalies within civilised society, but such literature is largely irrelevant to this study since, as prodigies and portents, monsters were not part of the geographical conception.[9] As Surekha Davies has discussed, there were three categories of monstrous prevalent in the sixteenth century: the monstrous as errors, the monstrous as prodigies and portents, and the monstrous peoples. It is this last category who belonged to, and indeed helped to create, the peripheries

---

and John Friedman has made an invaluable study of monstrous peoples in the Middle Ages. J. Romm, *Edges*; J. B. Friedman, *The Monstrous Races in Medieval Art and Thought* (New York: Syracuse University Press, 2000). Friedman, however, stops before the sixteenth-century period of transition. Mary Campbell has carried out an admirable analysis of the explorers' world, which provides some insight into the relationship between the monstrous peoples and the marvellous lands, and world geography, but her analysis has both a strong literary focus and an emphasis on explorers rather than overall geographical thought. M. B. Campbell, *The Witness and the Other World; Exotic European Travel Writing 400–1600* (Ithaca, NY: Cornell University Press, 1988). Jorge Magasich-Airola and Jean-Marc de Beer have shown how, in terms of monstrosity and wonder, the newly-discovered regions were not in fact entirely novel, but were thought to contain a variety of creatures derived from ancient myths both classical and biblical. J. Magasich-Airola, J.-M. de Beer, and M. Sandor, *America Magica: When Renaissance Europe Thought It Had Conquered Paradise*, 2nd edn (London: Anthem Press, 2007). Stephen Greenblatt has famously looked at the relationship between reactions of wonder and marvel, the linguistic creation of the monstrous, and the language of possession in European encounters with the New World. S. J. Greenblatt, *Marvelous Possessions: The Wonder of the New World: The Clarendon and the Carpenter Lectures 1988* (Oxford: Clarendon, 1991). Joyce Wright has written on monsters and the role they played in speculation about 'the other' in the period of the European discovery of mankind, and Christine Johnson has looked at monsters and marvels and the market they provided for geographical works. C. R. Johnson, 'Buying Stories: Ancient Tales, Renaissance Travelers, and the Market for the Marvelous', *Journal of Early Modern History*, 11, no. 6 (2007). J. M. Wright, 'True Peoples and Their Monsters: Speculations on the Other in the Age of Exploration', *Terrae Incognitae*, 37 (2005). Bartra and White have contributed a great deal to our understanding of the role monstrous peoples have played in European thought by tracing the portrayal of specific types of monstrous creatures through time. D. G. White, *Myths of the Dog-Man* (Chicago: University of Chicago Press, 1991); R. Bartra and C. T. Berrisford, *Wild Men in the Looking Glass: The Mythic Origins of European Otherness* (Ann Arbor: University of Michigan Press, 1994); S. Davies, 'Unlucky, Bad and Ugly'; J. J. Cohen, *Monster Theory*.

9 See for example L. Daston and K. Park, *Wonders and the Order of Nature, 1150–1750* (New York: Zone Books, 1998), particularly the central section, which further develops themes the same authors examined in 'L. Daston and K. Park, 'Unnatural Conceptions: The Study of Monsters in Sixteenth- and Seventeenth-Century France and England', *Past and Present*, 92 (1981); K. M. Brammall, 'Monstrous Metamorphosis: Nature, Morality and the Rhetoric of Monstrosity in Tudor England', *Sixteenth Century Journal*, 27 (1996). For an earlier examination of such monsters, see J. Céard, *La nature et les prodiges. L'Insolite au XVI siècle en France* (Geneva: Droz, 1977).

of the knowable world. Lorraine Daston and Katherine Park have discussed the geography of the monstrous peoples in the medieval and early modern periods in their study of *Wonders and the Order of Nature*, observing that 'the marvellous races were a phenomenon of the margins, and embellishment and completion of the natural order'.[10] They have, however, noted that in the medieval period, for the cosmographers, these peoples and wonders were without relevance for Europeans and 'occupied a space almost wholly discontinuous from that of the European center'.[11] They raise the concept of centre and periphery, but then abandon it for the early modern period, saying that once people had a true experience of the world as a sphere, the categories of margin and centre became meaningless. Monstrous peoples became just as much a feature of Europe as anywhere else in the world.[12] While I would not disagree with this statement, Daston and Park have assumed that this loss of the margins occurred earlier than it did. Throughout much of the sixteenth century, geographers and travellers continued to rely on a classical model of the world which indicated that monsters inhabited the regions of the world furthest from the temperate zones of Europe. Their purported existence helped Europeans to conceptualise space in terms of centre and periphery.

Classical authors believed that the Mediterranean (and more specifically Delphi and Rome in Greek and Roman geography respectively) lay at the centre of the world.[13] As one moved further out from this centre, the territories and their inhabitants became ever more bizarre. The regions bordering the Mediterranean but a little removed from the Greek and Roman heartlands, such as Persia and Egypt, produced people who, though alien, were nonetheless comprehensible to the Greeks and Romans. The ancient writers still had a language with which they could communicate the main features of these cultures. Beyond these lay regions of greater peculiarity and beyond them lay the *eschatai* (the furthest reaches) which were the realms of the monstrous immediately abutting the *eremoi* (waste-lands).[14] These regions marked the boundary between the true *oikoumene* and the uninhabitable realms of Ocean and of the frigid and torrid zones. They supported

10 L. Daston and K. Park, *Order of Nature*, 51.

11 *Ibid.*, 56.

12 *Ibid.*, 175.

13 Strabo 9.3.6; Pausanias 10.16.13.

14 'For a Greek city, *eschatai* meant the zone beyond the cultivated area; it was 'the region at the end', land with a poor yield, difficult to use and then only intermittently, out toward the mountains or in the mountains which always marked the boundaries of a territory; it bordered on the frontier and merged with it, the region of mountains and forests which separates two city states, abandoned to the use of shepherds, woodcutters and charcoal burners'. F. Hartog and J. Lloyd, *The Mirror of Herodotus: The Representation of the Other in the Writing of History* (Berkeley: University of California Press, 1988), 13. In terms of the *oikoumene*, the *eschatai* held the same

life, but only life in its most extreme and deformed type. It was to these regions that classical writers had restricted the reports of creatures of wonder and dread, confining the tales of the marvellous and the exotic to the outskirts of the world. [15]

In the Greek imagination these regions were more stable than anywhere else in the world. As James Redfield put it 'the mythical people [were] unchanging and being unreachable [were] immune to outside influence'.[16] Here, the climate, the resources, the animals and the people were all thought to be extreme in nature. These semi-mythical regions surrounded the Greek *oikoumene* on all sides. To the north, one-eyed Arimaspasians battled against griffins in fights for carefully guarded gold; pygmies waged war with cranes; goat-people and werewolves wandered in the northern hills.[17] To the south, giant Ethiopians dined with the gods; deadly snakes flew through the Arabian skies; pigmies thrived in the south of Libya.[18] To the east, in India, lay the realm of mystery and marvels, of dog-headed men, of sorcerers, of giant-gold guarding ants, of one-legged sciapods, while to the west lay the realms of heroes and gorgons.[19]

---

position marking the frontier territory between the civilised world and the world beyond.

15 Although, as Homer's *Odyssey* clearly demonstrates, monstrous creatures were initially portrayed closer to home, wondrous or monstrous peoples and species soon came to be more or less restricted to the *eschatai*. The Cyclopes of Homer, for instance, lived in the Mediterranean, but by the sixth century they had been relocated by Aristeas and Herodotus to the *eschatai*. Interestingly, however, as late as the 4th century BC Thucydides referred to Cyclopes as having been the earliest inhabitants of Sicily, located well within the *oikoumene*. Homer, *Odyssey*, book 9; Aristeas of Proconnesus, *Arimaspeia*, T.4 in A. o. Proconnesus, 'Arimaspeia of Aristeas of Proconnesus', ed. by John Bolton (Oxford: Oxford University Press, 1962). Herodotus, 3.116; Thucydides, 6.2.

16 J. Redfield, 'Herodotus the Tourist', *Classical Philology*, 90 (1985), 109.

17 Aristeas of Proconnesus, *Arimaspeia*, T.4 in A. o. Proconnesus, 'Arimaspeia.'Herodotus, 3.116. Pygmies were first mentioned by Homer, *Iliad*, 3.3–7, and given a clear location by Herodotus, 2.32. Herodotus places them in Libya, as does Aristotle in *Hist. Animal*, 7.12 (597a). By the time Pliny the Elder was writing, however, they had become associated with the north and the east rather than the south. Pliny firmly situates them in Thrace (*NH* 4.11). Many of the classical monstrous peoples had a tendency to shift locations around the peripheries of the earth, but pygmies did so more than most. As Pietro Janni points out, pygmies moved the entire length of the *oikoumene* from their original location in Libya to Scandinavia. See P. Janni, 'I Pigmei in Scandinavia: Vicende di una leggenda etnografica', in *Cultura classica e cultura germanica settentrionale*, ed. by Diego Poli Pietro Janni and Carlo Santini (Rome: Herder, 1985). Herodotus, 4.25, 4.105.

18 Herodotus, 3.23, 3.107, 2.32.

19 Ctesias F45 = (37) Photius *Bibl.* 72, 45a21–50a4, Herodotus, 3.106.2, Arrian, *Indica* 15.1-5; Strabo 15.1.44, Hecataeus Fr. 327; Pliny, *NH*, 7.2. See e.g. Diodorus Siculus 3.52–55 for such a location of the Gorgons.

Some classical writers, such as Herodotus, were wary of writing too much about monsters. He, for instance, wrote of gold-guarding griffins, gold-grubbing ants, Amazons and one-eyed men. He mentioned races of sorcerers and giants and talked of chest-headed people. But while all of these lay on the peripheries of his world, they also lay on the peripheries of his history and geography. Similarly, Strabo was extremely derogatory of those authors who included too many monstrous creatures, saying that Deimachus and Megasthenes, who wrote about men sleeping in their own ears, one-eyed men, mouthless men and others, ought to be distrusted.[20] Their very reluctance to include too much of the exotic and marvellous gave credence to the reports of those they did include. Other writers such as Ctesias and Megasthenes were less scrupulous.[21] They gloried in the unusual. The more flamboyant the report, the more they appreciated it. While both Ctesias and Megasthenes, even in classical times, were attacked as unmitigated liars, their reports were swept into the realm of public knowledge and picked up and reused by such influential writers as Pliny, Solinus, and Pomponius Mela.[22] Once given currency by constant reiteration, these monsters did not die. Recorded again by the most influential of the medieval geographers, Isidore of Seville among others, they lived on throughout the Middle Ages and into the early modern period.[23]

Frequently it was custom rather than appearance which defined monstrosity. Amazons and anthropophagi (cannibals) were common denizens of the *eschatai*. They did not necessarily look any different in form from the women and men with whom the Greeks were used to coming into contact, but their customs were monstrous to the ancients. The *eschatai* by its extreme environment forced people partially to lose their humanity in order to survive.[24] In character or custom, the people of the *eschatai* were part of a world of nightmare which few would wish or dare to visit. Until the sixteenth century their purported existence created a conceptual hurdle which reinforced the limits of the *oikoumene*.

It was not merely the anthropomorphic creatures of the farthest reaches who were distorted and strange; the animals too were unlike any to be found in the rest of the world. The gold-digging ants of India, first mentioned by Herodotus as

20 Strabo, 2.1.9.

21 Megasthenes was considerably more reliable than Ctesias, and some of his reports actually provided fairly accurate information about India. Nonetheless he loved tales of the monstrous, as the Strabo citation proves. For more on Megasthenes' contributions to a fact-based knowledge of India, see D. F. Lach, *Asia*, chapter 1.

22 See Aristotle, *Hist. Animal*, 8.28 and Theopompus (115 F 381) writing about Ctesias, and Strabo, 9.1.9 concerning Megasthenes.

23 R. Wittkower, 'Marvels of the East. A Study in the History of Monsters', *Journal of the Warburg and Courtauld Institutes* (1942).

24 This idea of the dehumanising effect of the *eschatai* remained prevalent in the sixteenth century. Spanish conquistadors were very worried about this weakening and dehumanising nature of the New World. See R. Earle, *Body of the Conquistador*, 19–54.

living in 'the *eschate* of the *oikoumene* towards the east',[25] became a feature of Greek and Roman narratives. Herodotus said that he received the story from some Persians. But it became accepted fact, so that Nearchus, an officer of Alexander the Great, whose journey from the Indus river to the Persian Gulf was recorded by Arian in the first century AD, is reported to have said that when he went to India, he had not seen any of these dog-sized, gold-digging ants living, but had seen their sloughed skins.[26] This may tell us as much about Arrian's perception as Nearchus's. Yet it is clear that already in the ancient world, tale and fable, expectation and empiricism, had become intertwined so that accepted narratives influenced the eyes of the explorer.

The giant gold-grubbing ants were perhaps the most famous of the monstrous animals to become incorporated into the geographies and exploration reports as real creatures, but there were others.[27] Aristeas's griffins had a long-lasting grip on peoples' credulity.[28] The list of monsters and marvels grew as each later writer added to the existing reports. Herodotus, for instance, included only the gigantic ants as one of the truly monstrous creatures of India. Ctesias, his successor in writing about Persia and India, dismissed much of what Herodotus said, calling him a teller of tales, yet added *mantichorae* among others to the catalogue of Indian monsters, creatures with scorpion tails and triple rows of teeth, who battled with griffins and dined on human flesh.[29] Ctesias' tales in turn were picked up by later writers. Pliny the elder spoke of both the lion-like *mantichora*, and the dog-sized ants.[30] By the time medieval travel-writers, encyclopaedists, and authors of bestiaries were writing, the amassed works of the classical authors had left the legacy of a plethora of weird creatures inhabiting the margins of the *oikoumene*. Such *mirabilia* were so beloved in the

25 Herodotus, 3.106.2.

26 Arrian, *Indica*, 14,4.

27 There has been a lengthy historiography of the gold-digging ants, much of it trying to find real origins for the myth. The search for authenticity crosses all genres. In a volume of F. D. Adams and A. S. Woodward, *The Birth and Development of the Geological Sciences* (Baltimore: The Williams & Wilkins company, 1938), 433–88, for instance, Frank D. Adams included a chapter on 'Quaint Stories and Beliefs' in which he devoted a large section to the gold-grubbing ants (483–488). I would argue that the reality of the story was not important in the context of geographical ideas. I am not concerned with the origin of the story, but the belief in it which coloured expectation and allowed for the perpetuation of the myth.

28 See for instance Solinus, 15.22 for a classical reference to Scythian griffins, made more than eight hundred years after Aristeas.

29 Ctesias on Herodotus: Ctesias, *F.Gr.Hist.* 3C, 688, F8, line 7. Ctesias on mantichorae *F.Gr.Hist.* 3C F45 = (15) Photius *Bibl.* 72 45a21–50a4.

30 Pliny, *NH*, 7.30 and 11.36.

Middle Ages that even those who took a more sceptical attitude towards them found that they had to include them.[31]

During the Middle Ages the number of fabulous beings multiplied hugely so that by the end of the period over fifty monstrous peoples had become part of the geographical and ethnographical tradition.[32] They also moved far from their original locations as the *oikoumene* encroached outwards into their regions. The only constant was that the nature of the creatures of the peripheries could be defined in contrast to a norm at the centre of the world. This centre slightly shifted so that it now embraced much of southern and central Europe and was not a strictly Mediterranean norm, but the idea of a contrast between centre and periphery still remained. The monstrous peoples therefore always occupied the zone furthest from the European heartland. They had always been allocated fixed locations, but once knowledge of the regions in which they were believed to dwell

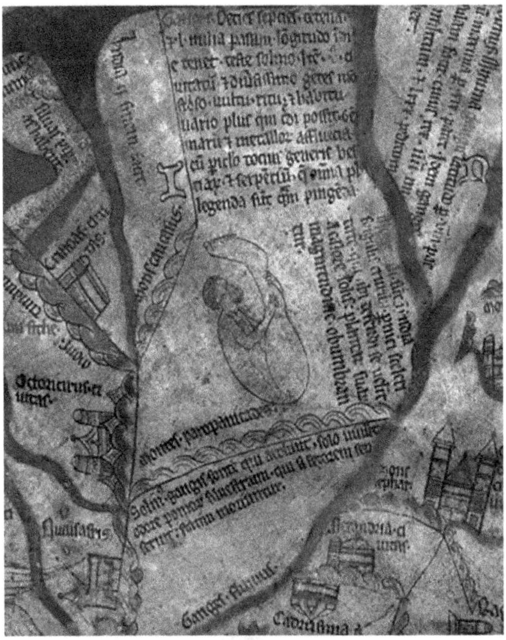

Fig. 3 Sciapods on the Hereford Map near the edge of earth and earthly paradise. Hereford Cathedral

[31] S. Davies, 'The Wondrous East in the Renaissance Geographical Imagination: Marco Polo, Fra Mauro and Giovanni Battista Ramusio', *History and Anthropology*, 23, no. 2 (2012).

[32] J. B. Friedman, *Monstrous Races*, 24.

became greater and their existence in these territories was disproved, they were not so much thought to be the products of the imagination or of idle or confused reports, as to be real but located in the wrong place. As southern Europeans became more familiar with Asia, for example, they relocated monstrous peoples from India to the far north of Europe which remained part of the *eschatai*.[33]

As the illustration from the Hereford map shows, the edges of the Ocean on the medieval *mappae mundi* had become the location of the realms of the monsters of Pliny and Solinus, many of which had their origins in Ctesias, Herodotus and even Homer.[34] Creatures such as pygmies and cynocephali became standardised in both verbal and visual depiction. But while they retained, almost unchanged, their classical appearance, they began to lose the sense of reality behind them. People still believed in them – there was no concept that they were mere fabrications – but there was a sense that they had little impact upon them. They subsided into the region of motif. As Mary Campbell has stated, being 'firmly located in a margin, the grotesque pose[d] little threat to the central order. It need not be integrated'.[35] Campbell's insight does not go far enough, however. Not only need the monsters not be integrated; in a sense they could not be integrated. They had become almost an irrelevance to the medieval world. They were creatures to be feared, but they provided no direct threat to the medieval *oikoumene*. St Augustine, for instance, might write about monsters, but his was an academic discussion.[36] The monstrous regions were too far away. Nobody need risk their own humanity by venturing into the realms of the monstrous. There were, of course, monstrous creatures closer at hand in the grotesque carvings on medieval churches and in the twisted creatures of medieval bestiaries, but though these could have a symbolic (and often moralising) function, they did not portray whole regions of monstrous peoples. These lived at the margins of the world and held sway there.

It was not until the late thirteenth or fourteenth centuries, as exemplified by the narratives of Marco Polo and Mandeville that these creatures began to be re-incorporated into the world, still at the fringes of the European horizon, but closer at hand than the shores of the all-embracing sea.[37] In so doing they became

33 *Ibid.*, 84.

34 M. B. Campbell, *Witness and the Other World*, 65.

35 *Ibid.*, 82.

36 Augustine, *City of God*, 5.43, 5.49.

37 There is always a problem in talking about the mediaeval concept of the world because two more or less contradictory themes run through it. The *mappae mundi* were never intended as realistic portrayals: they were maps intended to demonstrate a theological point. On the other hand they were also the most readily available maps and therefore helped to blur the distinction between spiritual and realistic mapping. Their portrayal of the Ocean as a seemingly finite boundary ran counter to the portrayal of mythical islands such as Hy-Brazil and St Brandon's isle which penetrated far into this boundary.

once again more tangible and more relevant. While Marco Polo scorned the tales of pygmies, for instance, he himself told tales of dog-headed men inhabiting Angaman and of an island of women five hundred miles south of Kesmacoran, who not only would not but could not live with men for more than a period of three months.[38] Marco Polo inherited his monsters indirectly from the classical authors, but unlike those overly credulous writers, he showed a scepticism towards the monstrous, so that when such creatures did creep into his reports, they became less stylised and more alive. They were part of the relevant world.

For the Renaissance writers the monstrous realms were more relevant than they had been even for Marco Polo. Not only were they convinced by thousands of years of accrued wisdom, and by revived interest in the classics and an idea of the world containing whole regions of strange and monstrous beings. Now there was the distinct possibility of actually entering the monstrous lands. And there would be repercussions for doing so: the idea of these peculiar marginal lands was so endemic in European thought that even the explorer's reports included tales of the bizarre. Ambassadors to Tartary mentioned dog-faced monsters and the diplomat, clergyman and ambassador to Russia, Carpini, 'wrote of men with one arm and one hande growing out the midst of their chest and but one foote' in a narrative that was still being reissued by Hakluyt three and half centuries after it was written. Travellers in Russia spoke of mermen. Chest-headed men were sighted in the New World. Dragons – half wolf, half eagle were reported in Africa.[39] Travellers knew that they were exploring the marginal territories. The monsters and exotic tales of classical literature and tradition had preconditioned the way in which explorers interpreted what they saw and heard. The horizon of the exotic was no longer at some distant place, but was at the very edge of the travelled region. It formed a relevant periphery.

This concept of relevant but peripheral regions inhabited by monstrosities continued to be an integral way of controlling and defining space, long after the Iberian voyages of exploration had theoretically destroyed the idea of an

[38] M. Polo, 'Viaggi', *NV*, vol. 3, 260, 65, 84. On Marco Polo and his disregard of monstrous peoples but location of fabulous creatures at the margins, see. C. v. Duzer, 'Hic Sunt Dracones: The Geography and Cartography of Monsters', in *The Ashgate Research Companion to Monsters and the Monstrous* (Routledge, 2013), 409–11.

[39] S. v. Herberstein, 'Commentari della Moscovia e della Russia composti gia latinamente per il signor Sigismondo libero barone in Herberstain, Neiperg e Guttenhag, tradotti di latino in lingua nostra volgare Italiana', *NV*, vol. 4, 824–5; G. P. de Carpine, 'The Voyage of Johannes de Plano Carpini unto the Northeast Parts of the World in the Year of Our Lord 1246', *PN*, vol. 1, 154; W. S. Raleigh, *The Discouerie of the Large, Rich, and Bevvtiful Empire of Guiana,* 2nd edn (Robert Robinson: London, 1596), 70. L. Africanus, 'Della descrizione dell'Africa e delle cose notabili che quivi sono per Giovani Lioni Africano', *NV*, vol. 1, 455. Leo does admit that he himself has not seen the dragon, but says it is commonly reported.

oikoumene. As Lestringant has pointed out, in the period that followed the Iberian expansion, 'the geography of the Renaissance no longer had a centre... [I]t represented a flat plain interspersed with hillocks and peculiar extrusions'.[40] At the beginning of the sixteenth century, however, these peculiar extrusions or singularities, as he calls them, were such as to create a centre to the world. They were features of almost every sixteenth-century geographical work, and were not, as he implies, randomly scattered throughout the world, but rather provided a sense that there was an underlying structure to the world in which there were zones of inhabitation ranging from the civilised in Europe to the monstrous in the farthest regions. They maintained the classical and late medieval model of a world fringed by exotic furthest reaches, and in effect they gave the world a centre again – a centre which was the normal. Ringing round that central heartland were concentric circles of ever more peculiar creatures. Even writers as widely differing and vehemently opposed as the French cosmographers, Belleforest and Thevet, drew weird and wonderful creatures and incidents out of the fabric of the ancient tales and wove them into their geographies.[41] The monsters and marvels whose existence was known and expected but not entirely comprehended formed a boundary in themselves between the human regions and the unknown outskirts, which while potentially habitable were also possibly disastrous and certainly somewhat ominous in character.

In large part, these regions were dangerous not just because they were inhabited by monsters, but also because these monstrous peoples indicated what Europeans might become should they enter the region. There was, and is, no single definition of monsters. For some, monsters were against nature, but for others they were part of nature.[42] The true danger lay in whether they were part of the natural balance of the world, made monstrous by the extreme environments which they inhabited. Humoral theories and environmental determinism were dominant theoretical approaches in European ethnography at the time.[43] As Giovanni Botero said, for example, northerners, beset by cold, were stupefied by excess of phlegm, while the southerners were 'rendered almost bestial by too much melancholy on account of the great heat'.[44] Environment could shape human nature. This meant that though during the course of the sixteenth century

40 F. Lestringant, 'Fortunes de la singularité à la Renaissance, le genre de l'Isolario', in *Écrire le monde à la Renaissance: Quinze études sur Rabelais, Postel, Bodin et la littérature géographique*, ed. by Frank Lestringant (Caen: Paradigme, 1993), 13.

41 See, for example, A Thevet, *Singularitez*, 112, who specifically cites Mela before recording sciapods, monosceli and headless men among others.

42 See C. v. Duzer, *Hic Sunt Dracones*, 387–88.

43 R. Earle, *Body of the Conquistador*; C. J. Glacken, *Traces* ; J.-M. Besse, *Les Grandeurs De La Terre*.

44 'Negli [populi di Mezzogiorno] abbondana ... la maninconia, che li rende quasi bestie.' G. Botero, *Ragion Di Stato*, 59.

in theory there was one single *oikoumene* embracing the whole world, in practice there was a centre and periphery in terms of human inhabitation. The temperate regions had the best climate, and therefore the least monstrous peoples. The *eschatai* contained extreme climates which formed monstrous peoples as part of the balance of nature. This theory in effect prolonged the existence of the old *oikoumene* as separate from the terrestrial earth. Throughout the sixteenth century, European geographical writers, regardless of their nationality, continued to present a literally Eurocentric view of the world in their writings. One of the most influential of such portrayals was the Cosmography of Sebastian Münster which circulated in German, Latin, Czech, English, and French. Münster ostensibly agreed with the idea that the whole world was knowable, but he created a geography which was dominated by the idea of a civilised centre and a monstrous periphery – a concept which really only fitted with an idea of an *oikoumene.*

Münster was firmly imbued with the classical tradition, and his concept of the world was as much defined by regions of marvels and monsters as was that of any Greek or Roman. His description of India is filled with citations from Pliny, Ctesias, Megasthenes and the Alexander stories of gryphons, two-headed snakes, man-eating river horses, triple-horned beasts, one-eyed people and many others.[45] He said that much of Sumatra was a desert and uninhabited because burned by the sun. Round this island circled monstrous man-eating *baleani*, so big they seemed like mountains with pointed, spined backs.[46] In the New World, he described Amazons and anthropophagi, and peculiar creatures such as one with a head like fox, a tail end like a monkey, human feet, ears like an owl's, and a pouch in which it carried its baby.[47] Many of those were descriptions of real animals and peoples but they appeared exotic and monstrous in Münster's narrative because of the difficulty of describing and transmitting the unknown to an unseeing readership. Such geographical descriptions ensured that the Greco-Roman system of classifying the world by rings of normality prevailed; the further one strayed from Central Europe, the more peculiar it became.[48]

Even writers looking at single regions, not at whole world geography, were imbued with the concept of an inhabitable centre and a marginal periphery. The zones of the monstrous helped to reinforce this idea, effectively continuing the idea of the *oikoumene* far into the sixteenth century. The *Historia de Gentibus Septentrionalibus* by the Swede, Olaus Magnus, written in 1555, clearly shows how integral monsters were to geographical perception. Northern Europe was very much part of the periphery of the world, and Olaus was the first person to provide specific historical and geographical information on Scandinavia to the

45 S. Münster, *Cosmographia*, 1070, 73–74, 79, 80, 82.

46 *Ibid.*, 1093.

47 *Ibid.*, 1100–02, 08.

48 M. Small, *Warring Traditions*.

rest of Europe.[49] His work is filled with geographical and historical details which frequently contradict the traditional image, and show his native land as both knowable and populated. Some parts of the traditional picture he clung to more closely, however, describing the Scandinavian regions as populated by monsters. Indeed, almost the only occasion when he contradicted a classical authority was when he asserted the existence of werewolves in the north, despite Pliny's doubts about their reality.[50] He inserted whole pages of quotations about griffins drawn directly from Pliny and Solinus.[51] He even recorded the existence of far northern pygmies, which had first been cited by Eustathios, Bishop of Thessalonike, in his twelfth-century commentary on the *Periegetes* of Dionysius.[52] Its seas were even more terrifying. According to Olaus the Northern Seas were filled with man-eating sirens, ship-crunching Leviathans, *baleani* (traditionally translated as whales but in Olaus's depiction truly ferocious), and similar horrid monsters drawn straight from the pages of classical texts. When writing about the seas off the Scandinavian coast, he declared:

> In its whirling water the vast Ocean offers a wondrous spectacle to all nations: it shows a variety of offspring and it brings them forth, not so much wonderful because of their size and the comparison with the stars as threatening in form, so that nothing seems to be concealed either in the heavens or on earth – whether in her bowels or in her household chattels – which is not held fast in [Ocean's] depths. For in that Ocean, so wide, supine, soft and fertile, with the air receiving regenerating seeds for growth, and with Nature always giving birth, there are found to be many monster-making things, which when things [i.e. Ocean and air] are confused and entwined hither and thither among themselves, now by wind and now by wave, seem to generate forms out of themselves and other elements, so that we hold that whatever is born in any

49 Magnus's work was the main source of information on Scandinavia for most European scholars throughout the sixteenth, and indeed, the seventeenth centuries. It went through six Latin editions before 1669, and was translated into Italian and German before the end of the century. It was also well known in England and is listed in the first catalogue of the Bodleian Library complied in 1605. See 'Introduction', in *Description of the Northern Peoples*, ed. by Peter Foote, vols 182, 187, 188 (London: Hakluyt Society, 1996–1998), lxx.

50 O. Magnus, *Historia*, 642; P. Foote, *Introduction*, lv.

51 P. Fisher, 'Ab Aquilone Aurum Venit; Olaus Magnus's Description of the North and His Feelings for His Homeland', *Journal of the Institute of Romance Studies*, 2 (1993), 672–73.

52 O. Magnus, *Historia*, 673. Eustathios of Thessalonike, *Commentary on Dionysius Periegetes*, (c. 1194), 39 in P. Dionysius and G. Bernhardy, *Dionysius Periegetes Graece Et Latine* (Leipzig: Weidmann, 1828), 93. For more on pygmies of the north, see P. Janni, *Pigmei*, 115.

part of nature is also pervasive for us in the sea, and that very many other things which can be encountered nowhere else are found there.[53]

Magnus explicitly gave the reason for his belief in monsters: the Ocean's 'wide, supine, soft and fertile nature' made it receptive to various seeds which

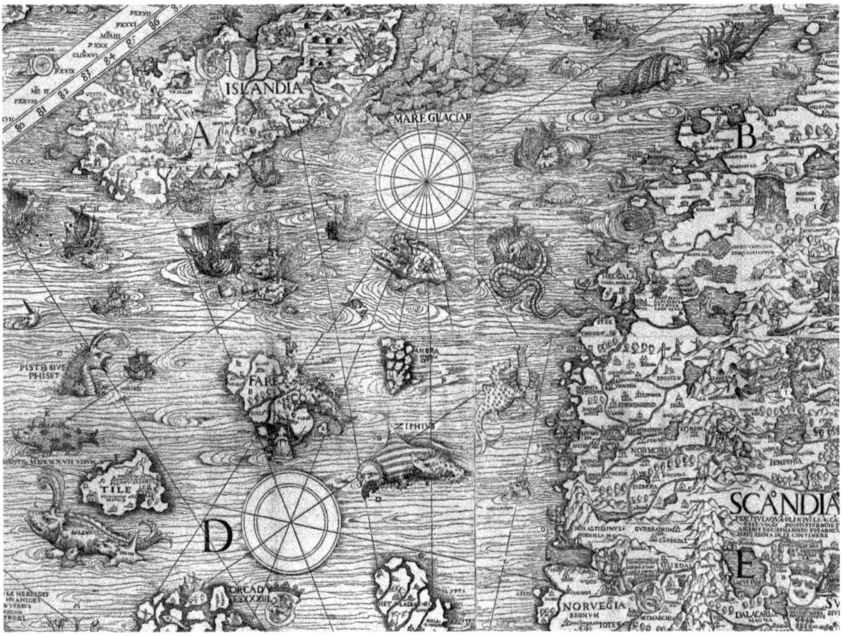

Fig. 4 From Carta Marina of Olaus Magnus, 1539. Uppsala University Library

[53] 'Admirabile spectaculum vastus Oceanus in suo gurgite cunctis nationibus offert: diversos partus ostendit, hosque non tam magnitudine et comparatione syderum mirabiles, quam forma minaces adducit, ut nec coelo, nec terra, aut eius visceribus, aut domesticis instrumentis aliquid abscondi videatur, quod non adsit in sua profunditate retentum. In eo namque Oceano tam lato, supino, molli, ac fertili, accremento accipiente semina genitalia, sublimi semperque pariente natura, pleraque esse monstrifica reperiuntur: quae perplexis, & insemet aliter atque aliter nunc flatu, nunc fluctu convolutis, ex se ac aliis principiis generare formas videntur, ut quicquid nascatur in ulla parte naturae, etiam in mari esse nobis pervasum habeamus et perquam multa alibi nusquam reperibilia, ibidem inveniri... Sunt et beluae in mari quasi hominis figuram imitantes, lugubres in cantu, ut nereides: etiam marini homines, toto corpore absoluta similitudine, qui ascendere navigia nocturnis temporibus videntur: statimque degravari quas insederint partes compertum est, et si diutius permanent, etiam ipsae naves merguntur.' O. Magnus, *Historia*, 729.

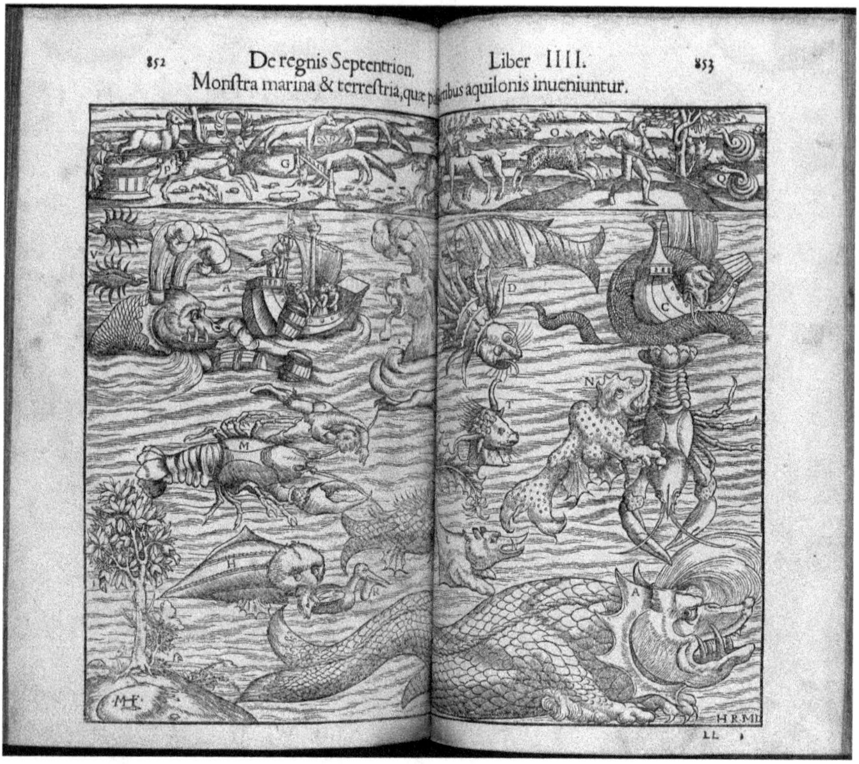

Fig. 5 Sea Monsters from Sebastian Münster, Cosmographiæ uniuersalis lib. VI. Basileae: apud Henrichum Petri mense Septemb. Anno salutis M. D. LIIII. Edinburgh University Special Collections, JY390

when combined with air and ocean were monstrous-making. Monsters of the northern sea were an inevitable product of Nature who is 'always giving birth'. Magnus also depicted these monsters on his *Carta Marina* of 1539. The images that he used were then reproduced by Münster in his *Cosmography*, who also provided an analysis of each monster, relating them back to classical sources (see Figs. 6 and 7).[54] Although English explorers had already begun searching for a northeast passage to Cathay three years before Olaus wrote his history, northern Europe remained the realm of mists and monsters for many years. Olaus's popular writings, given further support from Münster's *Cosmography*, did much to perpetuate the idea of it as a monstrous region. Realms of the monstrous could also act as interior boundaries within landmasses, effectively creating limits to the *oikoumene*. John Pory, in his sixteenth-

[54] S. Münster, *Cosmographia*, 849–50.

century English translation of Leo Africanus's (Hasan ibn Muhammad al-Wazzan) *Description of Africa*, whose book is the first work in Ramusio's compilation, added to the original text. Leo was a Berber travelling through North Africa and his work was considered the most authoritative on the region to be written in the early modern period, but this did not stop Pory from improving upon it to create monstrous regions. He stated that although Africa was much larger than Europe in a north-south direction, it was not so populous, in part because of the climate, but also because there was such an 'abundance of enormous and hurtfull creatures, that for feare of them the land in some places can very hardly, and in others by no meanes, be manured or inhabited be it never so fruitfull.'[55] Here it is clearly implied that although the exotic lands of Africa might in some ways be appealing, being 'fruitfull', they are outside the realm of human endeavour. Their creatures are not described as the objects of wonder or curiosity, but rather as 'enormous and hurtfull'. They are so terrifying and dangerous that they prevent cultivation and inhabitation. Leo had travelled through vast swathes of North Africa, and unsurprisingly had little patience with many of the more fanciful tales about the continent. Neither Leo nor Pory included any of the most extreme monstrosities of classical lore in his book, but while Leo did not mention any of the classical monstrous peoples such as Troglodytes, Ichthyophagi (fish-eaters) and Amazons, Pory clearly thought this was an unsupportable omission. He mentioned all of these. In particular he described Amazons whose characteristics were virtually identical to those recorded in antiquity. While many have attributed the original Greek stories of Amazons to reports of matriarchal societies, it seems far more likely that they were always mythical beings.[56] The step from myth to supposed fact was a small one, and later both Amazons came to be considered real beings.[57] Though they became accepted, however, it did not mean that they owed their original existence to report or observation. It is possible, though unlikely, that Pory's Amazons may have had more of a factual basis than the original Amazons, but they held a similar function in showing the twisting of norms which takes place at a remove from the European heartland of the world. Like the Amazons of the Greek world, they were not inhabitants of the regions which the Europeans knew autoptically, but were located further into the fringes of the lands unexplored by any European. The Amazons were now thought to live below Monomotapa, in the far south of Africa, yet despite their change in location they were, as Pory explicitly stated, similar to ancient Amazons:

55 A. Leo and J. Pory, *Historie of Africa*, 2.

56 *OCD*, 69–70.

57 The Cyclopes are less common in the classical histories and geographies than the Amazons but even they came to have some rather matter of fact mentions. See Thucydides, 6.2.

> These women, after the manner of the ancient Scythish or Asiatike Amazones, so much spoken of in histories of former times, seare off their left paps that they might not be an hindrance unto them in their shooting. They are most expert in war-like stratagems, and swifte of foote.[58]

These were Amazons straight out of classical mythology. The reports of Amazons were among the most prevalent tales of the monstrous at the fringes of the *oikoumene*. They resonated with Renaissance interests in sexuality.[59] Like the anthropophagi, or man-eaters, who also made appearances with startling frequency in the annals of both ancient and Renaissance travellers, they marked the part of the *oikoumene* which lay between the fantastical and the comprehensible.[60] They were neither supernatural nor misshapen, rather they were human but perverted. They belonged to a world where normal custom was turned inside out, to the extent that women could hound men from their homes and could defend themselves against all attack. Pory's Amazons, indeed, were believed to form the most valiant force in the army of the emperor of Monomotapa. Pory's expectation of Amazons was so great that he could not accept the lack of them in Leon's original narrative, and inserted them into the text. Ultimately his preconceptions distorted his belief in the autoptic writing, and fostered an ethnography derived from the classical myths of warrior women.

In the classical reports Amazons always lived on the edges of the *oikoumene*, just beyond the known world, but where they could interact with it.[61] Diodorus in his *Histories*, for instance, tells how they moved out from their island, and having attacked and subdued Atlantis, invaded and conquered much of the inhabited world, setting up colonies throughout Asia Minor, before 'returning to the continent' and continuing back to their home island.[62] Once in the known world, they intermingled with the other races and lost some of their distinctive identity. Their settlements might still have some cultural peculiarities, but they

[58] A. Leo and J. Pory, *Historie of Africa*, 33. In fact in classical accounts it was normally the right not the left breast which was cut off.

[59] The relationship between Amazons and early modern sexuality is not the concern of this book. Katherine Schwarz has written a fascinating study of this interest in the way that an interest in Amazons relates to early modern questions and ideas about sexuality. K. Schwarz, *Tough Love: Amazon Encounters in the English Renaissance* (Durham, NC: Duke University Press, 2000).

[60] See discussion below for more information on the anthropophagi.

[61] Both Herodotus and the author of the *Prometheus Bound*, writing in the fifth century BC, situated them in or near Themiscyra near Thermodon in Asia Minor. Diodorus Siculus, writing about 50 BC, thought their homeland lay a continent away on an island off the coast of Libya 'on the borders of the inhabited world.' Aeschylus, *Prometheus Bound*, 720ff; Herodotus, 4.110; ἐπι τοῖς πέρασι τῆς οἰκουμενης: Diodorus Siculus, 3.53.1.

[62] Diodorus Siculus 3.52–55.

did not contain the complete inversions of normal behaviour which typified the native lands of these aggressive people. Herodotus explains the characteristics of the Sauromatae as being the product of intermarriage between the Amazons and the Scythians, both of whom left their native regions and settled in another area.[63] As long as they lived without men they betrayed the characteristics of Amazons; but once men began to play a significant role in their lives, they retreated into the more traditional women's roles. This ability to move between monstrous and civilised was important. The Amazons lived at the margins, but being marginal creatures could not be entirely written off as the monstrous; instead they could become functioning members of the *oikoumene*.

Amazons are among the first monstrous people to be ascribed to the New World. Columbus described female warrior inhabitants of the island of Matenin who were visited by, and mated with, the cannibals who lived on the adjacent island of Charis, on the very fringes of the world the Columbus knew.[64] Columbus did not make explicit reference to Amazons, in his description of the warriors of Matenin, but the subsequent narratives ensure that it really was the classical monsters who became part of the geography of the New World. Peter Martyr, who embellished and fleshed out Columbus's narrative, made an emphatic comparison with the island Amazons described by Diodorus, who had attributed the foundation of Lesbos to the Amazons. Martyr said there was an island of Madanino or Matenin,

> which only women inhabit. The rumour of this island had come to our ears on our first journey. It is [has been] believed that the Cannibals go to them at certain times of the year, just as antiquity reported that the Thracians went across the straits to the Amazons of Lesbos, and in the same way they [i.e. the women of the Madanino island] send back their sons to their fathers after they have been weaned, but the female children they keep with themselves.[65]

63 Herodotus, 4.110. The Greeks generally believed that there were warrior women among these Sauromatae, but that they were only permitted to fight as long as they were virgins. When they lost their virginity, they took on what the Greeks would have thought of as the ordinary role of women, but they were not permitted to have sex until they had killed three men. Hippocrates, *Airs, Waters, Places*, 1.

64 C. Columbus, *Epistola Christofori Colom ... De Insulis Indie Supra Gangem Nuper Inve[N]Tis ... Ad ... D[Omi]N[U]M Gabrielem Sanchis ... Missa* (Rome: Stephan Plannck, 1493), no pagination.

65 'vocari insulam ab incolis Madanino affirmarunt quam sole mulieres inhabitant. Ad nostrorum aures primo itinere de hac insula fama devenerat. Ad eas haud secus Canibales certis anni temporibus concedere creditum est, atque ad Amazonas Lesbicas transfretasse Thraces retulit antiquitas, et eodem modo filios ad genitores mittere ablactatos, foeminas autem apud se retinere.' P. Martyr, *De Orbe Nouo Petri Martyris Ab Angleria Mediolanensis Protonotarij Cesaris Senatoris Decades* (Alcala: Michaele d'Eguia, 1530), 5v.

Sebastian Münster gave Columbus's description yet wider credence when he wrote about the island of Matinas, saying that women inhabit it in the manner of Amazons.[66] The Amazons of antiquity were being relocated to confined island spaces on the fringes of the known world where they could interact with the known communities.

Not all the Amazons were so clearly segregated from the more normal human societies. In the case of the island inhabitants, both water and custom had formed the divide between them and other populations. Many Amazon societies, however, were mainland communities: the Amazons of Asia, of Scandinavia, of southern Africa, and of the Amazon basin all clearly had a continental location. Nonetheless, they still acted as liminal peoples, marking the divide between the comprehensible and the truly monstrous, since although they were thought to be perverted, they were not visibly deformed by nature. Pory's Amazons, for instance, were mainland Africans, but belonged to the regions about which even the Abassins living in south-central Africa confessed to have little knowledge. Beyond their realm, as Pory pointed out, knowledge petered into nothingness.[67]

When Europeans did gain more autoptic knowledge of southern Africa, tales of Amazons persisted, still defining the liminal regions of the knowable world. The Franciscan friar, Francisco Alvares, having penetrated considerably further into central Africa than had Leon, wrote an account of his travels in 1520 in which he included a report of a matriarchal warrior society, but questioned the application of the name Amazon to these people.[68] He thought that they were misnamed Amazons, since they had, and lived with, husbands. Alvares had very little book learning and was able to look at Africa with an objective eye, untrammelled by classical scholarship.[69] Few of the expected classical monsters can be found in his work. Nonetheless, even he knew much about the features commonly attributed to Amazon peoples:

66 S. Münster, *Cosmographia*, 1101–2.

67 A. Leo and J. Pory, *Historie of Africa*, 32. It is not in the Ramusio version.

68 There was a tradition of warrior societies in Western Africa, particularly Benin, reports of which may have lain behind some of the tales of African Amazon women. Certainly by the eighteenth century the King of Dahomey had an elite guard of women and these do not seem to have been a new military force. For references to this guard, see R. Law, 'Dahomey and the Slave Trade: Reflections on the Historiography of the Rise of Dahomey', *Journal of African History*, 27 (1986), 240. R. A. Kea, 'Firearms and Warfare on the Gold and Slave Coasts from the Sixteenth to the Nineteenth Centuries', *The Journal of African History*, 12, no. 2 (1971), 213.

69 For information on Alvares, see the introduction to F. Alvares, 'The Prester John of the Indies: A True Relation of the Lands of the Prester John, Being the Narrative of the Portuguese Embassy to Ethiopia in 1520', ed. by Charles Fraser Beckingham, George Wynn Brereton Huntingford, and Lord Stanley of Alderley (Cambridge: Published for the Hakluyt Society at the Cambridge U.P, 1961, revised from 1881 edition), 133.

> Going towards the South [there] is a kingdom governed by women who could be called Amazons; according to what has been recounted and what is written in the book of the Infante Dom Pedro of Portugal, but these women (if it is the truth) all have husbands universally [i.e. in common] throughout the year and live with them. They have no Kings, but they have a Queen… They say that they are very strong women, and very warlike, and they fight on some sort of very swift animals that resemble cows, and they are great archers, and when they are little they have the left breast cut off so that it may not impede them in drawing arrows. They gather up an infinite amount of gold in this kingdom … The husbands of these women are not warriors.[70]

Although he denied their right to the title, there were many elements, such as the left breasts being cut off and their skill in archery in his report, which were surely derived more from the classical myth of the Amazon than actual fact. These are truly denizens of the *eschatai,* living in gold-rich exotic environments. Not even Alvares, although he had travelled much further into the heart of Africa than any previous European, had seen these Amazons. They conformed somewhat more to the values of the *oikoumene* as they were perceived by the Europeans, but they were still a fringe people, both literally and metaphorically.

Amazons were reported throughout the world, and the frequency of tales of Amazons led to a continued belief in their existence. There were certainly matrilineal societies and women warriors in the newly encountered territories, but for the most part the Europeans were doing more than recording their encounter. When Europeans expanded into new regions of the world, they anticipated that they would encounter societies of Amazons and other creatures of the classical heritage. Alvares was unusual in that he did not deliberately seek classical creatures in the lands which he was exploring. Yet, as the case of the Amazons demonstrates, the classical heritage was so strong that it still governed his interpretation.

The classical legacy also influenced interpretation of Amazon societies in the new world. Sir Walter Raleigh, for example, wrote of New-World Amazons and

70 'andando verso mezo giorno, è un regno governato da femine, che si potriano chiamare Amazoni, secondo che fu contato, & è scritto nel libro dell' infante don Pietro di Portogallo, ma queste femine, se è la verità, tutte tengonono i maritati universalmente tutto l'anno seco, e vivono con loro: non hanno Re, ma hanno una Reina, la quale non ha marito certo, ma con ogn'uno fa copia di se, e fa figliuoli, & la prima figliuola succede nel regno. Dicono esser Donne molto forti, & gran guerriere, & combattono sopra alcune sorti d'animali velocissimi, che somigliano vacche, & sono grand Arciere, & quando sono picciole, fanno lor secccare [sic] la mammella sinistra, accio non dia loro impedimento nel tirar delle saette. In questo regno raccolgono molto & infinito oro, il qual è portato poi nel regno di Damute, & indi in molte altre parti. Li mariti di queste donne non sono guerrieri.' F. Alvares, 'Viaggio Nella Aethiopia', *NV*, vol. 2, 237.

acephala (chest-headed men) which he had heard of in the course of his travels.[71] His Amazons were not a mere resurrection of classical deviants. They were more influenced by native reports and further distanced from classical legacy than even those of Alvares. They bore the traditional Amazon weapons, but they were not breastless. When writing about Amazons, Raleigh explicitly acknowledged earlier writers' discussions, however, saying:

> The nations of these women are on the south side of the riuer in the Provinces of Topago, and their chiefest strengths, and retracts are in the Ilands scituate on the south side of the entrance, some 60, leagues within the mouth of the saide riuer. The memories of the like women are verie ancient as well in Africa as in Asia: In Africa those that had Medusa for Queene: others in Scithia neere the riuers of Tanais and Thermadon: we finde also that Lampedo and Marthesia w ere Queens of the Amazones: in many histories they are verified to haue been, and in diuers ages and Provinces.

Raleigh was weaving together a version of the myth that was entirely derived from its South American context, with its own European-influenced preconceptions.[72]

Raleigh was writing, after all, several decades after the Spanish explorer Orellana had first associated the name Amazon with the South American river, because, when travelling towards its source, he had come across a tribe of warrior women who had attacked him ferociously. The full account of Orellana's voyage was not published until the nineteenth century, but a version was reported by Oviedo in 1543 and was republished by Ramusio. Therefore, a story of a specific European encounter with an Amazon community in the Americas was in common circulation. Orellana saw such great similarities between the women who attacked him and the people of classical lore, that he named not only the community of women, but the river itself, Amazon.[73] He ensured that classical ethnography left an indelible mark on the Americas.

[71] W. S. Raleigh, *Guiana*, 70; 23–24. Raleigh specifically relates the Ewaipanoma or chest-headed men back to monsters of the European tradition, saying that Mandeville had written of them. There does seem to have been a convergence of native and European tales in reference to the chest-headed men, however, since such creatures were also part of native lore. See N. Whitehead, 'The Historical Anthropology of Text: The Interpretation of Ralegh's 'Discoverie' of Guiana', *Current Anthropology*, 36 (1995), 56. Whitehead in fact argues that the tales of these men fit in more with the native politics of the region at the end of the sixteenth century. While this may be true, he under-emphasises how an education in the classics and Mandeville made Raleigh more ready to accept these stories.

[72] N. Whitehead, 'Anthropology of Text,' 58–59. For a discussion of Raleigh's Guianan monsters and how they had a basis in new world peoples and artefacts, see S. Davies, *Renaissance Ethnography*, Chapter 6.

[73] F. de Oviedo, 'La navigazione del grandissimo fiume Maragnon,' *NV*, vol. 6, 867.; see

Nobody was surprised to find Amazons in the Americas. In the location of classical monsters, the idea of symmetry and balance once again played their part. If a certain monstrous creature were to be found in several margins of the world, it must be found in them all. This idea again had a classical legacy. Herodotus, for instance, had been uncertain whether Hyperboreans existed, but, he argued, if they did do so, then there must also be Hypernoteans.[74] The former lived towards Boreas, the north wind, the latter toward Notos, the south wind. A similar attitude can be seen particularly clearly in André Thevet's *Singularitez de la France Antarctique,* first published in 1557, in which Thevet triumphantly proclaimed that a fourth sort of Amazon had been discovered. He listed the Amazons of antiquity, saying:

> The oldest were in Africa among which were the Gorgons who had Medusa for their Queen. The other Amazons were in Scythia near the river Tanais; who since then have reigned in a part of Asia near the river Thermodon. And the fourth sort of Amazons are those which we are talking about at the moment.[75]

Thevet argued that these newly discovered Amazons completed the description of antiquity. In the perfection of the design each continent had its own complementary Amazon peoples.[76] Otherwise nature would not be balanced.

Thevet was not the only one to be anxious to find traces of Amazon culture in the New World. Indeed New World explorers in general seem to have been so keen to discover this semi-monstrous people in their new-found territories that they record having discovered or heard of Amazons in eight distinct areas: in the Amazon basin, on the plateaux of Chaco and southwest Brazil, in the Antilles (Columbus mentions these), in Colombia (recorded by Fernan Perez de Quesada), in New Grenada, in Nicaragua, and on the Island of Women in Mexico discovered by Pedro de Grijalva.[77] Throughout the century they remained features of exploration reports. So prolific are the descriptions of these people and so closely do they conform to the ancient myths that some of the stay-at-home cosmographers became rather suspicious of the tales. Eventually even such a strong supporter of the reality and location of Amazons began to believe they were simply mythical creatures. He retracted his statements about New World Amazons in his

also M. Milanesi, 'Notes on Oviedo,' *NV*, vol. 6, 861.

74 Herodotus, 4.36.

75 'les plus anciennes ont esté en Afrique, entre lesquelles ont esté les Gorgones, qui avoyent Meduse pour Roine. Les autres Amazones ont esté en Scythie pres le fleuve de Tanaïs; lesquelles depuis ont regné en une partie de l'Asie, pres le fleuve Thermodoon. Et la quatrième sorte des Amazones, sont celle desquelles parlons presentement'. A. Thevet, *Singularitez*, 125.

76 F. Lestringant, *Mapping*, 78–79.

77 *Ibid.*, 17–80.

Cosmographie Universelle.[78] It was an incident when classical theories left more of a mark on exploration than they did upon academic scholarship.

Amazons were one of the few fringe peoples who could adapt to and be absorbed into society. The other key group, whose customs rather than physical characteristics segregated them from the civilised world, were the anthropophagi or man-eaters. Anthropophagi and cannibals have received considerable attention in academic scholarship so they will not be dealt with in great detail here, but they too were marginal people who in early reports had characteristics more derived from traditional tales than from reality.

Stories of cannibals were much more than a mere motif. More than any other single group of inhabitants, whose dehumanised and almost monstrous characteristics acted as boundaries to the *oikoumene*, they served to frame in the civilised and human societies and environments of Europe. There were reports of them in every fringe region of the world. They are a more complex case than the Amazons, however, since, as writers like Jean de Léry and Montaigne pointed out, cannibalism could exist within European societies and was not merely a feature of the margins.[79] In some accounts, however, cannibalism was the defining feature of certain societies in European reports and these societies were all at the margins of the world.

Cannibals pose a problem when one discusses the influence of classical thought on geographical perception. In many cases these people were real. The western Europeans who sailed east and west did witness fellow travellers being eaten by natives. The discovery of such people was exactly what classical report had led readers to expect; but the details of the specific characteristics of these people were the result of autopsy not reading. The gradual substitution of the new word cannibal for the classical word anthropophagi shows a distancing from a people of legend to a people of reality, as does the decreasing emphasis on the lurid and grotesque and the more casual acceptance of man-eating as a general state of affairs.[80] Nonetheless, to some extent these were still people of

[78] *Ibid.*, 80.

[79] M. de Montaigne, R. Dezeimeris, and H. A. Barckhausen, *Essais de Michel de Montaigne: texte original de 1580 avec les variantes des éditions de 1582 et 1587*, vol. 1 (Bordeaux: Féret & Fils, 1870), 181; J. de Léry, *Histoire d'un voyage fait en la terre de Brésil* (Geneva: Antoine Chuppin, 1585 (1st edition 1578)), 410.

[80] See, for instance, Maximilian of Transylvania, who tells how the islands once known as the Fortunate Islands are now known as the Canaries, and then moves on to talk of the Caribbean and man-eating people: 'Anthrophagi whom the Indians call cannibals (ab Anthropophagis quos Indi cannibales vocant)': M. Transylvanus, *De Moluccis*, A4. He specifically discusses the ancient terminology but then starts to use the more modern terms, acknowledging but distancing the reality from the classical literary heritage. By the time Montaigne was writing his essay on cannibals in 1580, the term had clearly become widely understood without need for correlation with the

classical creation. For writers such as Peter Martyr and initially Columbus, the discovery of New World cannibals disrupted their theory that the New World peoples belonged to a Hesiodic Golden Age. Columbus at first refused to believe all evidence of cannibalism because it did not fit with his image of the lands he had discovered.[81] Martyr described the cannibals as wild, mischievous and brutal, taking delight in eating human flesh.[82] For both writers, cannibals were an aberration. They were distanced from some of the other New World inhabitants and shown as truly monstrous, sometimes not just in behaviour but also in physical characteristics often being described as dog-headed, and were discussed very much in the context of the classical literary tradition.[83]

There can be no doubt that travellers to the New World or eastern Asia encountered, and were appalled by, man-eaters.[84] Some writers, such as Hernando Cortés who led the conquest of the Aztecs, appear almost resigned to the cannibalism both of their allies and their foes.[85] The natural historian Oviedo betrays an

---

Greek word anthropophagus. Montaigne was using them as a literary motif just as the classical writers had used anthropophagi, but it was a motif using the supposed reality of the New World cannibals to reflect European society back to the Europeans. It was a literature dependent on an understanding of the cannibals as real creatures of the New World. Gillies, in fact, argues the opposite, stating that 'The Canniba or Cannibales discovered on his first voyage, quickly took over the generic identity of Herodotus's 'Andropophagi' – to such an extent that the ancient term eventually passed out of usage'. J. Gillies, *The Geography of Difference*, 39. I would argue, however, that it is through a distancing from the ancient identity that the term cannibal came to replace that of anthropophagi.

81 See discussion in F. Lestringant and R. Morris, *Cannibals: The Discovery and Representation of the Cannibal from Columbus to Jules Verne* (Berkeley: University of California Press, 1997), 17.

82 P. Martyr, *De Orbe Nouo*, 3v.

83 F. Lestringant and R. Morris, *Cannibals*, passim. See also S. Davies, *Renaissance Ethnography*, Chapter 3. In it she discusses the creation of the Brazilian cannibal looking at the various factors which contributed to the identity and mapping of the Brazilian cannibal.

84 There have been two different trends in the study of cannibalism. The first studies it as a reality. The Europeans who encountered cannibalism focussed much upon it because it was a prevalent and horrific practice which forced itself on people's attention. The other trend sees cannibalism much more as a motif, a product of European imagination. The argument of this thesis is that cannibalism existed, but was embroidered by European travellers, so that many of the narratives of cannibalism were more fictional than real. For a good introduction to the literature on cannibalism, see P. Hulme, 'Introduction: The Cannibal Scene', in *Cannibalism and the Colonial World*, ed. by Francis Barker, Peter Hulme, and Margaret Iversen (Cambridge: Cambridge University Press, 1998). The book also contains a selection of essays which involve both attitudes towards the study of the discourse of cannibalism.

85 H. Cortés, *Carta de relación enviada a su majestad del emperador nuestro señor por el capitán general de la Nueva España* (Seville: Jacono Cromberger, 1522).

attitude of almost fatalistic acceptance, when he describes Caribs 'eating all men that they kill'.[86] Others, however, make cannibalism the key feature of the society, emphasising the bestiality of the people, dehumanising them and marking them out as monstrous peoples. Travellers to both east and west provided lurid descriptions of societies in the farthest reaches of the world where man-eating was so customary that people ate their own relatives, and smoked and salted human flesh was sold and bought at market stalls as a great delicacy.[87] In such tales cannibalism became the defining feature of the societies; such anthropophagi were described as living in the extreme climates, farthest from Europe, and were so abhorrent that they marked the limits of where it was advisable to explore. Ludovic of Varthema, for instance described the people of Java as killing each other like dogs and as 'the worst race that was ever created on earth'.[88] He declared that he had even seen shops where people sold the flesh of their own relatives, although this is almost certainly an invented tale. Repulsed, he claimed to have turned back, partly because he was afraid of being eaten, and partly because he felt that if he went any further he might die from the extreme cold, for he had heard tell that further south the sun only shone about four hours a day and that it was the coldest country on earth.[89] To his mind the anthropophagism and the extreme climate went hand in hand. These people, living at the edge of the *oikoumene*, gave credence to the reports of the yet more monstrous, the in fact entirely mythical creatures, who lived beyond them in the third zone of the *oikoumene*.

Throughout the sixteenth century the classical idea of *eschatai*, if not the term, persisted in the geographical literature, so that all the regions on the peripheries of the explored world contained monstrous creatures both animal and anthropomorphic. As people explored further, these creatures retreated still further, but they were always there at the peripheries marking off the *oikoumene*. They had fixed geographical locations, but really they lived in the realm of the unknown world, in the territory of the unexplored. Both their own existence and the geographical descriptions of the lands they inhabited were products of hearsay, fantasy and miscommunication. It was in this zone that the classical tales came into their own once again. Raleigh's chest-headed people could be found in Guiana, as Laurence Keymis also affirmed.[90] These creatures, first located by Herodotus in

86 'Tutti quelli che pigliano se gli mangiano.' F. de Oviedo, 'Sommario dell'Indie Occidentali', *NV,* vol. 5, 243.

87 A. Vespucci, 'Sommario di Amerigo Vespucci fiorentino, di due sue navigazioni, al Magnifico M. Pietro Soderini, Gonfalonier della magnifica republica di Firenze', *NV*, vol. 1, 662; L. of Varthema, 'Itinerario Di Lodovico Barthema', *NV*, vol. 1, 867.

88 L. of Varthema, 'Itinerario di Lodovico Barthema', *NV*, vol. 1, 867.

89 *Ibid.*, 868.

90 W. S. Raleigh, *Guiana*, 70; L. Keymis, 'A Relation of the Second Voyage to Guiana Performed and Written in the Yeere 1596', *PN*, vol. 10, 465.

the westernmost reaches of Libya, had drifted across the Atlantic to the farthest regions of Brazil.[91] Meanwhile dog-headed men, also located by Herodotus in Libya, had moved eastward to Asia; Marco Polo's ones were in Angaman in Azerbaijan; Sigismund von Herberstein, a German ambassador to Moscow in 1517 and 1526, located them beyond the Tachin river in Russia, and Giovanni da Pian del Carpine, a thirteenth-century papal ambassador to the Mongols, situated them in the farthest reaches of central Asia.[92] All around, the world was enclosed by the truly monstrous. They marked off the territory which Europeans could control and colonise from that in which they might become dehumanised.

Such monstrous realms proved to be only 'provisional boundaries for a knowledge in a perpetual state of progress'.[93] By the end of the century, although there was a far from complete knowledge of global geography, the *oikoumene* really was beginning to expand out to cover the sphere. Just as the Ocean was now seen as drawing the world together and bringing it into a single unit, so the lands themselves were seen as part of this unit. They were all inhabitable – specific monsters might exist, but there was no single zone of the monstrous. Once this was the case, the monsters themselves could become the subject of curiosity rather than terror. As John Friedman has pointed out, monsters were 'increasingly transported from the edges to the centre' so that no longer were they situated at the margins of the world, but rather could be found among us.[94] Amazons and anthropophagi naturally persisted since they had some basis in reality, but their territory ceased to mark the bounds of human exploration, and they began to conform more to ethnographical reality and less to classical myth. Some monsters, the product of new discovery and difficulty of description, continued to exist, but the creatures of classical miscommunication and imagination were no longer needed to create a margin for the world. They continued to hold a place in other branches of literature, but they began to disappear from the geographical texts. They were no longer needed as a means of mapping and managing space. They were the last of the classical boundaries to be transformed before the *oikoumene* truly could be said to coincide with the sphere, and the Renaissance theory of a world designed by God truly under human dominion could be possible.

91 Herodotus, 4.191.

92 G. P. de Carpine, 'Voyage of Johannes de Plano Carpini', *PN*, vol. 1, 147; S. v. Herberstein, 'Commentari della Moscovia', *NV*, vol. 4, 824; M. Polo, 'Viaggi', *NV*, vol. 3, 265. These eastern dog-headed men may have been part of their own Asian tradition, and were not identical to each other (according to Carpine, for instance only the males had dog-heads), but for Renaissance authors they all resonated with the monstrous dog-heads of antiquity. For more on dog-heads see D. G. White, *Dog-Man*.

93 F. Lestringant, *Mapping*, 4.

94 J. B. Friedman, 'Foreword,' in *The Ashgate Research Companion to Monsters and the Monstrous*, ed. Asa Simon Mittman (Farnham, Surrey, England: Ashgate, 2012), xxxiii.

Conclusion: A World Made for Humans

The sixteenth century witnessed a shift from a geography of limit to one of communication. The old idea of an *oikoumene* separate from the rest of the globe was destroyed. The margins of the *oikoumene* had shifted over the years, but the concept had remained for more than two millennia. Only a comparative few had ever discussed the world beyond, and even fewer had left any widely read, written legacy of voyages beyond the boundaries. These regions had remained unreachable and unknowable. Roger Bacon, John Mandeville, Marco Polo and other late medieval scholars and travellers had all spoken of an inhabitable torrid zone, and of the possibility of travelling to lands across the equator, yet the fear of its all-consuming heat was such that few dared attempt to break the southern bounds of the inhabitable world described by most of the classical geographical writers. Eventually, however, the pragmatic desire to find new trade routes to the East, coupled with an increasing number of scholarly works describing an inhabitable equatorial region, convinced enough men in command that the tropical zone might be entered and the equator crossed.[1] The Portuguese entered the torrid zone, and the myth of a climatic confine to human inhabitation was demonstrably proved to have no foundation. At more or less the same time, the other great boundary of the ancient world – the all-enclosing Ocean – was transformed, when Columbus, building on the work of Peter d'Ailly and his own readings of Ptolemy, rather than thinking of the Atlantic as a limit to exploration, saw it as a pathway to Asia, and set sail for India. His alteration of Ocean from a boundary to a means of communication initially harmed traditional perceptions of geography more than the discovery of the Americas. Thereafter geographers and explorers alike tended to accept the existence of Ocean, but it rapidly became altered from a barrier to a series of transportation routes. The humanist geographers transformed the concept of Ocean, retaining the idea of its circumambient nature, but not the idea of limit. In doing so, they paved the way for a completely modern geography, because by transforming the way Ocean was perceived into seeing it as the key means for making the whole world interconnected, it laid the foundations of

[1] See for example the reasoning attributed to Henry the Navigator by G. E. de Azurara, *Descobrimento da Guiné*, 56–57.

European overseas empires and global trade networks. Without a transportation route, there could be no globalisation.

As Lauren Benton states in her discussion on *Law and Geography in European Empires*, 'Oceans have become the quintessential metaphor for globalization. People, ideas, germs, plants and capital move in transnational and cross-regional currents'.[2] Without a change in the idea of Ocean from confine to communication there could have been no connected histories. The global trade networks of the seventeenth century were predicated on ideas of the world as a single body, united in a pattern of land and sea in which all lands were surrounded by Ocean, which made all regions of the world accessible. This idea of the terraqueous globe, with its long-lasting implications for European empires, developed out of the geographical vacuum of the sixteenth century when nobody knew the specific figure or arrangement of any continent, including Europe. The world was suddenly without shape or form. In seeking to gain an understanding of this now formless world, the humanist geographers and explorers did not turn away from the classical authors who had provided such faulty understanding of the earth. Instead they scoured the classical works to find precepts and ideas usable in their new geography of the world.

Their attitude to the ancient sources was equivocal. Authors from Vespucci to Thevet derided the ancients for their belief in the climatic boundaries to the *oikoumene*,[3] but nonetheless, virtually all sixteenth-century geographers turned to classical theory for a mould into which they could pour their empirical information. They also published their ideas. Humanist scholars trained in the classics began to print works of descriptive geography, and created an exchange of information that ensured a new pan-European understanding of world geography.

Sixteenth-century geographical writers rarely stated the theory on which they founded interpretations explicitly, but almost all their works betray a belief that there was proportionality and harmony in the structure of the world (and indeed of the universe), and a hidden balance between all its component parts. This proportionality was evident in the great counterbalancing relationships between the continents, and also in patterns visible within human geography. Their adherence to the doctrine of proportionality is most evident in their discussions of the existence of a Southern continent and of Northern passages to China. Mercator declared outright that there must be a Southern continent, to counterbalance the Northern one; George Best compared what we now know as the Frobisher straits to the straits of Magellan.[4] Both were clearly relying on an idea

2 L. Benton, *A Search for Sovereignty: Law and Geography in European Empires, 1400–1900* (Cambridge: Cambridge University Press, 2009), 104.

3 A. Vespucci, *Epistola*, cxiiii; A. Thevet, *Singularitez*, 88.

4 G. Mercator, *Atlas*, 20. G. Best, 'A True Discourse of the Three Voyages of Discoveire for the Finding of a Passage to Cathaya', *PN*, vol. 7, 280.

of an essential proportionality to the world. This idea had appeared in classical geographical writings as early as the sixth century BC, and had become part of the geographical and natural philosophical legacy of the Greeks. A classically-trained writer such as Mercator consciously drew on the writings of the Greeks and Romans for support for his theories of balance and proportionality in world geography, but for others the underlying principle was merely so embedded in their understanding of the world that they used the classical ideas without direct reference to the ancient authors.

In addition to such incidental use of classical ideas, even the less educated writers made explicit reference to classical authority. Thevet, whose very position as cosmographer to the king was granted because of his perceived authority as an eyewitness reporter of vast tracts of the world, still felt it necessary to shore up his own authority with reference to classical authors even though he himself had little classical education. John Davis, the pragmatic Arctic explorer, who also had had little formal education, having gone to sea as a young boy, peppered his tracts with references to Plato, Pliny and other Greek and Roman authorities as well as to contemporary reports.[5]

Whether implicit or explicit, the traditional concepts of balance, harmony and proportionality pervaded virtually all aspects of sixteenth-century geographical thought. The regions of the world which the Ancient Greeks would have called the *eschatai*, lying furthest from the old *oikoumene*, were still thought to be liberally endowed with the finest natural resources of the world, as had been assumed in ancient times.[6] Until nearly the end of the sixteenth century, Western Europeans, whether travellers or sedentary scholars, supposed that the inhabitants of these regions were deformed and monstrous, to an extent dehumanised. To survive in the harsh climates that prevailed in these potentially wealthy regions at the extremes of the world, people had had to adapt and become such creatures as seemed almost more mythical than real, but they had advantages of wealth and luxury goods which were unknown in the more temperate climates at the centre of the world. These regions of the monstrous also created internal boundaries within the world. The absolute limits of the *oikoumene* may have been destroyed, but, while the monstrous lands persisted in the geographical lore, there was in a sense a continuation of the idea of an *oikoumene*. It was only toward the end of the century when Europeans had travelled across a much greater proportion of the earth than they had done at the beginning of it that the monstrous realms began to disappear from Western European geography. Their

5 See for example where he refers to Cornelius Nepos and the potential circumnavigation of Africa. J. Davis, *Worldes Hydrographical*, pages unnumbered.

6 See N. Wey Gómez, *Tropics of Empire*, passim; R. Padrón, '(Un)Inventing America: The Transpacific Indies in Oviedo and Gómara', *Colonial Latin American Review*, 25, no. 1 (2016), 22.

legacy, however, was a geography of centre and periphery. At the very point of creating a new understanding of the world, geographical theory suggested that Europe lay at the unofficial 'heart of the world', with a climate, topography and environment which was the most auspicious for humans and best suited to civilised life. Arguments about construction and design, about Nature and Providence and about human beings' relationships and control of their environment continued to evolve throughout the seventeenth and eighteenth centuries. Men such as Matthew Hale in the 1670s continued to debate about what ordained the balance of nature. Others such as Humboldt and Linnaeus continued the debates into the eighteenth century.[7]

The ideas were more than academic, however. Ideas derived from the classics enabled the sixteenth-century European geographers to extrapolate from what was known to create their new human and physical geography for the world. They relied not just on the general ideas of balance and proportionality, but especially on the Platonic concept of *harmonia mundi* from which they inferred that everything within the universe was linked by harmonic proportions and hidden affinities. This was true as much for elements upon the earth as for those of the cosmos as a whole. Consequently many Europeans saw the New World as the counterpart to the Old World, which was linked to it by hidden affinities. They envisaged the inhabitants of the New World as corresponding to Europeans living in a golden age, and the New World cities as the counterparts of the known cities of the Old World. By seeking such hidden affinities, they reduced the unknown to the comprehensible.

These theories of balance were universal. They were fundamental to the understanding of both physical and human geography. They enabled sixteenth-century geographers and explorers to study humans throughout the world and examine the interconnection of humanity and the human environment. The world was not made to contain a region of human inhabitation that formed only a small part of a wider earthly sphere; instead it was all wholly designed for human inhabitation and use. Because of the natural balance in the world no land could have everything, but the result of this, in the Renaissance interpretation, was that the world was meant to be interconnected so that humans could trade with each other for the resources to be found in different parts of the world. In this interpretation a human element was imposed on the physical geography.

This interpretation then became key to many of the debates of subsequent centuries. Grotius, whose work on the legality of ocean ownership was famously sponsored by the Dutch East India Company after it had seized a Portuguese ship,

7 C. J. Glacken, *Traces* 481, 501–03.

used the language and ideas about an interconnected world to rationalise why there should be freedom of the seas. He declared:

> God Himself says this speaking through the voice of nature; and inasmuch as it is not His will to have Nature supply every place with all the necessaries of life, He ordains that some nations excel in one art and others in another... For do not the ocean, navigable in every direction with which God has encompassed all the earth, and the regular and the occasional winds which blow now from one quarter and now from another, offer sufficient proof that Nature has given to all peoples a right of access to all other peoples? Seneca thinks this is Nature's greatest service, that by the wind she united the widely scattered peoples, and yet did so distribute all her products over the earth that commercial intercourse was a necessity to mankind.[8]

The great legal authority, Grotius, who precipitated much of the debate about the right to control the oceans – which lasted throughout early modernity, being so vital an aspect of nascent globalisation – was still using the ideas of balanced distribution of goods, and indeed classical authors and citations, to lay the groundwork of the debates. One can see clearly in Grotius the shift in geographical understanding that had taken place in the sixteenth century. Ocean was still a vast expanse but it was not a boundary. Rather it had moved from being a 'terrifying wild wherein societies and nature interact' to an empty space which was criss-crossed by connected networks and sea-lanes.[9] Ocean had ceased to be a forbidden space, but was, rather, a benign one. But it was still really the classical Ocean. Until the existence of a northwest passage was proved by Roald Amundsen in the twentieth century, it was still only a theory that the ocean encompassed the earth. Likewise Grotius's idea that Nature's products were distributed all over the earth was a classical one (although really this idea is closer to Herodotus's statement that no one land can have everything than to Seneca's claim that god gave the winds so that the advantages of each region should be known to all).[10] The context, however, had changed. The understanding was that the balance evident in the world was part of the unity of the world and a mechanism for making all parts of the globe interact in this world of global commodities and global exchange.

8 H. Grotius, *The Freedom of the Seas, or the Right Which Belongs to the Dutch to Take Part in the East Indian Trade, Translated by Ralph Van Deman Magoffin, Introduction by James Brown Scott, Director of the Carnegie Endowment for International Peace* (New York: Oxford University Press, 1916) <https://oll.libertyfund.org/titles/552#Grotius_0049_203> (accessed 2 July 2019).

9 P. E. Steinberg, *The Social Construction of the Ocean* (Cambridge: Cambridge University Press, 2001), 105.

10 Herodotus, 1.32; Seneca, *Naturales Quaestiones,* 5.18.

Some writers implied that this balance was one of the laws of nature, but many also asserted that it was ordained by God. The problem of the connection between Nature and Divine Providence had permeated philosophical thought in the Hellenistic world. By the fourth century BC, philosophers came to view nature as an 'unchangeably stable, self-consistent and predictable order' with definable, comprehensible characteristics, although as we have seen they began to introduce the idea of *pronoia* or providence.[11] This understanding of nature had a strong influence in the sixteenth century. The writers with whom I have been concerned in this book were devoutly Christian, but though they accepted that God was omnipotent, they did not believe that he would interfere in the natural order which was part of his creation. They held therefore that the workings of the world system could be understood by studying natural laws within a geographical context. These laws could be determined both by philosophical interpretation and by empirical evidence. Geography was more a philosophy than a science as yet, frequently carried out by people whom we would consider to be natural or moral philosophers, but such a belief in natural law shows the way in which this reframed geography, while not yet a scientific study in the modern sense of the world, foreshadowed modern scientific ideas.

In relying on ideas of natural laws of balance, proportionality, and *harmonia mundi*, the sixteenth-century geographers and explorers were creating a geography that extended well beyond pure investigation. Investigation formed a guideline for studies and a basis for the rejection of classical ideas. But although many writers used investigative knowledge as a reason to refute existing geographical theories, in other areas, classical theory became the backbone of geography. The explorers used the work of classically trained geographers to gain some understanding about the world that they were to explore. The results of their investigations were then reported back to the stay-at-home geographers who then reinterpreted them in the light of their classical studies. The ancient thought and the new thought, the stay-at-home geography and the product of investigation, became intertwined. While recognising that classical geographical thought had proved hugely flawed, sixteenth-century geographers and explorers chose to reread the classics in a new light to provide a new framework for the world.

The reinterpretation of classical theories in the light of new information coming in from the sixteenth-century discoveries had long-lasting implications for ideas of race, ethnicity, European supremacy, empire and globalisation. It was in this period that humans became central to the understanding of geography as the study of the whole world which they believed was divinely appointed for their inhabitation. Thus Ocean was redefined as a global waterway, laws of balance and symmetry were reread in human terms, and a geography of centre and periphery

[11] A. C. Crombie, *Styles of Scientific Thinking*, 20.

with Europe at its heart was emphasised. This rereading of the classics ultimately created a world-view which allowed Europeans to believe in the possibilities of colonisation and trade anywhere in the world.

Despite the enormous diversity of their backgrounds, experiences and intellectual formation, the key geographers of the period were unified by this emerging concept of humans as central to understanding the structure and nature of the entire world. God had designed the world for humans to be a single, global, interconnected unit subject to the universal natural laws. Hence John Davis could argue that the ancient theories, which maintained that large portions of the world were uninhabitable, were sacrilegious, saying:

> If any man be persuaded to the contrary of this truth, he shall doe himselfe wrong in having so base an imagination of the excellency of God's creation, as to thinke that God creating the world for man's use, and the same being divided but into five parts, three of those parts should be to no purpose: but let this saying therefore of the Prophet Esayas be your full satisfaction, to confirme that which by experience I have truely spoken: For thus saith the Lord that created heaven, God himselfe that framed the earth, and made it, he that prepared it, he created it not in vaine, he framed it to be inhabited (Esay. 45.18).[12]

God had framed the world to be inhabited in entirety by humans: to this extent, the ancients' idea of the world had been in error. The error was not, however, in the structures of balance and proportionality of distribution of goods throughout the earth, or even in the idea of Ocean. It was in the idea of limited inhabitability. This book has argued that the inhabitable world, the *oikoumene*, was the fundamental aspect of early modern geography, but it was fundamental in that the world-view developed over the course of the sixteenth century so that the *oikoumene* became coterminous with the sphere. The reinterpretation of classical authority in the light of the European exploration and belief in a world created by God for humans allowed the sixteenth-century geographers to devise a new geography which was dominated by the idea that the whole world was a knowable, inhabitable, exploitable, connected space.

[12] J. Davis, *The Seaman's Secrets Devided into Two Parts Wherein Is Taught the Three Kindes of Sayling, Horizontall, Paradoxall, and Sayling Upon a Great Circle* (London: Thomas Dawson, 1594), C1 (16).

Bibliography of Classical Texts

Aristotle, *De Caelo. On the Heavens*, trans. by W. K. C. Guthrie, in *Aristotle in twenty-three volumes*, vol. 6, ed. by W. K. C. Guthrie (Cambridge, MA: Harvard University Press, 1939).

——, *Hist Animal.* 'Aristotle, History of Animals', ed. and trans. by D. M. Balme, in *Aristotle in twenty-three volumes*, vol. 3, ed. by Allan Gotthelf and D. M. Balme (Cambridge, MA, 1970).

——, *Meteorologica*, trans. by H. D. P. Lee, in *Aristotle in twenty-three volumes*, vol. 7, ed. by Henry Desmond Pritchard Lee (Cambridge, MA: Harvard University Press, 1952).

Athenaeus, *Athenaei Navcratitae Deipnosophistarvm Libri XV*, ed. by Georg Kaibel, 3 vols (Leipzig: Teubner, 1887).

Arrian, *Anabasis*, trans. by E. Iliff Robson (London: Heinemann, 1933).

Augustine, *The City of God.* 'Sancti Aurelii Augustini Episcopi De Civitate Dei Libri XXII', ed. by B. Dombart, 2 vols (Leipzig: Teubner, 1877).

Cicero, 'Dream of Scipio', *De Re Publica*, trans. by Clinton W. Keyes, in *Cicero: in twenty-eight volumes*, vol. 16, ed. by Clinton Walker Keyes (London: Heinemann, 1928).

——, *De Natura Deorum. On the Nature of the Gods*, trans. by H. Rackham (London: Heinemann, 1933).

Cleomedes, 'Cleomedis Caelestia (Meteora)', ed. by Robert B. Todd (Leipzig: Teubner, 1990).

Cornelius Nepos, *Vitae Att. Life of Atticus*, in *On Great Generals; on Historians*, with an English translation by J. C. Rolfe', ed. by John Carew Rolfe (Cambridge, MA: Harvard University Press, 1929).

Crates of Malos, *Sphairopoiia: Untersuchungen Zur Kosmologie Des Krates Von Pergamon: Mit Einem Anhang, Texte*, ed. by H. J. Mette (Munich: Beck, 1936).

——, *Cratete Di Mallo: I Frammenti*, ed. by M. Broggiato (La Spezia: Agorà, 2001).

Ctesias, *La Perse, l'Inde, autres fragments*, ed. by Dominique Lenfant (Paris: Belles Lettres, 2004).

——, *Ex Tôn Ktêsiou, Agatharchidou, Memnonos Historikôn Eklogai, Appianou Hibêrikê Kai Annibaikê* (Geneva: Henri II Estienne, 1557).

Diodorus, S., 'Diodorus of Sicily', in *Diodorus in 12 volumes* (London: Heinemann, 1939).

Diogenes Laertius, *Lives of Eminent Philosophers*, trans. by R. D. Hicks, vols VI–X, ed. by Robert Drew Hicks (London: Heinemann, 1925).

Geminus, *Introduction aux phénomènes*, ed. by Germaine Aujac (Paris: Belles lettres, 1975).

——, *Geminos's Introduction and Study of a Hellenistic Survey of Astronomy*, ed. by Berggren, J. L., and J. Evans (Princeton: Princeton University Press, 2006).

Dionysius of Halicarnassus, *On Literary Composition*, ed. by W. Rhys Roberts, xiii (London: MacMillan and Co., 1910).

——, *Dionysius Periegetes Graece Et Latine*. ed. by G. Bernhardy (Leipzig: Weidmann, 1828).

Hanno, *Periplus: Or, Circumnavigation of Africa*, ed. by Al N. Oikononides (Chicago: Ares, 1995).

——, 'La navigazione di Annone, capitano de' Cartaginesi nelli parti dell'Africa fuori delle colonne d'Ercole', in *Giovanni Battista Ramusio, Navigazioni e viaggi*, ed. by Marica Milanesi, 6 vols, vol. 1 (Turin: Giulio Einaudi, 1978), 547–50.

Harpocration, V, *Lexeis of the Ten Orators*, ed. by John J. Keaney (Amsterdam: Hakkert, 1991).

Hermogenes, *Hermogenis Opera*, ed. by Hugo Rabe (Leipzig: Teubner, 1913).

Herodotus, *Historiae, Vol 1 and 2*, ed. by C. Hude, 2 vols (Oxford: Clarendon Press, 1963–1975).

——, *Historiarum Liber Primus – Nonus*, ed. by Lorenzo Valla (Venice, 1474).

Homer, *The Odyssey*, 2 vols, ed. and trans. by A. T. Murray (London: Heinemann, 1919).

Horace, 'Odes and Epodes', ed. and trans. by Niall Rudd (Cambridge, MA: Harvard University Press, 2004.

Josephus, *Josephus*, vol. 1, *the Life; against Apion*, ed. and trans. by H. St. J. Thackeray (London: Heinemann, 1926).

Lucian, *Lucian in Eight Volumes*, ed. by K. Kilburn (London: Heinemann, 1968).

Macrobius, *Ambrosii Theodosii Macrobii Commentarii in Somnium Scipionis*, ed. by James Willis (Leipzig: Teubner, 1970).

——, *Commentary on the Dream of Scipio*, ed. by William Harris Stahl (New York: Columbia University Press, 1952).

Martianus Capella, *De nuptiis Philologiae et Mercurii*, ed. by James Willis (Leipzig: Teubner, 1983).

Ovid, *Fasti*, trans. and ed. by James George Frazer (London: Heinemann, 1931).

——, *Metamorphoses*, trans. by Frank Justus Miller, ed. by G. P. Goold and Frank Justus Miller (Cambridge, MA: Harvard University Press, 1977).

Pliny the Elder, *Natural History*, trans. by H. Rackham, in *Pliny in ten volumes*, ed. by H. Rackham (London: Heinemann, 1949).

Pomponius Mela, *De Situ Orbis*, 'De Chorographia Libri Tres: una cum Indice Verborum', ed. by Gunnar Ranstrand (Stockholm: Almqvist & Wiksell, 1971).

——, "Description of the World", ed. by F. E. Romer (Ann Arbor: University of Michigan Press, 1998).

Posidonius, *Posidonius. Vol. 1 the Fragments*, 2nd edn, ed. L. Edelstein and I. G. Kidd, (Cambridge: Cambridge University Press, 1989).

Ptolemy, *Ptolemy's Geography: An Annotated Translation of the Theoretical Chapters*, ed. by J. L Berggren and A. Jones (Princeton: Princeton University Press, 2000).

Pausanias, *Description of Greece in Four Books*, VIII (XCII)–X, trans. and ed. by W. H. S. Jones and Omerod, H. A. (London: Heinemann, 1935).

Photius, *The Bibliotheca: A Selection*, ed. by N. G. Wilson (London: Duckworth, 1994).

Plato, *Timaeus, Critias, Cleitophon, Menexenus, Epistles*, trans. by R. G. Bury, in *Plato: in twelve volumes*, vol. 9, ed. by Robert Gregg Bury (London: Heinemann, 1929).

——, *Timaeus*, trans. by H. D. P. Lee (Harmondsworth: Penguin Books, 1965).

Plutarch, *Oeuvres morales; opinions des philosophes*, ed. by Guy Lachenaud (Paris: Belles Lettres, 1993).

Polybius, *The Histories*, trans. by W. R. Paton, *Unattributed Fragments*, ed. and trans. by S. Douglas Olson, ed. by W. R. Paton, F. W. Walbank, Christian Habicht, and S. Douglas Olson (Cambridge, MA: Harvard University Press, 2010–2012).

Posidonius, *Posidonius. Vol. 1, The Fragments*, ed. by Ludwig Edelstein and I. G. Kidd (Cambridge: Cambridge University Press, 1972).

——, *Posidonius. Vol. 2 the Commentary: Testimonia and Fragments 1–149*, ed. by I. G. Kidd (Cambridge: Cambridge University Press, 1988).

Proclus, *The Commentaries of Proclus on the Timæus of Plato*, trans. by T. Taylor (London: Printed for the Author by A. J. Valpy, 1820).

Ptolemy, *Claudii Ptolemaei Geographia*, ed. by Karl Friedrich August Nobbe (Leipzig: Sumptibus succ. Ottonis Holtze, 1898).

Ptolemy. *La Geografia*, trans. by Pietro Andrea Mattiolo, ed. by Sebastian Münster, Giacomo Gastaldi, and Pietro Andrea Mattiolo (Venice: Gioa. Baptista Pedrezano, 1548).

Quintilian, *The Institutio Oratoria of Quintilian*, trans. by H. E. Butler, ed. by H. E. Butler (London: Heinemann, 1921).

Seneca, *Declamations, Volume 2: Controversiae, Books 7-10. Suasoriae. Fragments*, ed. by Michael Winterbottom (Cambridge, MA: Harvard University Press, 1974.

——, *Hercules, Trojan Women, Phoenician Women, Medea, Phaedra*, ed. znd trans. by John G. Fitch (Cambridge, MA: Harvard University Press, 2002).

——, *Naturales Quaestiones*, in *Seneca in ten volumes*, vol. 2, ed. by Thomas H. Corcoran (Cambridge, MA: Harvard University Press, 1971–1972).

Solinus, *C. Iulii Solini Collectanea Rerum Memorabilium*, ed. by Theodor Mommsen (Berolini: Weidmann, 1958).

——, *The Excellent and Pleasant Work of Iulius Solinus, Polyhistor*, trans. by Arthur Golding (London: Thomas Hacket, 1587).

Statius, P. P., *Thebaid*, vol. 1, books 1–7; vol. 2, books 8–12, *Achilleid*, ed. and trans. by D. R. Shackleton Bailey, vol. 33, ed. and trans. by D. R. Shackleton Bailey (Cambridge, MA: Harvard University Press, 2003).

Strabo, *The Geography of Strabo*, trans. by Horace Leonard Jones, in *Strabo in eight volumes*, ed. by Horace Leonard Jones (London: Heinemann, 1935).

Tacitus, *De Vita Agricola* and *Germania*, trans. by M. Hutton, rev. by R. M. Ogilvie; *Germania*, trans. by M. Hutton, rev. by E. H. Warmington; *Dialogus*, trans. by Sir W. Peterson, rev. by M. Winterbottom, in *Tacitus in 5 volumes* (London: Heinemann, 1970).

——, 'De Vita Agricola', ed. by R. M. Ogilvie and I. Richmond (Oxford: Oxford University Press, 1967).

Virgil, *Eclogues, Georgics, Aeneid*: Books 1–6, ed. by H. Rushton Fairclough, rev. by G. P. Goold (Cambridge, MA: Harvard University Press, 1999).

——, *Vergili Maronis Opera*, ed. by R. A. B. Mynors (Oxford: Clarendon Press, 1969).

Bibliography of Medieval and Early Modern Sources

——., 'The Second Voyage to Guinea Set out by Sir George Barne, Sir John Yorke, Thomas Lok, Anthonie Hickman and Edward Casstelin, in the Yere 1554.' In *The Principal Navigations, Voiages, Traffiques and Discoveries of the English Nation*, ed. by Richard Hakluyt, 2nd edition, 12 vols, vol. 6 (Glasgow: Maclehose and Sons, 1903, first published 1598-1600. Reprint, 2nd), 165–74.

——., 'Vita Sanctissimi Confessoris Christi Brendani', Transcription Du Manuscrit D'alençon À La Bibliothèque Municipale D'alençon, Codex 14, F° 1 R À 11 V., Chapter Xiv, Xième Siècle.

Abbot, G. *A Brief Description of the Whole Worlde Whereby Are Particularly Described Al the Monarchies, Empires and Kingdomes of the Same with Their Severall Titles and Situations Thereunto Adjoining (*London: T Iudson for Iohn Browne, 1599).

Acosta, J. de, *Historia natural y moral de las Indias* (Seville, 1590; repr. Madrid, 1894).

——., *A Naturall and Morall Historie of the East and West Indies, Reprinted from Translation of Edward Grimston, 1604*, ed. by C. R. Markham (London: Hakluyt Society, 1880; first published in Latin and Spanish, Seville, 1590).

Africanus, L. Della descrizione dell'Africa e delle cose notabili che quivi sono per Giovani Lioni Africano, in *Giovanni Battista Ramusio, Navigazioni e viaggi*, ed. by Marica Milanesi, 6 vols (Turin: Giulio Einaudi, 1978), I, 19–460.

Ailly, d', P., E. J. P. Buron, and Columbus, *Ymago Mundi: Texte latin et traduction française des quatre traités cosmographiques de d'Ailly et des notes marginales de Christophe Colomb. Étude sur les sources de l'auteur* (Paris: Maisonneuve frères, 1930), II, 454–55.

Albertus (Magnus), Habes in hac pagina *De natura locorum librum mira eruditione, & singulari fruge refertum, & iam primum summa diligentia reuisum, in lucem aeditum, quem leges diligentius, vel si cosmographia vel physica profecisse te volueris* (Strasbourg: Atlantsee, 1515).

Alvares, F., The Prester John of the Indies: A True Relation of the Lands of the Prester John, Being the Narrative of the Portuguese Embassy to Ethiopia in 1520, trans. by Lord Stanley of Alderley, ed. by Charles Fraser Beckingham, George Wynn Brereton Huntingford (London: Hakluyt Society, 1881; repr. Cambridge: for Hakluyt Society at the Cambridge UP, 1961).

——., 'Viaggio nella Aethiopia', in *Giovanni Battista Ramusio, Navigazioni e viaggi*, ed. by Marica Milanesi, 6 vols, vol. 2 (Turin: Giulio Einaudi, 1980), 75–386.

Azurara, G. E. de, *Crónica do descobrimento e conquista da Guiné* (Paris: J. P.Aillaud, 1841; repr. Mem-Martins: Publicações Europa-Améric, 1989).

Bacon, R., *Opus Maius*, ed. by John Henry Bridges (Oxford: Clarendon Press, 1897–1900).

Barbaro, I., 'Di messer Iosafa Barbaro, gentiluomo veneziano, il viaggio della Tana e nella Persia', in *Giovanni Battista Ramusio, Navigazioni e viaggi*, ed. by Marica Milanesi, 6 vols, vol. 3 (Turin: Giulio Einaudi, 1980), 481–576.

Barros, J. de, and D. de Couto, *Da Asia de João de Barros e de Diogo de Couto* (Lisboa: Na Regia Officina Typografica, 1778).

Belleforest, F. de, *La Cosmographie universelle de tout le monde: augmentée, ornée et enrichie* (Paris: Nicolas Chesneaus, 1575).

Bembo, P., *Della historia vinitiana libri xii* (Venice: Appresso Gualtero Scotto, 1552).

Benzoni, G., 'Insignis et admirandis historia de reperta primum Occidentali India à Christophoro Columbo', trans. by Urbain de Chauveton, in *Americae Pars Quarta*, ed. by Theodor de Bry (Frankfurt: I Feyrabend, 1594).

——., *La historia del Mondo Nuovo laqual tratta dell'isole & mari nuovamente ritrovati & delle nuove città da lui proprio vedute, per acqua & per terra in quattordeci anni* (Venice: Francesco I Rampazetto, 1565).

Best, G., 'A True Discourse of the Three Voyages of Discoveire for the Finding of a Passage to Cathaya', in *The Principal Navigations, Voiages, Traffiques and Discoveries of the English Nation*, ed. by Richard Hakluyt, 2nd edn (repr.), 12 vols, vol. 7 (Glasgow: Maclehose and Sons, 1903; first published 1598), 250–83.

Blundeville, T., *A Brief Description of Universal Mappes and Cardes* (London: Roger Ward for Thomas Cadman, 1589).

Bodin, J., *De Republica Libri Sex, Latine Ab Autore Redditi, Multo Quam Antea Locupletiores. Cum Indice Copiosissimo* (Lyons: Jacques du Puys, 1586).

——., *Methodus Ad Facile Historiarum Cognitionem* (Paris: M. Iuuenem, 1566).

Boemus, J. *Omnium Gentium Mores Leges et Ritus ex Multis Clarissimia Rerum Scriptoribus, a I.B.A. Sacerdote Tentenicae Militiae Devoto Nuper Collectos: & in Libros Tris Distinctos Aphricam, Asiam, Europam, Optime Lector Lege* (Augsburg: Augustae Vindelicorum, 1520).

Botero, G., *Della ragion di stato libri dieci, con tre libri delle cause della grandeza e magnificenza della città* (Venice: i Gioliti, 1589).

——., *Relationi del mare* (Padua: Andréa Viani, 1598).

——., *Relationi universali devise in quattro parti. Novamente reviste, corrette e ampliate dall' istesse auttore* (Brescia: Compagnia Bresciana, 1598).

——., *The Worlde or a Historicall Description of the Most Famous Kingdomes and Common-Weales Therein. Relating Their Situations, Manners, Customes,*

*Civill Government and Other Memorable Matters, Translated into English and Enlarged by R. Johnson (*London, 1691).

Buonaccioli, M. A., *La prima parte della Geografia di Strabone, di greco tradotta in volgare italiano da M. Alfonso Buonaccioli* (Venice, 1562).

——., M. A., *La seconda parte della geografia di Strabone, di greco tradotta in volgare italiano da M. Alfonso Buonaccioli* (Ferrara, 1565).

Carpine, G. P. de, 'The Voyage of Johannes De Plano Carpini Unto the Northeast Parts of the World in the Year of Our Lord 1246', in *The Principal Navigations, Voiages, Traffiques and Discoveries of the English Nation,* ed. by Richard Hakluyt, 2nd edn (repr.), 12 vols, vol. 1 (Glasgow: Maclehose and Sons, 1903; first published 1598–1600), 134–78.

Columbus, C., *Epistola Christofori Colom ... De Insulis Indie Supra Gangem Nuper Inve[N]Tis ... Ad ... D[Omi]N[U]M Gabrielem Sanchis ... Missa* (Rome: Stephan Plannck, 1493).

——., 'Raccolta di documenti e studi pubblicati dalla R. Commissione colombiana', Part I, Vol. 2 (Ministero della pubblica istruzione: Rome, 1892).

——., The Libro De Las Profecías of Christopher Columbus: An En Face Edition, ed. by D. C. West and A. Kling (Gainesville: Univ. of Florida P, 1991).

Columbus, C., D. C. West, and A. Kling, *The Libro De Las Profecías of Christopher Columbus: An En Face Edition* [in English] (Gainesville: University of Florida Press, 1991).

Contarini, G., *La Republica, Ei Magistrati Di Vinegia* (Venice: Girolamo Scotto, 1544).

Coronado, V. di, 'An Extract of the Letter of Captain Vasquez di Coronado', in *The Principal Navigations, Voiages, Traffiques and Discoveries of the English Nation*, ed. by Richard Hakluyt, 2nd edn (repr.), 12 vols, vol. 9 (Glasgow Maclehose and Sons, 1903; first published 1598–1600), 116–18.

——., 'Relazione che mandò Francesco Vazquez di Coronado, capitano generale della gente che fu mandata in nome di sua Maestà al paese nuovamente scoperto: quel che successe nel viaggio, dalli ventidue d'aprile di questo anno MDXL, che partì da culiacan per innanzi, e di quel che trovò nel paese dove andava', in *Giovanni Battista Ramusio, Navigazioni e viaggi*, ed. by Marica Milanesi, 6 vols, vol. 6 (Turin: Giulio Einaudi, 1980), 607–28.

Corsali, A., 'Di Andrea Corsali fiorentino allo illustrissimo signor duca Giuliano de' Medici lettera scritta in Cochin, terra dell'india, nell'anno MDXV, alli VI di gennaio', in *Giovanni Battista Ramusio, Navigazioni e viaggi*, ed. by Marica Milanesi, 6 vols, vol. 2 (Turin: Giulio Einaudi, 1979), 13–74.

Cortés, H., *Carta de relación enviada a su majestad del emperador nuestro señor por el capitán general de la Nueva España* (Seville: Jacono Cromberger, 1522).

Ctesias, *Ex Tôn Ktêsiou, Agatharchidou, Memnonos Historikôn Eklogai, Appianou Hibêrikê Kai Annibaikê* (Geneva: Henri II Estienne, 1557).

Davis, J., *The Seaman's Secrets Devided into Two Parts Wherein Is Taught the Three Kindes of Sayling, Horizontall, Paradoxall, and Sayling Upon a Great Circle (*London: Thomas Dawson, 1594).

——., *The Worldes Hydrographical Description* (London: Thomas Dawson, 1595).

Dee, J., 'Concerning a New Location of the Island of Estotitland and the Province of Drogio', in *The Limits of the British Empire*, ed. by Ken Macmillan and Jennifer Abeles (Westport, CT: Praefer, 2004), 37–38.

——., *Mathematicall Praeface to the Elements of Geometrie of Euclid of Megara (1570)*, ed. by Allen G. Debus (New York: Science History Publications, 1975).

——., The Limits of the British Empire, ed. by Macmillan, K. and J. Abeles (Westport, Connecticut and London: Praefer, 2004).

Edward VI, 'Copy of Letter Sent by Edward VI to 'the Kings, Princes, and Other Potentates, Inhabiting the Northeast Partes of the Worlde, toward the Mighty Empire of Cathay, at Such Time as Sir Hugh Willoughby Knight, and Richard Chancelor, with Their Company, Attempted Their Voyage Thither in the Yeere of Christ 1553', in *The Principal Navigations, Voiages, Traffiques and Discoveries of the English Nation,* ed. by Richard Hakluyt, 2nd edn (repr.), 12 vols, vol. 2 (Glasgow: Maclehose and Sons, 1903; first published 1598–1600), 210.

Ficino, M., 'Compendium Marsilii Ficini in Timaeum', in *Omnia Divini Platonis Opera, Tralatione M. Ficini, Emendatione et ad Graecum Codicem Collatione S. Grynæi*, ed. by Simon Grynaeus and Marsilio Ficino (Basel: In officina Frobeniana, 1546), 672–700.

——., 'De divino furore, lettere I', in *Epistolarum familiarum liber I*, ed. by S. Gentile (Florence, 1990; first published 1495), 24.

Fracastoro, G., 'De Sympathia et Antipathia Liber I', in *Opera omnia* (Venice: Giunti Press, 1555).

——., 'Risposta dello eccellentissimo messer Ieronimo Fracastoro del crescimento del Nilo a messer Gio. Battista Ramusio', in *Giovanni Battista Ramusio, Navigazioni e viaggi*, ed. by Marica Milanesi, 6 vols, vol. 2 (Turin: Giulio Einaudi, 1978), 406–37.

Gascoigne, G., 'To the Reader', in *A Discourse of a a Discoverie for a New Passage to Cataia, Reprinted in Voyages and Colonising Enterprises of Sir Humphrey Gilbert*, ed. by David B. Quinn, 2nd edn (London: Hakluyt Society, 1940), 132.

Gastaldi, G., 'Tierra Nueva', in *La geografia di Ptolemeo* (Venice: G. B. Pedrezano, 1548), 1 map: b&w; 12 x 16 cm (plate 13.5 x 17.5 cm.), on sheet 17 x 21 cm.

Gilbert, H., 'A Discourse Written by Sir Humphrey Gilbert Knight, to Prove a Passage by the Northwest to Cathaia and the East Indies', in *The Principal*

Navigations, Voiages, Traffiques and Discoveries of the English Nation, ed. by Richard Hakluyt, 2nd edition (repr.), 12 vols, vol. 7 (Glasgow: Maclehose and Sons, 1903; first published 1598–1600), 158–90.

Giorgio, F. *Francisci Georgii ... De harmonia mundi totius cantica tria* (Venice: Bernardini de Vitalibus), 1525.

Giovio, P., 'Paolo Iovio istorico delle cose della Moscovia, a monsignor Giovanni Rufo, arcivescovo di Cosenza." In *Giovanni Battista Ramusio, Navigazioni e viaggi*, ed. by Marica Milanesi, 6 vols, vol 3 (Turin: Giulio Einaudi, 1980), 667–95.

Girava, G., *Dos libros de cosmographia* (Milan: Christoval Caron, 1556).

Giunti, T., 'Tommaso Giunti ai lettori', in *Giovanni Battista Ramusio, Navigazioni e viaggi*, ed. by Marica Milanesi, 6 vols, vol. 3 (Turin: Giulio Einaudi, 1978), 3–5.

Góis, D. de, *Chronica di Principe Dom Joam, Rei que foi destes regnos segundo do nome* (Lisbon: Francisco Correa, 1567).

Gomara, L. de, *La Istoria de las Indias, y Conquista de Mexico* (Çaragoça: Agustín Millán, 1552).

Grosseteste, R., *Die Philosophischen Werke des Robert Grosseteste*, ed. by L. Baur (Münster: Beiträge zur Geschichte der Philosophie des Mittelalters, 1912).

Grotius, H., *The Freedom of the Seas, or the Right Which Belongs to the Dutch to Take Part in the East Indian Trade, trans. by Ralph Van Deman Magoffin, intro. by James Brown Scott, Director of the Carnegie Endowment for International Peace (*New York: Oxford University Press, 1916).

Grynaeus, S., *Die new Welt, Der Landschaften unnd Insulen, so bis hie her allen Altweltbeschrybern unbekant* (Strasbourg: G. Ulricher, 1534).

—— ed., *Novus orbis regionum ac insularum veteribus incognitarum una cum tabula cosmographica, et aliquot alii consimilis argumenti libellis, quorum omnium catalogus sequenti patebit pagina* (Basle: Jo. Hervagium, 1532).

Guagnino, A., 'La descrizione della Sarmazia europea del magnifico cavalliere Alessandro Guagnino veronese, tradotta dalla lingua latina nel volgare italiano dal reverendo messer Bartolomeo Dionigi da Fano', in *Giovanni Battista Ramusio, Navigazioni e viaggi*, ed. by Marica Milanesi, 6 vols, vol. 4 (Turin: Giulio Einaudi, 1980), 319–606.

Hakluyt, R. (the elder)., 'Letter from Richard Hakluyt (the elder) to Abraham Ortelius, (c.1591) in Abraham Ortelius Correspondence (Kraus 57)'. Digitised and hosted online by the Harry Ransom Centre, The University of Texas at Austin, last accessed January 2020, https://hrc.contentdm.oclc.org/digital/collection/p15878coll99/id/156%20.

Hakluyt, R. (the younger), 'Discourse of Western Planting', in *Hakluyt Society Extra Series*, ed. by David Beers Quinn and Alison M. Quinn (London, 1584; facsimile edition 1993).

——., 'Divers Voyages Touching the Discoverie of Ameria, and the Ilands Adiacent Unto the Same, Made First of All by Our Englishmen, and Afterward by the Frenchmen and Britons: And Certaine Nots of Adverstisements for Observations, Necessarie for Such as Shall Heereafter Make the Like Attempt, with Two Mappes Annexed Heereunto for the Plainer Understanding of the Whole Matter', ed. by David B. Quinn (Amsterdam: Theatrum Orbis Terrarum, 1967; first published London: Thomas Woodcocke, 1582).

——., 'The Epistle Dedicatorie in the Third Volume of the Second Edition', in *The Principal Navigations, Voiages, Traffiques and Discoveries of the English Nation*, ed. by Richard Hakluyt, 2nd edn (repr.), 12 vols, vol. 1 (Glasgow: Maclehose and Sons, 1903; first published 1598–1600), lxxiv–lxxx.

——., ed. The Principal Navigations, Voiages, Traffiques and Discoveries of the English Nation, 2nd edn, 12 vols. Vol. 1 (Glasgow: Maclehose and Sons, 1903, first published 1598–1600).

Herberstein, S. v., 'Commentari della Moscovia e della Russia composti gia latinamente per il signor Sigismondo libero barone in Herberstain, Neiperg e Guttenhag, tradotti di latino in lingua nostra volgare Italiana', *NV*, vol. 4, 824–5

Herrera y Tordesillas, A. de, and C. Barlaeus. *Novus Orbis, Sive Descriptio Indiae Occidentalis...Metaphraste C. Barlaeo. Accesserunt & Aliorum... Descriptiones, & Navigationis...Australis Jacobi Le Maire...& Navigationum Omnium Per Fretum Magellanicum...Narratio* [in Undetermined] (Amsterdam: 'apud Michaelem Colinium, 1622).

Interiano, G., 'Giorgio Interiano genovese a messer Aldo Manuzio romano della vita de' Zichi, chiamati Ciarcassi', in *Giovanni Battista Ramusio, Navigazioni e viaggi*, ed. by Marica Milanesi, 6 vols, vol. 4 (Turin: Giulio Einaudi, 1978), 29–36.

Ionas, A., 'A Briefe Commentarie of Island: wherin the errors of such as have written concerning this Island, are detected and the slanders, and reproches of certaine strangers, which they have used over-boldly against the people of Island, are confuted', in *The Principal Navigations, Voiages, Traffiques and Discoveries of the English Nation*, ed. by Richard Hakluyt, 2nd edn (repr.), 12 vols, vol. 4 (Glasgow Maclehose and Sons, 1903; first published 1598–1600), 89–168.

Keymis, L., 'A Relation of the Second Voyage to Guiana Performed and Written in the Yeere 1596', in *The Principal Navigations, Voiages, Traffiques and Discoveries of the English Nation, ed.* by Richard Hakluyt, 2nd edn (repr.), 12 vols, vol. 10 (Glasgow: Maclehose and Sons, 1903; first published 1598–1600), 441–501.

Leo, A., and J. Pory, *A Geographical Historie of Africa*, Early English Books, 1475–1640 / 262:12 (George Bishop: London, 1600).

Léry, J. de, *Histoire d'un voyage fait en la terre de Brésil* (Geneva: Antoine Chuppin, 1585 ; first edn 1578).

Letts, M., ed. Mandeville's Travels; Texts and Translations, 2nd, vol. 102 (London: Hakluyt Society, 1953).

Magnus, O., *Historia de Gentibus Septentrionalibus (*Rome: J. M. de Viottis, 1555).

Mandeville, J. 'The Voyage of John Mandevil Knight in Latin, Begun in the Raigne of Edwarde the 2. Anno. 1322 Continued for the Space of 33. Yeeres, and Ended in the Raigne of Edwarde the 3. From England to Iudea, and from Thence to India, China, Tartaria, and as Farre as 33. Degrees to the South of the Aequinoctiall', in *The Principall Navigations, Voiages and Discoveries of the English Nation*, ed. by Richard Hakluyt, David B. Quinn and R. A. Skelton, 2 vols. Vol. 1 (London, 1589, 1965 facsimile edition, 25–79).

Martyr, P., *De Nouo Orbe, or the Historie of the West Indies ... Comprised in Eight Decades ... Three Formerly Translated into English, by R. Eden ... The Other Fiue by M. Lok, trans. by Michael Lok* (London: Thomas Adams, 1612).

——., *De Orbe Nouo Petri Martyris Ab Angleria Mediolanensis Protonotarij Cesaris Senatoris Decades* (Alcala: Michaele d'Eguia, 1530).

——., *The Decades of the Newe Worlde or West India, trans. by R. Eden* (London: in æd. G. Powell, 1555).

Medina, P. de, *Arte De Nauegar* (Francisco Fernandez de Cordoua: Valladolid, 1545).

Mela, P., 'Concerning the Situation of the World', ed. by Arthur Golding (London: Thomas Hacket, 1585).

Mendoza, J. G. de, *Historia de las cosas mas notables, rites y costumbres del gran Reyno de la China con un itinerario del Nuevo Mundo* (Rome: V. Accolti, 1585).

Mercator, G., *Atlas Sive Cosmographicae Meditationes De Fabrica Mundi Et Fabricati Figura* (Amsterdam: Albertus Busius, 1595).

——., 'The Letter of Gerard. Mercator to Richard Hakluyt of Oxford Touching the Intended Discoverie of the Northwest Passage', in *The Principal Navigations, Voiages, Traffiques and Discoveries of the English Nation*, ed. by Richard Hakluyt, 2nd edn (repr.), 12 vols, vol. 3 (Glasgow Maclehose and Sons, 1903; first published 1598–1600), 275–82.

Micheow, M. de, 'Trattato Il di Matteo di Micheovo, Dottor Fisico e Canonico Cracoviense, delle due Sarmazie'. in *Giovanni Battista Ramusio, Navigazioni e viaggi*, ed. by Marica Milanesi, 6 vols. Vol. 4 (Turin: Giulio Einaudi, 1980, 607–81).

Montalboddo, F. da, *Paesi novamente retrovati* (Vicenza: Henrico Vicentin, 1507).

Montaigne, M. de, R. Dezeimeris, and H. A. Barckhausen. *Essais De Michel De*

Montaigne: Texte Original De 1580 Avec Les Variantes Des Éditions De 1582 Et 1587. Vol. 1 (Bordeaux: Féret & Fils, 1870).

Münster, S., *Claudius Ptolemaeus: Geographia, First Edition 1540*, 4 editions (Basel: Heinrich Petri, 1552).

——., *Cosmographiae Universalis Lib VI* (Basel: Henric Petri, 1544).

——., *Cosmographiae Universalis Lib VI* (Basel: Henric Petri, 1550).

——., 'Introductio ad cosmographiam', in *Novus orbis regionum ac insularum veteribus incognitarum una cum tabula cosmographica, et aliquot alii consimilis argumenti libellis, quorum omnium catalogus sequenti patebit pagina*, ed. by Simon Grynaeus (Basel: Jo. Hervagium, 1532).

——., ed., 'Reverendo in Christo Patri ac Domino Philippo à Gundelsheim, Episcipo Basiliensi', in Ptolemy, *Geographia* (Basel, 1540; facsimile edn Amsterdam, 1966).

Navagero, A., *Andreae Nagerii Opera omnia quae quidem magna adhibita diligentia colligi potuerant*, ed. by Gaetano Cristoforo Volpi and Giovanni Antonio Volpi (Padua: J. Coominus, 1718).

Nebrija, A., *Introductorium Cosmographiae*, ed. by V. B. Sanchez (Seville: Lebrija, 2000, first published 1503).

Ortelius, A., *Abrahami Ortelli... Thesaurus Geographicus recognitus et actus* [in Latin] (Antverpiae: ex officina Plantiniana, 1596).

Oviedo, F. de, 'La navigazione del grandissimo fiume maragnon', in *Giovanni Battista Ramusio, Navigazioni e viaggi*, ed. by Marica Milanesi, 6 vols, vol. 6 (Turin: Giulio Einaudi, 1980), 859–70.

——., 'Sommario dell'Indie Occidentali', in *Giovanni Battista Ramusio, Navigazioni e viaggi*, ed. by Marica Milanesi, 6 vols, vol. 5 (Turin: Giulio Einaudi, 1980), 19–206.

——., 'Della naturale e generale istoria dell'Indie a' tempi nostri ritrovate, libro primo', in *Giovanni Battista Ramusio, Navigazioni e viaggi*, ed. by Marica Milanesi, 6 vols, vol. 6 (Turin: Giulio Einaudi, 1985), 345–956.

Pacioli, L. a. *Sūma De Arithmetica, Geometria, Proportioni & Proportionalita* (Venice: Paganino de Paganini, 1494).

Pires, T. 'The Suma Oriental of Tomé Pires : An Account of the East, from the Red Sea to Japan, Written in Malacca and India in 1512-1515, and the Book of Francisco Rodrigues, Rutter of a Voyage in the Red Sea, Nautical Rules, Almanack and Maps, Written and Drawn in the East before 1515', ed. by Armando Cortesão, 2 vols (London: The Hakluyt society, 1944).

Polo, M., 'Dei viaggi di messer Marco Polo, gentiluomo veneziano', in *Giovanni Battista Ramusio, Navigazioni e viaggi*, ed. by Marica Milanesi, 6 vols, vol. 3 (Turin: Giulio Einaudi, 1980).

Postel, G., *Cosmographicae disciplinae compendium* (Basel: Per Ioannem Oporinum, 1561).

Preciado, F., 'Ulloa's Voyage to Gulf of California', in *The Principal Navigations,*

Voiages, Traffiques and Discoveries of the English Nation, ed. by Richard Hakluyt, 2nd edn, 12 vols. Vol. 9 (Glasgow: Maclehose and Sons, 1903, first published 1598–1600. Reprint), 207–78.

Ptolemy, *La Geografia*, trans. by Pietro Andrea Mattiolo, ed. by Sebastian Münster, Giacomo Gastaldi and Pietro Andrea Mattiolo (Venice: Gioa. Baptista Pedrezano, 1548).

Purchas, S., *Hakluytus Posthumus or, Purchas His Pilgrimes*. Vol. 17 (Glasgow: Maclehose and Sons, 1905).

Raleigh, W. S., *The Discouerie of the Large, Rich, and Bevvtiful Empire of Guiana,* 2nd edn (Robert Robinson: London, 1596).

——., *The History of the World* [in English] (London: printed by William Iaggard for Walter Burre, 1617; 1st edition 1614).

Ramusio, G. B. ed., *Navigazioni e viaggi*, ed. by Marica Milanesi, 6 vols (Turin: Giulio Einaudi, 1978, first published Venice: Giunti, 1550–1559).

——., 'All'eccellentiss. M. Ieronimo Fracastor. Gio. Battista Ramusio', in *Giovanni Battista Ramusio, Navigazioni e viaggi*, ed. by Marica Milanesi, 6 vols, vol. 1 (Turin: Giulio Einaudi, 1978), 1–6.

——., 'Alli Lettori', in *Giovanni Battista Ramusio, Navigazioni e viaggi*, ed. by Marica Milanesi, 6 vols, vol. 2 (Turin: Giulio Einaudi, 1978), 1043–44.

——., 'Dichiarazione d'alcuni luoghi ne' libri di messer Marco Polo, con l'istoria del reubarbaro', in *Giovanni Battista Ramusio, Navigazioni e viaggi*, ed. by Marica Milanesi, 6 vols, vol. 3 (Turin: Giulio Einaudi, 1980), 56–71.

——., 'Discorso di messer Gio. Battista Ramusio sopra il crescer del fiume Nilo, allo eccellenitssimo messer Ieronimo Fracastoro', in *Giovanni Battista Ramusio, Navigazioni e viaggi*, ed. by Marica Milanesi, 6 vols, vol. 2 (Turin: Giulio Einaudi, 1978), 391–405.

——., 'Discorso di messer Gio. Battista Ramusio sopra il terzo volume delle Navigazioni e Viaggi nella parte del Mondo Nuovo', in *Giovanni Battista Ramusio, Navigazioni e viaggi*, ed. by Marica Milanesi, 6 vols, vol. 5 (Turin: Giulio Einaudi, 1984), 3–18.

——., 'Discorso sopra il libro di Odoardo Barbosa e sopra il sommario dell Indie Orientali, in *Giovanni Battista Ramusio, Navigazioni e viaggi*, ed. by Marica Milanesi, 6 vols, vol. 2 (Turin: Giulio Einaudi, 1978), 541–3.

——., 'Discorso sopra la Navigazione di Annone Cartaginese, fatto per un pilotto portoghese', in *Giovanni Battista Ramusio, Navigazioni e viaggi*, ed. by Marica Milanesi, 6 vols, vol. 1 (Turin: Giulio Einaudi, 1978), 551–61.

——., 'Discorso sopra la navigazione di Iambolo mercante antichissimo', in *Giovanni Battista Ramusio, Navigazioni e viaggi*, ed. by Marica Milanesi, 6 vols, vol. 1 (Turin: Giulio Einaudi, 1978), 903–8.

——., 'Navigazione di Sebastiano Caboto: prefazione nella seguente navigazione', in *Giovanni Battista Ramusio, Navigazioni e viaggi*, ed. by Marica Milanesi, 6 vols, vol. 4 (Turin: Giulio Einaudi, 1978), 105–8.

——., 'A Speciall Note Gathered by the Excellent Venetian Cosmographer M. John Baptista Ramusius out of the Arabian Geographie of Abilfada Ismael, Concerning the Trending of the Ocean Sea from China Northward Along the Coast of Tartarie and Other Unkowne Lands, and Then Running Westwards Upon the Northerne Coasts of Russia, and So Farther to the Northwest', in *The Principal Navigations, Voiages, Traffiques and Discoveries of the English Nation*, ed. by Richard Hakluyt, 2nd edn (repr.), 12 vols, vol. 3 (Glasgow: Maclehose and Sons, 1903; first published 1598–1600), 412–19.

Recorde, R. *The Castle of Knowledge* (London: Reyner Wolfe, 1556).

Roy, L. le, *De la vicissitude ou variété des choses en l'univers, et concurrence des armes et des lettres par les premieres et plus illustres nations du monde, depuis le temps où à commencé la civilité, et memoire humane iusques à present,* 2nd edition (Paris: Pierre l'Huilier, 1576).

Sévérac, J. C. de, 'Mirabilia Descripta, les merveilles de l'Asie', ed. by H. Cordier (Paris: Geuthner, 1925).

Signot, J., *La division du monde, contenant la déclaration des provinces et régions d'Asie, Europe et Aphricque* (Paris: Lotrian, 1539).

Sylvano, B., 'Dedicatory Letter to Acquaviva Sylvanus', in *C. Ptolemaeus, Cosmographia*, ed. by Bernardo Sylvano and R. A. Skelton (Venice, 1511; facsimile edition Amsterdam: Theatrum Orbis Terrarum, 1969), VII.

Thevet, A., *Cosmographie universelle illustrée de diverses figures des choses plus remarquables vevés par l'auteur, & incognues de noz anciens & modernes* (Paris: Pierre L'Huillier, 1575).

——., *Les singularitez de la France Antarctique, autrement nommée Amérique, & de plusieurs terres et isles découvertes de nostre temps* (Paris: Maurice de La Port, 1558).

——., *The New Founde Worlde of Antarctike*, trans. by Thomas Hacket (London: Printed for Thomas Hacket, 1568).

——., *Pourtraits et vies des hommes illustres grecz latins et payens*, 2 vols, vol. 2 (Paris: La vefue I. Keruert & G. Chaudière, 1584).

Thorne, R., 'The Booke Made by the Right Worshipful M. Robert Thorne in the Yeere 1527 in Sivil to Dr. Ley, Lord Ambassadour for King Henry the Eight, to Charles the Emperour, Being an Information of the Parts of the World, Discovered by Him and the King of Portingal: And Also of the Way to the Moluccas by the North', in *The Principal Navigations, Voiages, Traffiques and Discoveries of the English Nation*, ed. by Richard Hakluyt, 2nd edn (repr.), 12 vols, vol. 2 (Glasgow: Maclehose and Sons, 1903; first published 1598–1600), 164–78.

——., 'A Declaration of the Indies and Lands Discovered and Subdued Unto the Emperour, and the King of Portugal: And Also of Other Partes of the Indies and Rich Countries to Be Discovered Which the Worshipfull M. Robert Thorne Merchant of London (Who Dwelt Long in the Citie of Sivil) in Spaine

Exhorted King Henrie the Eight to Take in Hand (Ad 1527)', in *The Principal Navigations, Voiages, Traffiques and Discoveries of the English Nation*, ed. by Richard Hakluyt, 2nd edn (repr.), 12 vols, vol. 3 (Glasgow: Maclehose and Sons, 1903; first published 1598–1600), 259–61.

Torquemada, A. de, *Jardin de flores curiosas, en que se tratan algunas materias de humanidad, Philosophia, Theologia, Y Geographia, con otras cosas curiosas, etc* (Salamanca: J. B. de Terranova, 1570).

Transylvanus, M., *De Moluccis Insulis* (Cologne: Eucharius Cervicornus, 1523).

Van Durme, M., *Correspondance Mercatorienne* (Anvers: de Nederlandsche Boekhandel, 1959).

Varthema, L., 'Itinerario di Lodovico Barthema', in *Giovanni Battista Ramusio, Navigazioni e viaggi*, ed. by Marica Milanesi, 6 vols, vol. 1 (Turin: Giulio Einaudi, 1978), 763–892.

Verazzano, G., 'Al cristianissimo re di Francia Francesco Primo, relazione di Giovanni da Verrazzano fiorentino della terra per lui scoperta in nome di sua Maestà, scritta in Dieppa, adì 8 luglio MDXXIIII', in *Giovanni Battista Ramusio, Navigazioni e viaggi*, ed. by Marica Milanesi, 6 vols, vol. 6 (Turin: Giulio Einaudi, 1980), 887–906.

Vespucci, A., 'Sommario di Amerigo Vespucci fiorentino, di due sue Navigazioni, al magnifico M. Pietro Soderini, gonfalonier della magnifica republica di Firenze." In *Giovanni Battista Ramusio, Navigazioni e viaggi*, edited by Marica Milanesi, 6 vols, vol. 1 (Turin: Giulio Einaudi, 1980), 659–65.

——., 'Epistola De Alberico Vesputio', chap. 5, in *Paesi novamente retrovati*, ed. by Fracanzano da Montalboddo (Vicenza, 1507).

'Vita Sanctissimi Confessoris Christi Brendani' (Transcription du manuscrit d'Alençon à la bibliothèque municipale d'Alençon, Codex 14, fol. 1 r à 11 v., chapter xiv, 11th century' <https://www.hs-augsburg.de/~harsch/Chronologia/Lspost10/Brendanus/bre_navi.html> (accessed 30 November 2019).

Willes, R., 'Certaine Other Reasons, or Arguments to Prove a Passage by the Northwest, Learnedly Written by M. Richard Willes Gentleman', in *The Principal Navigations, Voiages, Traffiques and Discoveries of the English Nation*, ed. by Richard Hakluyt, 2nd edn (repr.), 12 vols, vol. 7 (Glasgow: Maclehose and Sons, 1903; first published 1598–1600), 191–203.

Bibliography of Secondary Literature

Abulafia, D., *The Discovery of Mankind; Atlantic Encounters in the Age of Columbus* (New Haven: Yale University Press, 2008).

Adams, F. D., and A. S. Woodward, *The Birth and Development of the Geological Sciences* (Baltimore: The Williams & Wilkins company, 1938).

Anastos, M. V., 'Pletho, Strabo and Columbus', ch. 17, in *Studies in Byzantine Intellectual History*, 1979.

Armitage, D., *The Ideological Origins of the British Empire*, Ideas in Context (Cambridge: Cambridge University Press, 2000).

Ashe, G., *Land to the West: St. Brendan's Voyage to America* (London: Collins, 1962).

Atkinson, G., *Les Nouvelles horizons de la Renaissance française* (Paris: Droz, 1935).

Baert, A., *Le Paradis terrestre: un mythe espagnol en Océanie: les voyages de Mendaña et e Quirós, 1567–1606* (Paris: L'Harmattan, 1999), 193–213.

Baker, J. N. L., 'Academic Geography in the Seventeenth and Eighteenth Centuries', in *The History of Geography: Papers by J.N.L. Baker* (Oxford, 1963), 14–32.

Barnes, J. R., 'Giovanni Battista Ramusio and the History of Discoveries: An Analysis of Ramusio's Commentary, Cartography and Imagery, in *Delle Navigationi et viaggi* (University of Texas, 2007).

Bartra, R., and C. T. Berrisford, *Wild Men in the Looking Glass: The Mythic Origins of European Otherness* (Ann Arbor: University of Michigan Press, 1994).

Beagon, M., *Roman Nature: The Thought of Pliny the Elder* (Oxford: Clarendon Press, 1992).

Beecher, D., 'The Legacy of John Frampton: Elizabethan Trader and Translator', *Renaissance Studies*, 20, no. 3 (2006), 320–39.

Benton, L., *A Search for Sovereignty: Law and Geography in European Empires, 1400–1900* (Cambridge: Cambridge University Press, 2009).

Berggren, J. L., and J. Evans, *Geminos's Introduction and Study of a Hellenistic Survey of Astronomy* (Princeton: Princeton University Press, 2006).

Berggren, J. L., and A. Jones, *Ptolemy's Geography: An Annotated Translation of the Theoretical Chapters* (Princeton: Princeton University Press, 2000).

Besse, J.-M, *Les Grandeurs de la terre; aspects du savoir géographique à la Renaissance* (Lyon: ENS editions, 2003).

Biraschi, A. M., P. Maribelli, G. D. Massaro, and A. Pagnotta, eds, *Strabone: saggio di bibliografia 1469–1978* (Perugia, 1981).

Boardman, J., *The Greeks Overseas: Their Early Colony and Trade*, 4th edn (London: Thames and Hudson, 1999).

Bolton, J. D. P., Aristeas of Proconnesus (Oxford: Clarendon Press, 1962).

Boumans, R., 'The Religious Views of Abraham Ortelius', *Journal of the Warburg and Courtauld Institutes,* 17, no. 3/4 (1954), 374–77.

Bowen, M., *Empiricism and Geographical Thought from Francis Bacon to Alexander Von Humboldt* (Cambridge: Cambridge University Press, 1981).

Bowen, W. R., 'Ficino's Analysis of Musical Harmonia', in *Ficino and Renaissance Neoplatonism*, ed. by Konrad Eisenbichler and Olga Zorzi Puglies (Ottowa: Dovehouse editions, 1986), 18–28.

Bowersock, G. W., 'Tacitus and the Province of Asia', in *Tacitus and the Tacitean Tradition*, ed. by T. J. Luce and A. J. Woodman (Princeton: Princeton University Press, 1993), 3–10.

Boxer, C. R., and J. de Barros, *João de Barros: Portuguese Humanist and Historian of Asia* (New Delhi: Concept Publishing Company, 1981).

Brammall, K. M., 'Monstrous Metamorphosis: Nature, Morality and the Rhetoric of Monstrosity in Tudor England', *Sixteenth Century Journal*, 27 (1996), 3–21.

Braudel, F., and S. Reynolds, *The Mediterranean and the Mediterranean World in the Age of Philip II* (Berkeley: University of California Press, 1995).

Broc, N., *La géographie de la Renaissance (1420–1620)* (Paris: Bibliothèque nationale, 1980).

Broggiato, M., ed., *Cratete di Mallo: I Frammenti* (La Spezia: Agorà, 2001).

Bucher, B., *Icon and Conquest: A Structural Analysis of the Illustrations of De Bry's Great Voyages*, trans. by Basia Miller Gulati (Chicago: Chicago University Press, 1981).

Bunbury, E. H., *A History of Ancient Geography among the Greeks and Romans From the Earliest Ages Till the Fall of the Roman Empire*, 2 vols, vol. 2 (New York: Dover Publications, 1959; first printed 1879).

Burgaleta, C. M., *José De Acosta, S. J., 1540–1600: His Life and Thought* (Chicago: Jesuit Way, 1999).

Burmeister, K., *Sebastian Münster: Versuch Eines Biographischen Gesamtbildes*, 2nd edn (Basel: Basler Beiträge zur Geschichtswissenschaft, 1969).

Burnet, J., *Early Greek Philosophy*, 4th edn (London: Black, 1958).

Büttner, M., 'The Significance of the Reformation for the Reorientation of Geography in Lutheran Germany', *History of Science*, 17 (1979), 151–69.

Butzer, K. W., 'From Columbus to Acosta: Science, Discovery and the New World', *Annals of the Association of American Geographers*, 82 (1992), 543–65.

Campbell, M. B., *The Witness and the Other World; Exotic European Travel Writing 400–1600* (Ithaca and London: Cornell University Press, 1988).

Carlton, G., *Worldly Consumers: The Demand for Maps in Renaissance Italy* (Chicago: The University of Chicago Press, 2015).

Carvalho, J. B. de, *La traduction espagnole du 'De Situ Orbis' de Pomponious Mela par Maitre Joan Faras et les notes marginales de Duarte Pacheco Pereira* (Lisbon, 1974).

Cassidy, V. H. de P., 'The Voyage of an Island', *Speculum*, 38 (1963), 595–602.

Cawley, A. C., 'Down Under: A Possible New Source for Mandeville's Travels', in *Iceland and the Mediaeval World, Studies in Honour of Ian Maxwell*, ed. by G. Turville-Peter and J. Stanley Martin (Melbourne, Australia, 1974), 144–49.

Céard, J., *La nature et les prodiges; L'insolite au XVI siècle en France* (Geneva: Droz, 1977).

Checkin, L. S., 'Mappae Mundi and Scandinavia', *Scandinavian Studies*, 65, no. 3 (1993), 487–521.

Chiappelli, F., ed., *First Images of America: The Impact of the New World on the Old* (Berkeley: University of California Press, 1976).

Chibnall, M., 'Pliny's Natural History and the Middle Ages', in *Empire and Aftermath*, ed. by T. A. Dorey (London: Routledge, 1975), 57–79.

Ciccolella , F., 'The Greek Donatus and the Study of Greek in the Renaissance', *International Journal of the Classical Tradition*, 12 (2005), 1–24.

Clarke, K., *Between Geography and History: Hellenistic Constructions of the Roman World* (Oxford: Oxford University Press, 1999).

——., 'An Island Nation: Re-Reading Tacitus' *Agricola*, *Journal of Roman Studies*, 91 (2001), 94–112.

Clay, D., 'Columbus' Senecan Prophecy', *American Journal of Philology*, 113 (1992), 117–20.

Clough, C. H., 'The New World and the Renaissance', in *The European Outthrust and Encounter, the First Phase, c. 1400–c. 1700: Essays in Honour of David Beers Quinn on His 85th Birthday*, ed. by C. H. Clough and P. E. H. Hair (Liverpool: Liverpool University Press, 1994), 291–328.

Codazzi, A., *Le Edizioni quattrocentesche e cinquecentesche della 'Geografia', di Tolomeo* (Milan: Goliardica, 1949).

Cohen, J. J., *Monster Theory: Reading Culture* (Minneapolis: University of Minnesota Press, 1996).

Colon, C., *Historia del Almirante Don Cristobal Colon*, Colección de libros raros o curiosos que tratan de America , 2 vols, vol. 2 (Madrid: Tomas Minues, 1892).

Conley, T., 'Early Modern Literature and Cartography: an Overview', in *Cartography in the European Renaissance; the History of Cartography*, ed. by David Woodward, vol. 3.1 (Chicago: University of Chicago Press, 2007), 401–10.

——., *The Self-Made Map: Cartographic Writing in Early Modern France* (Minneapolis: University of Minnesota Press, 2010).

Cook, H., 'Ancient Wisdom, the Golden Age, and Atlantis: The New World in Sixteenth-Century Cosmography', *Terrae Incognitae*, 10 (1978), 25–43.

Cordano, F., *La Geografia Degli Antichi* (Roma: Laterza, 1992).

Cormack, L. B., 'Flat Earth or Round Sphere: Misconceptions of the Earth and the Fifteenth-Century Transformation of the World', *Ecumene*, 1(1994), 363–86.

——., 'Good Fences Make Good Neighbours; Geography in Early Modern England', *Isis* (1991), 631–61.

——., *Charting an Empire: Geography at the English Universities 1580–1620* (Chicago: Chicago University Press, 1997).

——., 'Ptolemy at Work: The Role of the *Geography* in Geography and Mathematics Teaching in Early Modern England', in *Ptolemy's Geography in the Renaissance*, ed. by Zur Shalev and Charles Burnett, Warburg Institute Colloquia, Vol. 17 (London: The Warburg Institute, 2011), 207–24.

Cornford, F. M., *From Religion to Philosophy: A Study in the Origins of Western Speculation* (New York, 1957).

Cosgrove, D. E., 'Mapping New Worlds: Culture and Cartography in Sixteenth–Century Venice', *Imago Mundi: The International Journal for the History of Cartography*, 44 (1992), 65–89.

——., *Apollo's Eye: A Cartographic Genealogy of the Earth in the Western Imagination* (London: Johns Hopkins University Press, 2001).

Coulter, J. A. *The Literary Microcosm: Theories of Interpretation of the Later Neoplatonists* [in English; Greek, Ancient (to 1453) (summary)] (Leiden: Brill, 1976).

Couprie, D. L., Heaven and Earth in Ancient Greek Cosmology: From Thales to Heraclides Ponticus (New York: Springer, 2011).

Crane, N., *Mercator: The Man Who Mapped the Planet* (London: Weidenfeld and Nicholson, 2002).

Crombie, A. C., *Styles of Scientific Thinking in the European Tradition: The History of Argument and Explanation Especially in the Mathematical and Biomedical Sciences and Arts* (London: Duckworth, 1994).

Cuningham, W., *The Cosmographical Glasse* (London: in officina Ioan Daij, 1559).

Cunliffe, B., *The Extraordinary Voyage of Pytheas the Greek: The Man Who Discovered Britain.* (London: Penguin Books, 2001).

Dalché, P. G., 'L'œuvre géographique du Cardinal Fillastre (1428). Représentation Du monde et perception de la carte à l'aube des découvertes', in *Humanisme et culture géographique à l'époque du concile de Constance: Autour de Guillaume Fillastre, Actes du colloque de l'université de Reims*, ed. by Didier Marcotte (Turnhout: Brepols, 2002), 293–355.

——., *La Géographie De Ptolémée En Occident* (Turnhout: Brepols, 2009).

——., 'Principes et modes de la représentation de l'espace géographique durant le Haut Moyen Âge', in *L'espace géographique au Moyen Âge (*Florence: Edizioni del Galuzzo, 2013; first published 2003), 119–50.

——, 'The Reception of Ptolemy's Geography (End of the Fourteenth to Beginning of the Sixteenth Century), in *Cartography in the European Renaissance: the History of Cartography*, ed. by David Woodward, vol. 3.1 (Chicago: University of Chicago Press, 2007), 285–364.

——., *L'espace géographique au Moyen Âge (*Florence: Edizioni del Galuzzo, 2013; first published 2003).

Daston, L., and K. Park, 'Unnatural Conceptions: The Study of Monsters in Sixteenth- and Seventeenth-Century France and England', *Past and Present,* 92 (1981), 20–54.

Daston, L., and K. Park, *Wonders and the Order of Nature, 1150–1750* (New York: Zone Books, 1998).

Davies, S., 'America and Amerindians in Sebastian Münster's Cosmographiae Universalis Libri VI (1550)', *Renaissance Studies*, 25 (2011), 351–73.

——., *Renaissance Ethnography and the Invention of the Human: New Worlds, Maps and Monsters* (Cambridge: Cambridge University Press, 2016).

——., 'Unlucky, Bad and Ugly: Categories of Monstrosity from the Renaissance to the Enlightenment', in *The Ashgate Research Companion to Monsters and the Monstrous*, ed. by Asa Simon Mittman (Farnham, Surrey, England: Ashgate, 2012), 49–77.

——., 'The Wondrous East in the Renaissance Geographical Imagination: Marco Polo, Fra Mauro and Giovanni Battista Ramusio', *History and Anthropology*, 23, no. 2 (2012), 215–34.

Debus, A. G., 'Introduction', in *Mathematicall Praeface to the Elements of Geometrie of Euclid of Megara (1570)*, ed. by John Dee (New York: Science History Publications, 1975).

Depuydt, J., 'Le Cercle d'amis et de correspondants autour d'Abraham Ortelius', in *Abraham Ortelius 1527–1598 Cartographe et humaniste* (Turnhout: Brepols, 1998), 117–40.

Diehl, E., *Anthologia Lyrica Graeca*, vol. 1 (Leipzig: Teubner, 1925).

Dilke, O. A. W., *Greek and Roman Maps* (London: Thames and Hudson, 1985).

Dilke, O. A. W., and M. Dilke, 'The Adjustment of Ptolemaic Atlases to Feature the New World', in *The Classical Tradition and the Americas: European Images of the Americas and the Classical Tradition*, ed. by Wolfgang Haase and Meyer Reinhold, vol. 1 (Berlin: De Gruyter, 1994), 117–34.

Diller, A., 'The Oldest Manuscripts of Ptolemaic Maps', *Transactions and Proceedings of the American Philological Association*, 71 (1940), 62–67.

——., *The Textual Tradition of Strabo's Geography with Appendix: The Manuscripts of Eustathius' Commentary on Dionysius Periegetes* (Amsterdam: A. M. Hakkert, 1975).

Donattini, M., 'Giovanni Battista Ramusio e le sue 'Navigationi', appunti per una biografia', *Critica Storica, Rivista Trimestrale*, 17 (1980–1981), 55–100.

Dueck, D., *Strabo of Amasia: A Greek Man of Letters in Augustan Rome* (London and New York: Routledge, 2000).

Dueck D., and K. Brodersen, 'Introduction', in *Geography in Classical Antiquity, Key Themes in Ancient History* (Cambridge: Cambridge University Press, 2012).

Duzer, C. v., 'Hic Sunt Dracones: The Geography and Cartography of Monsters', in *The Ashgate Research Companion to Monsters and the Monstrous* (London: Routledge, 2013), 387–435.

Earle, R., *The Body of the Conquistador: Food, Race, and the Colonial Experience in Spanish America, 1492–1700* (Cambridge: Cambridge University Press, 2012).

Eastwood, B., 'Mediaeval Empiricism: The Case of Grosseteste's Optics', *Speculum*, 43 (1968), 306–21.

Edelstein, L., and I. G. Kidd. *Posidonius. Vol. 1 the Fragments*, 2nd edn, vol. 1 (Cambridge: Cambridge University Press, 1989).

Edgerton, S. Y. J., 'From Mental Matrix to "Mappamundi" to Christian Empire: The Heritage of Ptolemaic Cartography in the Renaissance', in *Art and Cartography*, ed. by David Woodward (Chicago: Chicago University Press, 1987), 10–51.

Edson, E., *The World Map, 1300–1492: The Persistence of Tradition and Transformation* (Baltimore: Johns Hopkins University Press, 2007).

Elliott, J. H., *The Old World and the New*, 1st Canto edn (Cambridge: Cambridge University Press, 1970)

Eratosthenes, and D. W. Roller, *Eratosthenes' Geography* (Princeton: Princeton University Press, 2010).

F. Preciado, 'Ulloa's Voyage to Gulf of California', in *The Principal Navigations, Voiages, Traffiques and Discoveries of the English Natione*, ed. by Richard Hakluyt, 2nd edn (repr.), 12 vols, vol. 12 (Glasgow: Maclehose and Sons, 1903; first published 1598–1600).

Fauvel, J., 'Platonic Rhetoric in Distance Learning: How Robert Record Taught the Home Learner', *For the Learning of Mathematics*, 9, no. 1 (1989), 2–6.

Fernández-Armesto, F., *Before Columbus: Exploration and Colonisation from the Mediterranean to the Atlantic 1229–1492* (Basingstoke: Macmillan Education, 1987).

——., *Columbus and the Conquest of the Impossible* (London: Phoenix Press, 1974).

Fisher, P., 'Ab Aquilone Aurum Venit; Olaus Magnus's Description of the North and His Feelings for His Homeland', *Journal of the Institute of Romance Studies*, 2 (1993), 181–90.

——., 'Olaus Magnus and the Limits of Credence', in *Acta Conventus Neo-Latini*

Hafniensis (Binghamton, NY: Medieval & Renaissance Texts & Studies, 1995), 409–19.
Flint, V., '*The Imaginative Landscape of Christopher Columbus* (Princeton Princeton University Press, 1992).
Foote, P., 'Introduction', in *Description of the Northern Peoples*, ed. by Peter Foote, vols 182, 187, 188 (London: Hakluyt Society, 1996–1998).
Friedman, J. B., 'Foreword', *The Ashgate Research Companion to Monsters and the Monstrous*, ed. by Asa Simon Mittman (Farnham, Surrey, England: Ashgate, 2012).
——., *The Monstrous Races in Medieval Art and Thought* (New York: Syracuse University Press, 2000).
Fryde, E. B., *Humanism and Renaissance Historiography* (Hambledon Press: London, 1983).
Fuller, M. C., *Voyages in Print: English Travel to America, 1576–1624* (Cambridge: Cambridge University Press, 1995).
Ganong, W. F. *Crucial Maps in the Early Cartography and Place-Nomenclature of the Atlantic Coast of Canada* Toronto: University of Toronto Press, 1964.
Gerbi, A., and J. Moyle, *Nature in the New World: From Christopher Columbus to Gonzalo Fernandez De Oviedo* (Pittsburgh: University of Pittsburgh Press, 1985).
Gersh, S., 'The Medieval Legacy from Ancient Platonism', in *The Platonic Tradition in the Middle Ages: A Doxographic Approach*, ed. by S. Gersh and J. F. M. Maarten (New York: De Gruyter, 2002), 3–30.
Gilbert, H., 'A Discourse Written by Sir Humphrey Gilbert Knight, to Prove a Passage by the Northwest to Cathaia and the East Indies', in *The Principal Navigations, Voiages, Traffiques and Discoveries of the English Nation*, ed. by Richard Hakluyt, 2nd edition (repr.), 12 vols, vol. 7 (Glasgow: Maclehose and Sons, 1903; first published 1598–1600), 158–90.
Gillies, J., *Shakespeare and the Geography of Difference* (Cambridge: Cambridge University Press, 1994).
Glacken, C. J., *Traces on the Rhodian Shore: Nature and Culture in Western Thought from Ancient Times to the End of the Eighteenth Century (*Berkeley: University of California Press, 1967).
Goldstein, T., 'Geography in Fifteenth-Century Florence', in *Merchants and Scholars: Essays in the History of Exploration and Trade: Collected in Memory of James Ford Bell*, ed. by John Parker (Minneapolis: University of Minnesota Press, 1965), 9–32.
Gormley, C. M., M. A. Rouse, and R. H. Rouse, 'The Medieval Circulation of the De Chorographia of Pomponius Mela', *Medieval Studies*, 46 (1984), 266–320.
Goss, J. J. S., *The Mapmaker's Art: A History of Cartography* (London: Studio Editions, 1993).

Grafton, A., *Joseph Scaliger: A Study in the History of Classical Scholarship* vol. 2 (Oxford: Clarendon Press, 1993).
Grafton, A., A. Shelford, and N. Siraisi, *New Worlds, Ancient Texts: The Power of Tradition and the Shock of Discovery* (Cambridge, MA: Harvard University Press, 1992).
Grande, S., 'Le relazione geografiche fra P. Bembo, G. Fracastoro, G. B. Ramusio, G. Gastaldi." *Memorie della società geografica italiana,* 12 (1956), 93–197.
Greenblatt, S. J., *Marvelous Possessions: The Wonder of the New World: The Clarendon and the Carpenter Lectures 1988* (Oxford: Clarendon, 1991).
Grendler, P. F., *Schooling in Renaissance Italy, Literacy and Learning, 1300–1600 (*Baltimore: John Hopkins University Press, 1989).
——. *The Universities of the Italian Renaissance* (Baltimore: Johns Hopkins University Press, 2001).
Groesen, M. v., 'The De Bry Collection of Voyages (1590–1634): Early America Reconsidered', *Journal of Early Modern History*, 12, no. 1 (2008), 1–24.
——., *The Representations of the Overseas World in the De Bry Collection of Voyages (1590–1634)* (2008).
Gudger, E. W., 'Pliny's 'Historia Naturalis, the Most Popular Natural History Ever Published', *Isis*, 6 (1924), 269–81.
Haase, W., and M. Reinhold, *The Classical Tradition and the Americas* (Berlin: De Gruyter, 1994).
Hair, P. E. H., 'The Early Sources on Guinea', *History in Africa*, 21 (1994), 87–126.
——., 'The "Periplus of Hanno" in the History and Historiography of Black Africa', *History in Africa*, 14 (1987), 56–58.
Hamilton, A., *The Family of Love* (Cambridge: James Clarke and Co. Ltd., 1981).
Hamilton, A., and L. Garbett, *The Apocryphal Apocalypse: The Reception of the Second Book of Esdras (4 Ezra) from the Renaissance to the Enlightenment* (Oxford: Oxford University Press, 1999).
Hanke, L., *All Mankind Is One: A Study of the Disputation between Bartolomé De Las Casas and Juan Ginés De Sepúlveda in 1550 on the Intellectual and Religious Capacity of the American Indians* ((DeKalb, Ill: Northern Illinois University Press, 1994; 1st edition 1974).
——., *'All the Peoples of the World Are Men'*, James Ford Bell Lectures, 8 (Minnesota: University of Minnesota Press, 1970).
——., *Aristotle and the American Indians: A Study in Race Prejudice in the Modern World* (Chicago: Henry Regnery, 1959).
——., *The First Social Experiments in America a Study in the Development of Spanish Indian Policy in the Sixteenth Century (*Cambridge, MA.: Harvard University Press, 1935).
Hankin, J., 'Galileo, Ficino and Renaissance Platonism', in *Humanism and Early*

Modern Philosophy, ed. by Jill Kraye and M. W. F. Stone (London: Routledge, 2000), 209–37.

——., 'The Study of the Timaeus in Early Renaissance Italy', in *Natural Particulars: Nature and the Disciplines in Renaissance Europe*, ed. by Antony Grafton and Nancy Siraisi (Cambridge, MA: MIT Press, 1999), 77–120.

Hanson, E. P., *New Worlds Emerging* (London: Victor Gollancz Ltd, 1950).

Harley, J. B., 'Silences and Secrecy: The Hidden Agenda of Cartography in Early Modern Europe', *Imago Mundi; The International Journal for the History of Cartography*, 40 (1988), 57–76.

Harpocration, *Lexeis of the Ten Orators*, ed. by J. Keaney (Amsterdam: A. M. Hakkert, 1991).

Hartog, F., and J. Lloyd, *The Mirror of Herodotus: The Representation of the Other in the Writing of History* (Berkeley: University of California Press, 1988).

Harvey, P. D. A., 'Medieval Maps: An Introduction', in *The History of Cartography*, ed. by J. B. Harley and D. Woodward, vol. 1 (Chicago and London: University of Chicago Press, 1987), 283–85.

Hatzimichali, M., 'Strabo's Philosophy and Stoicism', in *The Routledge Companion to Strabo*, ed. by Daniela Dueck (London: Routledge, 2017), 9–21.

Hawkes, C. F. C., 'Pytheas: Europe and the Greek Explorers', in *The Eighth J. L. Myres Memorial Lecture* (Oxford: Blackwell, 1975).

Headley, J. M., 'The Sixteenth-Century Celebration of the Earth's Total Habitability: The Issue of the Fully Habitable World for Renaissance Europe', *The Journal of World History*, 8 (1997), 4–18.

Hemmerdinger, B., *Les manuscrits d'Hérodote et la critique verbale* (Genoa: Istituto di filologia classica publicaziavi, 1981).

Hiatt, A., 'Blank Spaces on the Earth' *The Yale Journal of Criticism*, 15 (2002), 223–50.

——., *Terra Incognita; Mapping the Antipodes before 1600* (London: British Library, 2008).

Hirsch, R., 'Printed Reports on the Early Discoveries and Their Reception', in *First Images of America: The Impact of the New World on the Old*, ed. by Fredi Chiappelli, 2 vols, vol. 2 (Berkeley: University of California Press, 1976), 537–58.

——., 'Printing and the Spread of Humanism in Germany', in *The Printed Word: Its Impact and Diffusion (Primarily in the Fifteenth and Sixteenth Centuries)*, vol. 3 (London: Ashgate, 1978), 24–37.

Horden, P., and N. Purcell, *The Corrupting Sea: A Study of Mediterranean History* (Oxford: Blackwell, 2000).

Hornblower, S., and A. J. S. Spawforth, *The Oxford Classical Dictionary*, rev. 3rd edn (Oxford: Oxford University Press, 2003).

Huffman, C. A., *Philolaus of Croton: Pythagorean and Presocratic, a Commentary on the Fragments and Testimonia with Interpretive Essays* (Cambridge: Cambridge University Press, 1993).

Hulme, P., 'Introduction: The Cannibal Scene', in *Cannibalism and the Colonial World*, ed. by Francis Barker, Peter Hulme, and Margaret Iversen (Cambridge: Cambridge University Press, 1998), 1–39.

Hulton, P., ed., *The Work of Jacques Le Moyne De Morgues: A Huguenot Artist in France, Florida and England* (London: British Museum Publications, 1977).

Ingstad, A. S., and E. S. Seeberg, *The Discovery of a Norse Settlement in America* (Oslo: Universitetsforlaget, 1977).

Ingstad, H., and A. S. Ingstad, *The Viking Discovery of America: The Excavation of a Norse Settlement in L'anse Aux Meadows, Newfoundland* (New York: Checkmark Books, 2001).

Irby, G. I., 'Mapping the World: Greek Initiatives from Homer to Eratosthenes', in *Ancient Perspectives: Maps and Their Place in Mesopotamia, Egypt, Greece, and Rome*, ed. by Richard J. A. Talbert (Chicago and London: Chicago University Press, 2012).

Jacob, C., *Géographie et ethnographie en Grèce ancienne* (Paris: Armand Colin, 1991).

Jane, L. C., ed., 'Introduction', *Select Documents Illustrating the Four Voyages of Columbus*, 2 vols, vol. 2 (London: Hakluyt Society, 1933).

Janni, P., 'I Pigmei in Scandinavia: Vicende di una leggenda etnografica', in *Cultura classica e cultura germanica settentrionale*, ed. by Diego Poli Pietro Janni and Carlo Santini (Rome: Herder, 1985), 113–23.

Johnson, C. R., 'Buying Stories: Ancient Tales, Renaissance Travelers, and the Market for the Marvelous', *Journal of Early Modern History*, 11, no. 6 (2007), 405–46.

——., 'German Cosmographers and the Naming of America', *Past and Present*, 191 (2006), 3–43.

——., *The German Discovery of the World: Renaissance Encounters with the Strange and Marvelous* (Charlottesville: University of Virginia Press, 2008).

Joost-Gaugier, C. L., 'Ptolemy and Strabo and Their Conversation with Appelles and Protogones: Cosmography and Painting in Raphael's School of Athens', *Renaissance Quarterly*, 51 (1998), 761–87.

Kahn, C. H., *Anaximander and the Origins of Greek Cosmology* (New York, 1960).

——., *The Art and Thought of Heraclitus: A New Arrangement and Translation of the Fragments with Literary and Philosophical Commentary* (Cambridge: Cambridge University Press, 1979).

Kaiser, C. B., *Towards a Theology of Scientific Endeavour: The Descent of Science* (Aldershot: Ashgate, 2007).

Kea, R. A., 'Firearms and Warfare on the Gold and Slave Coasts from the

Sixteenth to the Nineteenth Centuries', *The Journal of African History*, 12, no. 2 (1971), 185–213.

Kelly, D. R., 'Tacitus Noster: The *Germania* in the Renaissance and Reformation', in *Tacitus and the Tacitean Tradition*, ed. by T.J. Luce and A. J. Woodman (Princeton: Princeton University Press, 1993), 152–67.

Keuning, J., 'The History of Geographical Map Projections until 1600', *Imago Mundi: The International Journal for the History of Cartography*, 12 (1955), 1–24.

Kirk, G. S., ed., *Heraclitus: The Cosmic Fragments*, 1st edn (Cambridge: Cambridge University Press, 1954).

——., 'Sense and Common Sense in the Development of Greek Philosophy', *Journal of Hellenic Studies*, 81 (1961), 105–18.

Kirk, G. S., J. E. Raven, and M. Schofield, *The Presocratic Philosophers: A Critical History with a Selection of Texts*, 2nd edn (Cambridge: Cambridge University Press, 1983).

Kirk, G. S., J. E. Raven, and M. Schofield, *The Presocratic Philosophers: A Critical History with a Selection of Texts*, 2nd edn (Cambridge: Cambridge University Press, 1983).

Klaus A. Vogel, 'Cosmography', in *The Cambridge History of Science*, ed. by Katherine Park and Lorraine Daston (Cambridge: Cambridge University Press, 2006), Chapter 20.

Klebs, A. C., 'Incunable Editions of Pliny's Historia Naturalis', *Isis*, 24 (1935), 120–21.

Kohl, J. G., *A Descriptive Catalogue of Those Maps, Charts and Survey Relating to America: Which Are Mentioned in Vol. III of Hakluyt's Great Work* (Washington: Henry Polkinhorn, 1857).

Kristeller, P. O., 'Ficino and His Work', in *Marsilio Ficino e il retorno di Platone*, ed. by G. C. Garfagnini (1986), 15–96.

——., 'The First Printed Edition of Plato's Works and the Date of Its Publication (1484)', in *Studies in Renaissance Thought and Letters*, vol. 3 (Rome, 1993), 135–46.

——., *Renaissance Thought: The Classic, Scholastic, and Humanist Strains; a Revised and Enlarged Edition of the Classics and Renaissance Thought* (New York: Harper, 1961).

Labarre, A., 'Diffusion de L'Historia Naturalis de Pline au temps de la Renaissance', in *Festschrift für Claus Nissen* (Wiesbaden: Guido Pressler, 1973), 451–571.

Lach, D. F., *Asia in the Making of Europe*, 3 vols, vol. 1 (Chicago and London: University of Chicago Press, 1965).

Larner, J., 'The Church and the Quattrocento Renaissance in Geography', *Journal of Renaissance Studies*, 12 (1998), 26–39.

——., *Marco Polo and the Discovery of the World* (New Haven: Yale University Press, 1999).

Law, R., 'Dahomey and the Slave Trade: Reflections on the Historiography of the Rise of Dahomey', *Journal of African History*, 27 (1986), 237–67.

Law, R. C. C., 'The Garamantes and Trans-Saharan Enterprise in Classical Times', The Journal of African History, 8 (1967), 181–200.

Lee, H. D. P., 'Introduction', *Timaeus,* trans. with an intro. by H. D. P. Lee (Harmondsworth: Penguin Books, 1965).

Lestringant, F., *Andre Thevet: Cosmographe des derniers Valois* (Geneve: Droz, 1991).

——., 'Fiction de l'espace brésilien à la Renaissance; l'exemple de Guanabara', in *Arts et legendes d'espaces: figures du voyage et rhetoriques du monde, ed. by Christian Jacob et Frank Lestringant* (Paris: Presses de l'Ecole normale supérieure, 1981), 205–57.

——., 'Fortunes de la singularité à la Renaissance, le genre de l'Isolario', in *Écrire le monde à la Renaissance: Quinze études sur Rabelais, Postel, Bodin et la littérature géographique*, ed. by Frank Lestringant (Caen: Paradigme, 1993), 17–48.

——., *Mapping the Renaissance World : The Geographical Imagination in the Age of Discovery* (Berkeley: University of California Press, 1994).

——., and R. Morris, *Cannibals: The Discovery and Representation of the Cannibal from Columbus to Jules Verne* (Berkeley: University of California Press, 1997).

Lewis, M. W., and K. E.Wigen, *The Myth of Continents, a Critique of Metageography* (Berkeley: University of California Press, 1997).

Leydi, S., *Sub Umbra Imperialis Aquilae: Immagini del potere e consenso politico nella Milano di Carlo V* (Firenze: Olschki, 1999).

Lilley, K. D., ed., *Mapping Medieval Geographies: Cartography and Geographical Thought in the Latin West and Beyond: 300–1600* (Cambridge: Cambridge University Press).

Little, G. A., *Brendan the Navigator: An Interpretation* (Dublin: M. H. Gill and Son, Ltd, 1945).

Lloyd, G. E. R. *Polarity and Analogy: Two Types of Argumentation in Early Greek Thought* (Cambridge: Cambridge University Press, 1966).

Lovejoy, A. O., *The Great Chain of Being: A Study of the History of an Idea* (Cambridge, MA: Harvard University Press, 1936).

Lozovsky, N., *The Earth Is Our Book: Geographical Knowledge in the Latin West ca. 400–1000* (Ann Arbor: University of Michigan Press, 2000).

Lutz, C., 'Martianus Capella', in *Catalogus Translationum Et Commentariorum: Mediaeval and Renaissance Latin Translations and Commentaries; Annotated Lists and Guides*, ed. by Paul Oskar Kristeller (Washington D.C: The Catholic University of America Press, 1971), 398.

MacCormack, S., *On the Wings of Time: Rome, the Incas, Spain, and Peru* (Princeton: Princeton University Press, 2007).

——., *Religion in the Andes: Vision and Imagination in Early Colonial Peru* (Princeton: Princeton University Press, 1991).

Mackley, J. S., The Legend of St. Brendan: A Comparative Study of the Latin and Anglo-Norman Versions (Leiden: Brill, 2008).

Magasich-Airola, J., J.-M. de Beer, and M. Sandor, *America Magica: When Renaissance Europe Thought It Had Conquered Paradise*, 2nd edn (London: Anthem Press, 2007).

Mancall, P., *Hakluyt's Promise: An Elizabethan's Obsession for an English America* (New Haven and London: Yale University Press, 2007).

Mangani, G., 'Abraham Ortelius and the Hermetic Meaning of the Cordiform Projection', *Imago Mundi; The International Journal for the History of Cartography*, 50 (1998), 59–83.

Martin, C., 'Experience of the New World and Aristotelian Revisions of the Earth's Climates During the Renaissance', *History of Meteorology*, 3 (2006), 1–16.

Mauny, R., 'Trans-Saharan Contacts and the Iron Age in West Africa', in *The Cambridge History of Africa*, ed. by J. D. Fage, vol. 2 (Cambridge: Cambridge University Press, 1978), 272–335.

McConica, J., 'Humanism and Aristotle in Tudor Oxford', *The English Historical Review*, 94, no. 371 (1979), 291–317.

Meganck, T., *Erudite Eyes: Friendship, Art and Erudition in the Network of Abraham Ortelius (1527–1598)* (Leiden: Brill, 2017).

Metzler, I., 'Perceptions of Hot Climate in Medieval Cosmography and Travel Literature', *Reading Medieval Studies*, 23 (1997), 69–105.

Meurer, P., *The Strabo Illustratus Atlas: A Unique Sixteenth-Century Composite Atlas for the House of Bertelli in Venice* (Bedburg Hau: Antiquariat Gebr. Haas, 2004).

Milanesi, M., 'Notes on Oviedo's *Navigazione del grandissimo fiume*', in *Giovanni Battista Ramusio, Navigazioni e viaggi*, ed. by Marica Milanesi, 6 vols, vol. 6 (Turin: Giulio Einaudi, 1978).

——., *Tolomeo sostituito: studi di storia delle conoscenze geografiche nel XVI secolo* (Milan: Unicopli, 1984).

Milham, M., 'An Introduction to the Renaissance Tradition of Pomponius Mela', Acta Conventus Neo Latini Amstelodamensis: Proceedings of the Second International Congress of Neo-Latin Studies (Munich, 1979), 786–93.

Momigliano, A., 'The Place of Herodotus in the History of Historiography', *Historia*, 43 (1958), 1–13.

Moretti, G., *Gli Antipodi: Avventure letterarie di un mito scientifico* (Parma: Nuova Pratiche Editrice, 1994).

——., 'Nec Sit Terris Ultima Thule (La Profezia Di Seneca Sulla Scoperta Del Nuovo Mondo)', *Columbeis*, 1 (1986), 95–106.

Mund-Dopchie, M., 'Hanno', in *Catalogus Translationum Et Commentariorum:*

Mediaeval and Renaissance Latin Translations and Commentaries; Annotated Lists and Guides, ed. by Virginia Brown, vol. VIII (Washington DC: The Catholic University of America Press, 2003), 49–55.

Murphy, T. M., *Pliny the Elder's Natural History: The Empire in the Encyclopedia (*Oxford: Oxford University Press, 2004).

Myers, M. Y., 'The Frontiers of the Empire and the Edges of the World in the Augustan Poetic Imaginary', Stanford University, 2008.

Nansen, F., *In Northern Mists, Arctic Exploration in Early Times*, 2 vols, vol. 1 (London: William Heinemann, 1911).

Nauert, C., 'Humanists, Scientists, and Pliny: Changing Approaches to a Classical Author', *American Historical Review,* 84 (1979), 72–85.

Nowell, C. E., *Magellan's Voyage around the World: Three Contemporary Accounts* (Evanston, IL: Northwestern University Press, 1962).

Nunn, G. E., 'Marinus of Tyre's Place in the Columbus Concepts', *Imago Mundi*, 2, no. 1 (1937), 27–35.

O'Doherty, M., *The Indies and the Medieval West: Thought, Report, Imagination* (Turnhout: Brepols, 2013).

Oikonomides, A. N., ed., Hanno the Carthaginian: Periplus or Circumnavigation of Africa (Chicago: Ares Pub, 1977).

Oxford English Dictionary, '*Philosophy, n.*' (Oxford: Oxford University Press).

Paassen, C. v., *The Classical Tradition of Geography* (Gröningen: J. B. Wolters, 1957).

Padrón, R., '(Un)Inventing America: The Transpacific Indies in Oviedo and Gómara', *Colonial Latin American Review*, 25, no. 1 (2016), 16–34.

Pagani, L., *Claudii Ptolemaei Cosmographia* (Torriana: Stella Pollare, 1990).

Pagden, A., *European Encounters with the New World* (New Haven: Yale University Press, 1993).

——., *The Fall of Natural Man: The American Indian and the Origins of Comparative Ethnology*, repr. with corrections and additions (Cambridge: Cambridge University Press, 1982).

——., *Peoples and Empires: Europeans and the Rest of the World, from Antiquity to the Present* (London: Phoenix Press, 2002).

Pantin, Isabelle, 'Oronce Fine's Role as Royal Lecturer', in *The Worlds of Oronce Fine: Mathematics, Instruments and Print in Renaissance France*, ed. by Alexander Marr (Donington: Shaun Tyas, 2009), 13–30.

Parker, J., *Books to Build an Empire: a Bibliographical History of English Overseas Interests to 1620* (Amsterdam: Nico Israel, 1965).

Parker, R., 'The Origins of Pronoia: A Mystery', in *Apodosis: Essays Presented to Dr. W. W. Cruickshank to Mark His Eightieth Birthday* (London: St.Paul's School, 1992), 84–94.

Parks, G. B., 'The Contents and Sources of Ramusio's Navigationi', in *Terza edizione delle Navigationi e viaggi, Facsimile reprint*, ed. by R. A. Skelton, G.

B. Parks and G. B. Ramusio, 3rd edn, 3 vols, vol. 1 (Amsterdam: Theatrum Orbis Terrarum, 1967–70), 2–37.
——., 'Ramusio's Literary History', *Studies in Philology*, 52 (1955), 127–49.
Parmelee, L. F., 'French Propaganda in Late Elizabethan England', *Sixteenth-Century Journal*, 25 (1994), 853–72.
Parry, J. H., *Age of Reconnaissance*, 2nd edn (London: Praeger, 1969; 1st edition 1963).
Pearce, S., 'Nature and Supernature in the Dialogues of Girolamo Fracastoro', *The Sixteenth Century Journal*, 27, no. 1 (1996), 111–32.
Pédech, P., *La Méthode Historique De Polybe* (Paris: Belles Lettres, 1964).
Philips, W. D. Junior, and C. R. Philips, *The Worlds of Christopher Columbus* Cambridge (Cambridge University Press, 1992).
Portuondo, M. M., *Secret Science: Spanish Cosmography and the New World* (Chicago: University of Chicago Press, 2013).
Powell, W. S., *John Pory 1572–1636: the Life and Letters of a Man of Many Parts* (Chapel Hill: The University of North Carolina Press, 1977).
Quinn, D. B., 'The Early Cartography of Maine in the Setting of Early European Exploration of New England and the Maritimes', in *American Beginnings: Exploration, Culture and Cartography in the Land of Norumbega*, ed. by Emerson W. Baker (Lincoln, NE: University of Nebraska Press, 1995), 37–59.
——., 'Introduction', in *The Principall Navigations, Voiages and Discoveries of the English Nation*, ed. by Richard Hakluyt, David B. Quinn, and R. A. Skelton, 2 vols, vol. 1 (London, 1589; facsimile edition 1965).
Ramachandran, A., *The Worldmakers: Global Imagining in Early Modern Europe* (Chicago: The University of Chicago Press, 2015).
Randles, W. G. L., 'Classical Models of World Geography and Their Transformation Following the Discovery of America', in *The Classical Tradition and the Americas: European Images of the Americas and the Classical Tradition,* ed. by Wolfgang Haase and Meyer Reinhold, vol. 1 (Berlin: De Gruyter, 1994), 5–76.
——., *De la Terre plate au globe terrestre; une mutation épistémologique rapide (1480–1520)* (Paris: Librarie Armand Colin, 1980).
——., 'The Evaluation of Columbus' "India" Project by Portuguese and Spanish Cosmographers in the Light of the Geographical Science of the Period', *Imago Mundi: The International Journal for the History of Cartography*, 42 (1990), 50–64.
——., 'The Recovery of Ptolemy's Geography in Renaissance Italy and Its Impact in Spain and Portugal in the Period of the Discoveries', in *Geography, Cartography and Nautical Science in the Renaissance: the Impact of the Great Discoveries (*Aldershot: Ashgate, 2000), Chapter V.
Redfield, J., 'Herodotus the Tourist', *Classical Philology*, 90 (1985), 97–118.

Reeve, M. D., 'The Rediscovery of Classical Texts in the Renaissance', in *Itinerari Dei Testi Antichi*, ed. by O. Pecere (Rome, 1991), 115–59.

Reynolds, L. D., *Texts and Transmission: A Survey of the Latin Classics* (Oxford: Oxford University Press, 1983).

Reynolds, L. D., and N. G. Wilson, *Scribes and Scholars: A Guide to the Transmission of Greek and Latin Literature*, 3rd edn (Oxford: Oxford University Press, 1991).

Roa-de-la-Carrera, C. A., and S. Sessions, *Histories of Infamy: Francisco López De Gómara and the Ethics of Spanish Imperialism* (Boulder: University Press of Colorado, 2005).

Roberts, S. E., *Printing a Mediterranean World: Florence, Constantinople, and the Renaissance of Geography* (Cambridge: Harvard University Press, 2013).

Rogers, E. M., *Diffusion of Innovations* (New York: Free Press, 2003; first published 1963).

Roller, D. W., *Through the Pillars of Herakles: Greco-Roman Exploration of the Atlantic* (London: Routledge, 2006).

——., ed., Eratosthenes' Geography, (Princeton: Princeton University Press, 2010).

Rombai, L., 'Tolomeo e Toscanelli: fra Medioevo ed età moderna; Cosmografia e cartografia nella Firenze del XV secolo', in *Il Mondo di Vespucci e Verrazzano: Dalla Terrasanta all'America*, ed. by Leonardo Rombai (Florence: Leo S. Olshki, 1992), 29–69.

Romm, J., *The Edges of the Earth in Ancient Thought: Geography, Exploration, and Fiction (*Princeton: Princeton University Press, 1992).

——., 'Mythes, cartes et histoire: L'utilisation par Ortelius de l'Atlantide dans le développement de la théorie de la dérive des continents', in *Abraham Ortelius 1527–1598: Cartographe et humaniste* (urnhout: Brepols, 1998), 104–16.

——., 'New World and "Novos Orbes": Seneca in the Renaissance Debate over Ancient Knowledge of the Americas', in *The Classical Tradition and the Americas: European Images of the Americas and the Classical Tradition,* ed. by Wolfgang Haase and Meyer Reinhold, vol. 1 (Berlin, 1994), 77–113.

Roseman, C. H., ed., *Pytheas of Massalia: On the Ocean* (Chicago: Ares, 1994).

Rubies, J. P., 'New Worlds and Renaissance Ethnology', in *Travellers and Cosmographers: Studies in the History of Early Modern Travel and Ethnology*, Variorum Collected Studies (Aldershot: Ashgate, 2007), 157–97.

——., *Travel and Ethnology in the Renaissance; South India through European Eyes, 1250–1625* (Cambridge: Cambridge University Press, 2000).

Russell, J., *Inventing the Flat Earth: Columbus and His Modern Historians* (New York: Praeger, 1991).

Ryan, L. V., 'Richard Hakluyt's Voyage into Aristotle', *The Sixteenth Century Journal*, 12, no. 3 (1981), 73–84.

Sbordone, F., 'Exerpta ed epitomi della Geografia di Strabone', *Studi Bizantini e Neollenici,* 7 (1953), 201–6.

——., F., 'La Tradizione umanistica della Geografia di Strabone', *Bollettino del Comitato per la preparazione dell'edizione nazionale dei classici greci e latini*, 4 (1961), 11–32.

Schellhase, K. C., *Tacitus in Renaissance Political Thought* (Chicago: Chicago University Press, 1976).

Schlesinger, R., and A. P. Stabler, eds, *André Thevet's North America: A Sixteenth-Century View* (Kingston and Montreal: McGill University Press, 1986).

Schmidt, B., *Innocence Abroad: The Dutch Imagination and the New World, 1570–1670* (Cambridge: Cambridge University Press, 2001).

Schmitt, C. B., Q. Skinner, E. Kessler, and J. Kraye, eds, *The Cambridge History of Renaissance Philosophy* (Cambridge: Cambridge Unviersity Press, 1988).

Schoonheim, P. L., *Aristotle's 'Meteorology' in the Arabico-Latin Translation: Critical Edition of Both Texts, Introduction, Index and Registers (Aristoteles Semitico-Latinus)* (Leiden: Brill, 2000).

Schwarz, K., *Tough Love: Amazon Encounters in the English Renaissance* (Durham, NC: Duke University Press, 2000).

Seaver, K., 'Norumbega and Harmonia Mundi in Sixteenth-Century Cartography', *Imago Mundi: The International Journal for the History of Cartography*, 50 (1998), 34–58.

Seed, P., 'Navigating the Mid-Atlantic, or What Gil Eanes Achieved', in *The Atlantic in Global History, 1500–2000*, ed. by Jorge Cañizares-Esguerra and Erik R Seema (London: Pearson, 2007), 77–89.

Shalev, Z., and C. Burnett, *Ptolemy's 'Geography' in the Renaissance*, Warburg Institute Colloquia, 17 (2011).

Sherman, W., *John Dee: The Politics of Reading and Writing in the English Renaissance (*Amherst: University of Massachussetts Press, 1995).

Shrimpton, G. S., *Theopompus the Historian* (Montreal: McGill-Queen's University Press, 1991).

Sitwell, N. H., *The World the Romans Knew* (London: Hamish Hamilton, 1984).

Skelton, R. A., 'Hakluyt's Maps', in *Hakluyt Handbook*, ed. by D. B. Quinn, vol. 1 (London: Hakluyt Society, 1974), 48–73.

Skelton, R. A., and B. Sylvano, *Ptolemaeus Cosmographia with an Introduction by R. A. Skelton and Dedicatory Letter to Acquaviva Sylvanus* (Venice, 1511; facsimile edition, Amsterdam: Theatrum Orbis Terrarum, 1969).

Small, M., 'Displacing Ptolemy? The Textual Geographies of Ramusio's Navigazioni e viaggi', in *Mapping Medieval Geographies: Cartography and Geographical Thought in the Latin West and Beyond, 300–1600*, ed. by Keith D. Lilley (Cambridge: Cambridge University Press, 2014), 152–72.

——., 'From Jellied Seas to Open Waterways: Redefining the Northern Limit of the Knowable World', *Renaissance Studies* ,21, no. 3 (2007), 315–39.

——., 'From Thought to Action: Gilbert, Davis, and Dee's Theories Behind the Search for the Northwest Passage', *The Sixteenth Century Journal*, 44, no. 4 (2013), 1041–58.

——., 'Real and Theoretical Boundaries: Human Geography in Herodotus' (University of Victoria, 1998).

——., 'Warring Traditions: Ptolemy and Strabo in the Geography of Sebastian Münster', in *Ptolemy's Geography in the Renaissance*, ed. by Zur Shalev and Charles Burnett, Warburg Institute Colloquia, vol. 17 (London: The Warburg Institute, 2011), 167–86.

——., 'A World Seen through Another's Eyes: Hakluyt, Ramusio and the *Navigazioni e viaggi*', in *Richard Hakluyt and Collected Travel Writing in Early Modern Europe*, ed. by Daniel Carey and Claire Jowett (London: Hakluyt Society and Ashgate, 2011).

Smith, W., ed., *A Dictionary of Greek and Roman Geography by Various Writers*, vol. 1 (London: Walton and Maberly, 1878).

Spitzer, L., and A. G. Hatcher, *Classical and Christian Ideas of World Harmony: Prolegomena to an Interpretation of the Word 'Stimmung'* (Baltimore: Johns Hopkins Press, 1963).

Stahl, W. H., 'Astronomy and Geography in Macrobius', *Transaction and Proceedings of the American Philological Association*, 73 (1942), 232–58.

——., 'Dominant Traditions in Early Medieval Latin Science', *Isis*, 50, no. 2 (1959), 95–124.

Steinberg, P. E., *The Social Construction of the Ocean* (Cambridge: Cambridge University Press, 2001).

Steinberg, S. H., *Five Hundred Years of Printing, rev. by J. Trevitt,* 4th edn (London: Oak Knoll Press, 1996; first printed 1959).

Strauss, G., *Sixteenth-Century Germany: Its Topography and Topographers* (Madison: University of Wisconsin Press, 1959).

Swanson, J. T., 'The Myth of Trans-Saharan Trade During the Roman Era', *The International Journal of African Historical Studies* (1975), 582–600.

Tafuri, M., *Venice and the Renaissance*, trans. by Jessica Levine (Cambridge MA: MIT Press, 1989).

Talbert, R. J. A., *Ancient Perspectives: Maps and Their Place in Mesopotamia, Egypt, Greece, and Rome (*Chicago: Chicago University Press, 2012).

Taub, L. C., *Ptolemy's Universe: The Natural, Philosophical and Ethical Foundations of Ptolemy's Astronomy* (Chicago: Open Court Press, 1993).

Taylor, E. G. R., 'The Cosmographical Ideas of Mandeville's Day', in *Mandeville's Travels: Texts and Translations*, ed. by M Letts (London: Hakluyt Society, 1953).

—— ed., *The Original Writings and Correspondence of the Two Richard Hakluyts*, 2 vols, vol. 1 (London: Hakluyt Society, 1935), 77–83.

——., *Tudor Geography 1485–1583* (London: Methuen, 1930).

Thorndike, L., ed., *The Sphere of Sacrobosco and Its Commentators* (Chicago: University of Chicago Press, 1949).

Thrower, N. J. W., 'The Rediscovery of Ptolemy and Cartography in Renaissance Europe', in *Maps and Civilization: Cartography in Culture and Society*, ed. by Norman J. W. Thrower (Chicago: University of Chicago Press, 1999), 58–75.

Tibbetts, G. R., 'The Beginnings of a Cartographic Tradition', in *History of Cartography: Cartography in the Traditional Islamic and South Asian Societies*, ed. by J. B. Harley and D. Woodward, vol. 2.1 (Chicago: University of Chicago Press,1992), 96–104.

Tierney, J. J., 'The Map of Agrippa', *Proceedings of the Royal Irish Academy, Section C: Archaeology, Celtic Studies, History, Linguistics, Literature*, 63 (1962), 151–66.

Tillinghast, W. H., 'The Geographical Knowledge of the Ancients Considered in Relation to the Discovery of America', in *Narrative and Critical History of America*, ed. by Justin Winsor, 8 vols, vol. 1 (Boston, 1884; reprinted New York, 1967), 1–58.

Todd, R. B., 'Cleomedes', in *Catalogus Translationum Et Commentariorum: Mediaeval and Renaissance Latin Translations and Commentaries; Annotated Lists and Guides*, ed. by Virginia Brown, vol. VII (Washington DC: The Catholic University of America Press, 1992), 1–6.

——., 'Geminus and the Pseudo-Proclan Sphaera', in *Catalogus Translationum Et Commentariorum: Mediaeval and Renaissance Latin Translations and Commentaries; Annotated Lists and Guides*, ed. by Virginia Brown, vol. VIII (Washington DC: The Catholic University of America Press, 2003), 8–24.

Tymowski, M. "Death and Attitudes to Death at the Time of Early European Expeditions to Africa ". *Cahiers d'Études Africaines* 54, no. 215 (2014): 787–811.

Veneri, T., 'Giovanni Battista Ramusio, molto piu di uno spettatore. le quinte delle "Navigationi et viaggi', *Italica*, 89, no. 2 (2012), 162–201.

Vignaud, H., *Toscanelli and Columbus. The Letter and Chart of Toscanelli on the Route to the Indies by Way of the West, Sent in 1474 to the Portuguese Fernam Martins, and Later on to Christopher Columbus. A Critical Study of the Authenticity and Value of These Documents [&C.]* (London: Sands & Co., 1902).

Vogt, J., *Orbis* (Freiburg: Herder, 1960).

Walbank, F. W., *Polybius* (London and Berkeley: University of California Press, 1990).

Waterfield, R., 'Introduction', in *The Theology of Arithmetic: On the Mystical, Mathematical and Cosmological Symbolism of the First Ten Numbers by Iamblichus* (Grand Rapids, MI: Phanes Press, 1988), 23–33.

Weiss, B., 'The Geography in Print: 1475–1530', in *Ptolemy's Geography in the Renaissance*, ed. by Zur Shalev and Charles Burnett, Warburg Institute Colloquia, vol. 17 (London: The Warburg Institute, 2011), 91–120.

Wey Gómez, N., *The Tropics of Empire: Why Columbus Sailed South to the Indies* (Cambridge, MA: MIT Press, 2008.

Whitaker, I., 'The Problem of Pytheas' Thule', *The Classical Journal*, 77, no. 2 (1981), 148–64.

White, D. G., *Myths of the Dog-Man* (Chicago: University of Chicago Press, 1991).

Whitehead, N., 'The Historical Anthropology of Text: The Interpretation of Raleigh's 'Discoverie' of Guiana', *Current Anthropology,* 36 (1995).

Wittkower, R., 'Marvels of the East. A Study in the History of Monsters', *Journal of the Warburg and Courtauld Institutes* (1942), 159–97.

Wittkower, R., and I. Warburg, *Architectural Principles in the Age of Humanism* (London, 1949).

Woodward, D., *Cartography in the European Renaissance*, The History of Cartography, vol. 3, parts i and ii (Chicago: University of Chicago Press, 2007).

——., *Maps as Prints in the Italian Renaissance: Makers, Distributors and Consumers* (London: The British Library, 1996).

——., 'Roger Bacon's Terrestrial Coordinate System', *Annals of the Association of American Geographers*, 80, no. 1 (1990), 109–22.

Wright, J. M., 'True Peoples and Their Monsters: Speculations on the Other in the Age of Exploration', *Terrae Incognitae*, 37 (2005), 1–15.

Wroth, L., *The Voyages of Giovanni da Verrazzano 1524–1528* (New Haven: Yale Univeristy Press, 1970).

Yates, F., *The Occult Philosophy in Elizabethan England* (London: Routledge and Kegan Paul, 1979).

Zerubavel, E., *Terra Cognita: The Mental Discovery of America* (New Jersey: Rutgers University Press, 1992).

Zeyl, D. J. , 'Introduction', in *Plato, Timaeus* (Indianapolis: Hackett Pub. Co., 2000), xiii–xc.

Index

Printed in the USA
CPSIA information can be obtained
at www.ICGtesting.com
LVHW010754221023
761596LV00007B/1261

9 781783 275205